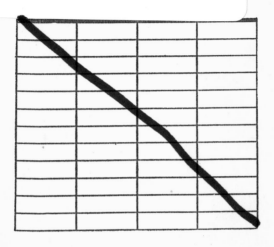

DEMCO

NATURE, TECHNOLOGY, AND SOCIETY

Cultural Roots of the Current Environmental Crisis

Victor Ferkiss

NEW YORK UNIVERSITY PRESS

New York and London

NEW YORK UNIVERSITY PRESS
New York and London

Library of Congress Cataloging-in-Publication Data
Ferkiss, Victor C.
    Nature, technology, and society : cultural roots of the current
environmental crisis / Victor Ferkiss.
        p.   cm.
    Includes bibliographical references and index.
    ISBN 0-8147-2611-9
    1. Human ecology—Philosophy—History.   2. Human ecology—
Religious aspects—History.   3. Technology and civilization.
4. Nature—Religious aspects—History.   I. Title.
GF21.F47   1993
304.2—dc20   92-27184

New York University Press books are printed on acid-free paper,
and their binding materials are chosen for strength and durability.

Manufactured in the United States of America

c 10 9 8 7 6 5 4 3 2 1

CONTENTS

PREFACE

The planet Earth, our common home, is in a state of crisis. Pollution menaces our health if not our lives. Resources dwindle as population soars. The technologies that many count on to solve our environmental problems create new ones, including threats to our personal and political freedoms. This book is about such dire problems, but only incidentally. It assumes we are in grave but not unsurmountable difficulties, but it focuses primarily on the question of how we got to be where we are. It asks, What are the cultural and intellectual roots of the attitudes toward nature and technology that have created modern technological civilization? Unless we can understand what causes us to act as we do, there appears to be little hope that we can change our ways and solve our problems.

In the first part of this book we will examine how various civilizations and cultures, including Western civilization, have historically looked at how humanity should relate to nature and technology. Given the vastness and complexity of the subject matter, I have been forced to rely almost exclusively on secondary sources. Nonetheless, I have tried to choose them judiciously and keep them as up to date as possible, being as fair as the truth allows to differing viewpoints on controversial matters. The second part of the book deals with the conflicting attitudes of contemporary thinkers vis-à-vis the issues involved in the human relationship to nature through the means of technology. Here I am able to make much more use of primary sources, but they are less exhaustive than I would ike.

As in any book that deals with such matters, there are bound to be problems of fairness. Virtually all of the sources are in English, which reflects not simply my own linguistic limitations but the fact that most of the controversy about environmental matters in the contemporary world has been carried out in English. As time goes on this shortcoming will change, as it is already beginning to do. In any case, the book reflects who I am: a Westerner and an American.

All of us, try as we may, are in large measure the creatures of our upbringing and environment. But that need not prevent us from searching for truths and values that in some way may transcend our origins and situations. Even the casual reader will detect in the pages that follow my bias toward certain positions: a reverence for nature, an unwillingness to allow technology and its owners to set the terms on which human societies use it, and a concern for human freedom and dignity. For this I offer no apology. I can only hope that even those who will disagree with my interpretations will be moved to ask the all-important question of where we go from here. Whether we like it or not, there will be a future in which our relations with nature and technology will be even more important than they are today. I firmly believe we should play a conscious role in shaping that future.

I wish to give special thanks to my editor at New York University Press, Niko Pfund, not only for his faith in the book but for his many suggestions on improving it and, above all, I thank him for his patience.

Nature, Technology, and Society

Chapter One

TO SERVE MAN OR TO SERVE NATURE? NATURE AND TECHNOLOGY IN ANCIENT AND CLASSICAL CIVILIZATIONS

From the beginnings of human society on Earth, human beings have struggled to survive and flourish, and they have used technology as an aid in that struggle. But early humanity's attitudes toward nature and technology were fundamentally different from our own. Nature was accepted as an all-encompassing environment, and many—in some cases all—of its manifestations were worshiped and looked upon with awe. Thunder and lightning reflected the wrath of the gods. Certain physical places were regarded as sacred. The earth that bore fruit was considered divine by agricultural peoples, and hunters paid tribute to the spirits of the animals they pursued and killed. Humanity had to struggle to survive, but it was a struggle within nature as much as against it.

Throughout history, people have argued about what distinguishes us from the other animals. For most philosophers and their cousins the theologians, we have been *homo sapiens*, distinguished from other animals by having the ability to reason. For Lewis Mumford, for example, man was first of all a dreamer who can remember his dreams.[1] For others he has been *homo ludens*, or "man the playing animal."[2] For still others, man is *homo faber*, a tool-making animal, first and foremost a producer. The argument on this subject is not simply an academic one. The ancients made an important distinction between contemplation and action. The former, they argued, was the higher and ultimately the only proper occupation of men. If man is to be distinguished from the other animals by his ability to reason or by being a dreamer who can remember his dreams, or even by being devoted to play rather than "work," then contemplation might be his primary goal. If, on the other hand, man is primarily a producer, then action is what counts.

Alert readers may have noted that when referring to human beings I have used the word "man." I am not simply being chauvinistic or following conventional usage. As we shall see, a growing body of feminist thought argues that the technological conquest of nature has been essentially a masculine enterprise. Certainly the period in which Western man has dominated nature through technology has coincided with rigid and perhaps even growing masculine control of society, so that all of modern technology can be viewed as an expression of male sexism.[3]

Human beings have always been users of technology. Fire may have been the first technology, devised by humans, used to keep hostile animals at bay in the night and to cook food. But some say that the first human tool was a crudely shaped stone used as a hand ax when man scavenged for animals, before he became a hunter. Still others have argued that the purpose of this first tool was to kill fellow humans, thus making technology coeval with murder.[4] In any case, there is no question that prehistoric human beings used tools to affect their environment and each other. Indeed, for many anthropologists the use of tools has become the primary means of distinguishing proto-humans from other species.

There is little direct evidence about how our remote ancestors actually lived. Early anthropologists assumed that the lives of "savage" peoples with whom Europeans were coming in contact as they spread their power over the globe could tell us something about the nature of "primitive man" and give us clues about our own past. Though this approach is still used, with some legitimacy, far more attention is now given to actual fossil and archaeological records. But time has been kinder to bones, tools, and physical artifacts than to ideas and institutions, and only rarely do we find material evidence such as a cave painting that is testimony to early humans' artistic as well as hunting abilities.

Early civilizations tried in various ways to deal intellectually with nature and technology. Their ideas, however, must often be disentangled from their total cultures and social systems. Ideas can usually not be observed directly but must be inferred from what was done rather than from what was said, simply because there are no records of the latter.

Religion played a major role in human attitudes toward technology and, above all, toward nature. The most ancient tradition apparently looked upon the earth as a female entity, a Mother alternately nurturant or terrible, and attitudes toward "Mother Nature" often varied according to the shifting fortunes of belief and disbelief in the old religious traditions and the later patriarchal religions that succeeded them.[5] On the whole, however, the female religions such as the cults of the Great Mother looked upon the earth as benign as well as actively fecund, while patriarchal religions considered the earth both dangerous yet, paradoxically, generally passive. Later philosophical systems that dualistically divided the world and human individuals into higher and lower parts and faculties can be traced back to this primitive distinction between triumphant male sky gods and defeated earth goddesses.

Even though technology consisted of masculine tools of domination, especially weapons, agriculture, probably the most momentous advance in human technology until modern times, was almost certainly an invention of women, who were the first to notice that seeds carelessly left behind at campsites sprouted into plants. But this important new technology involved cooperation with nature rather than domination over it. Whatever advantages control over nature may have given to women was somehow lost as men took over power in society.[6]

By the time the peoples of the Near East established their place in history, the masculine principle of struggling against nature had triumphed.[7] Both the Egyptians and the Mesopotamians regarded nature as basically nurturant, and because the deities of the latter could appear "more unruly and harsher" than those of the former has been ascribed to the fact that the Nile was more predictable and uniformly benign than the rivers of Mesopotamia.[8] Though a source of nurture, nature was above all subject to human ends. Nature was represented in Mesopotamian mythology as "monstrous chaos," and it was the proper task of mankind to tame wild things.[9]

These Near Eastern urban societies were the first to abandon the ancient religious views of the oneness of humanity and nature and to form the new view of separation between the two. According to one observer, "the dominant myth and reality in Mesopotamian society was the conquest of nature by divine-human order."[10] The major distinctive technologies of Mesopotamian civilization centered around irrigation, which, by its very nature, demands a high degree of governmental centralization and control over citizens—a pioneering example of what Karl Marx would many millennia later call "oriental despotism."[11] Beyond such despotism, some would sometimes cut down trees simply to underscore their power. The *Epic of Gilgamesh*, dating from over three thousand years before Christ, tells of how one "heroic" king destroyed a forest in order to "establish his name."[12] The upshot was that irrigation and deforestation led to salinization, flooding, and the eventual destruction of the ecological base of the Tigris-Euphrates river civilization, a forerunner of the environmental disasters that would plague humanity in subsequent centuries.

What role did politics play in all this? Certainly the political order underwrote the basic cultural norms of these societies. At base all were theocracies, with no distinction made between the political and religious obligations of the subjects. The decision to treat nature and technology in certain ways was ultimately a political one.

The Egyptians were clearly more dependent on their natural environment than their neighbors. Egypt was "the gift of the Nile." The regular and predictable rising of the river was so basic to Egyptian life that the Egyptians became pioneers in the calculation of time, and concurrently in the development of astronomy. Their gods were usually nature deities. To the Egyptians, "all nature was animate and filled with gods."[13] These beliefs worked well for them. It was not until modern times and the construction of the Aswan High Dam that Egypt has faced major ecological disaster.

In the nearby Persian Empire, which also had great influence on the ideas of early Western man, reverence for nature was a basic element in the religious tradition as well. Even the coming of Zoroaster in the sixth century B.C., with his cosmic dualism of lightness and darkness, did not completely obliterate this reverence. Yet there is little reason to think that the Persians were especially kind to nature in practice.

In all these ancient civilizations, religion provides the major focus for concepts about nature. Actual practice usually reflects these concepts, though not always. Technologies can arise as a result, but are usually taken for granted. It is only with the Greeks that attitudes toward nature and, to a slightly lesser extent, technology become self-conscious and are endowed with rationality. Insofar as the Greeks discovered philosophy, they also discovered the philosophy of nature and of technology.

The Greeks

The Greeks discovered nature as a concept, as part of their discovery of philosophy. They were the first to realize that while human customs differed from city to city, from society to society, and from time to time, and were therefore conventional, some things about humans—and in the nonhuman world as well—were universal.[14]

However, just as the Greeks disagreed with each other about philosophical ideas generally, they especially disagreed about what "nature" meant. In the third century B.C. Aristotle was already able to catalog fourteen definitions of the nature of nature, and from his day to ours we have been plagued by the fact that an idea so basic to human thought is so difficult to define.[15]

But virtually all Greeks agreed on one thing: what was natural was in some sense more basic, more ancient, and therefore more venerable and superior to what was newer and/or made by humans. The universe of nature, moreover, was not a chaos but a cosmos, governed by laws of nature that people must obey: "Nature had become the sum of all that was divine."[16]

Nature was for the Greeks an "intellectual organism,"[17] sometimes personified and spoken of as Gaia.[18] Although it is not clear what Aristotle himself understood the word to mean (the nearest Greek equivalent to the Latin *natura* is *physis*), in his *Physics* he seems to regard nature as "the totality of natural objects, i.e., of objects which are capable of change and bringing it to an end, of objects which have an inner tendency to change."[19] For the Greeks, nature could not be opposed, but it could be ennobled and used for human ends, for example, for medicine.[20]

This philosophical approach to nature was supplemented by a popular view that centered on a lively sense of the world as being "the theatre of the gods,"[21] an arena in which spirit had a major role to play. "Greek religion," according to one observer, "was in large part the worship of nature, and the old Greek gods were essentially nature deities."[22] Think of Poseidon, god of the sea, and Zeus, god of lightning. These gods entered the realm of man from the wilderness, which was their proper home. Shrines and temples were established in places of natural beauty, with trees considered especially sacred. Greek religion thus affirmed the oneness of man and nature.[23] Greek architecture also took its cue from natural surroundings, and buildings were designed to complement the basic meaning of the landscape.[24] Finally, Greek sculptors and poets expressed themselves widely in the glorification of nature.[25]

The Greeks, unlike their predecessors, sought to understand nature in a rational way. Even so, because of their belief in human unity with nature, they often did not make a rigid distinction between what we would today regard as the living and the nonliving, thinking of the stars as alive and the soul as material. They were much concerned about the interaction between people and their natural environment. The famous physician and savant Hippocrates discussed in his classic *Airs, Waters, Places* the influence of the environment on human health, the historian Thucydides examined the influence of nature on history, and Sophocles and Anaxagoras analyzed the effect of humans on the natural environment.

But this interaction was not always benign. In the ancient world, including Greece, the landscape was affected by mining, canal building, and agricultural practices that caused soil erosion.[26] Even though the germs of an ecological perspective existed in the Greek philosophy of nature, in practice the Greeks paid scant attention to it or to its implications. The destruction of the land by Greek agriculture was not simply the work of the omnivorous goat (which they introduced) but of farming activity as a whole.[27] Sophocles' play *Antigone* is rife with passages deploring man's destruction of the land. And in his

Critias, the philosopher Plato describes the devastating effect of Greek agriculture on the ecological balance.[28]

Despite their concepts of oneness with nature and their belief that the gods had their origins in the wilderness, the Greeks always preferred cultivated fields to the wild. And despite their belief that nature involved change, they essentially abhorred change, preferring a static situation to an altered one. In their view, changelessness was next to godliness. Even Plato's *Republic* is a veritable handbook for a utopia whose central feature is immobility. Of the important Greek philosophers, only Heraclitus accepted, and indeed glorified, change. Basically, for the Greeks, space was all important in shaping their worldview and time was essentially an illusion.[29]

Though the word "ecology" is Greek in its roots, the ancient Greeks did not invent the concept, which is a product of the nineteenth century. But it is the Greek philosophers, especially Aristotle, who have shaped many of our traditional views of the subject. In Plato's *Timeaus* the whole world is endowed with life and likened to an animal. Plato believed in a world-soul or demiurge, which could impart ends to nature, and he taught that nature was to be understood not in terms of its beginnings but of its ends.[30] By Plato's time, "the cosmos had become a teleological system, and God is 'the teacher of the whole world,' "[31] a teleologist being someone who believes that everything has a purpose, an end.

Even more than Plato, Aristotle, a biologist by training, was a teleologist. Biology taught him that certain parts of the body served specific purposes: the lungs were for breathing, the heart for pumping blood. He believed that everything in the universe had a final cause, a purpose determined by nature, an end inherent in nature itself. For Aristotle, humans, being rational, were higher than the animals, and since hierarchy was a basic characteristic of the universe—a derivation from the dualistic categorization of things as higher and lower—the purpose of all lower things was to serve the higher. Therefore, he reasoned logically, the purpose of the natural universe was to serve mankind. As he put it, "Now if Nature makes nothing incomplete, and nothing in vain, the inference must be that she has made all animals for the service of man."[32] His great pupil, Theoprastus, often called the father of botany, dissented and held that things in nature had their own ends as well.[33] But Aristotle's view became the dominant one and was later reinforced by the Christian belief in the unique and favored status and destiny of human beings. In addition, for Aristotle matter was associated with the feminine and was passive, while ideas were masculine and active, bolstering the basic attitude that man should rule over nature.

Almost alone among the Greek philosophers was Protagoras. In the declining years of Greek culture, he taught the basic unity of all species and the universe, even extending to the transmigration of souls. He espoused a special veneration of nature not unlike that found in some religious systems in India and the Far East.[34]

Though the Greeks had long speculated about nature, the mainstream of Greek philosophical thought, beginning with Socrates, took a turn from the contemplation of the physical universe surrounding human beings to the contemplation of the social scene among them. Socrates concluded that speculations about nature were dogmatic, unprovable, and useless,[35] and that man should instead study how to live properly as an individual and with his fellows.

The Greeks had no real interest in or concept of technology as we think of it today, though they took it for granted that man's use of tools and technologies distinguished him from the beasts.[36] The reasons were both socioeconomic and cultural. Ancient Greece was hardly a land flowing with milk and honey: famine was often a threat, and infanticide, especially of female children, was commonly used to keep the population down. The real dirty work of the economy was performed by slaves, so there was no incentive in Greek society for inventions that made labor less or easier.[37] In addition, because change was abhorred, uncontrolled technological innovation was considered a danger to social stability.[38]

In addition to these social considerations, Greek philosophical thought did not separate physical technologies from other kinds of disciplined expertise. The Greek word *techne* applied not only to the use of the craftsman's tools but to cultural techniques as well.[39] Rhetoric, all-important to the Greek freeman/citizen as a means of public discussion and debate, was the prototype of technology.

In large measure this disdain for the practical arts was not simply based on abstract philosophical considerations conjoined with the association of physical work with slavery in the narrow sense, but on a general rejection of all that our present society takes for granted as good in the way of efficiency, economic rationality, and increased productivity. Xenophon, for example, did not see production as creating something new, but simply as rearranging what already existed.[40] Aristotle echoed this view; for him, nature alone produced wealth, and man could only organize and manage it. In *Politics*, he states that he will not discuss acquisition since it is not in good taste.[41] This aristocratic contempt for the economic and commercial worlds had foreseeable results. The Greeks' lack of enthusiasm for technology, or even an interest in it, was reflected in very little technological progress.

The Romans

Although the Romans defeated the Greeks physically, they were largely conquered by the Greeks spiritually. This amalgam became the world of Hellenic civilization, throughout which Greek culture flourished under Roman imperial protection and patronage—with Greek tutors often the slaves of their Roman students—and Greek ideals became the standard of Roman literature and ideas.[42]

Like the Greeks, the Romans loved nature, or at least certain aspects of it. Their most ancient religious traditions were centered on the family hearth and the soil. The word "patriotism" comes from the Latin *patria*, fatherland. The Romans began their existence as a nation with strong feelings of "geopiety," the reverence for the sacred soil where one's ancestors are buried. In the legendary founding of Rome, Romulus is portrayed as performing "various rites in order to be purified of all irreverence," necessary because the site of the city had been moved.[43] But soon these feelings became attenuated as republic was transformed into empire, and nature and soil alike became simply new areas of domination.

Greek literature was the model and standard for the Romans. Originally, much of the Roman love of nature, especially as expressed in literature, came from their Greek mentors. When one sees references to nature in Roman literature—above all in Virgil, Lucretius, Ovid, and Cato—it becomes easy to accept the idea that Roman literature is really "a late chapter in the history of Greek thought."[44]

But this Roman love of nature was a selective one. For Romans, the wilderness held no charms, nor did the mountains or the open seas.[45] They preferred their nature tamed. The great Roman passion was to get away from the city—though they were great city builders[46]—to their country villas, which reminded them of their cherished rural past and were far more healthful and pleasant than city dwellings. While Roman religion, like that of the Greeks, was originally highly animistic, it was concerned above all with ritual and with finding some way to make use of nature. It was predominantly agricultural,[47] but "the Roman mind was marked by its practicality, and Roman attitudes toward nature were distinctly utilitarian."[48]

Though the explicit expression of ideas did not change very much, behavior did. Even more than the Greeks, the Romans made the increased veneration of nature a standard.[49] A major role here was played by the philosopher and statesman Cicero, though he added little that was really new to the corpus of Greek or Roman thought. His greatest contribution was to synthesize and to some extent popularize it. He was above all significant in being the major vehicle through which the classical heritage was passed on to the Middle Ages and beyond; when many of the original Greek texts were temporarily lost, his work was still accessible. His exhortations to follow "nature," often loosely defined, did much to create a tradition that was to influence later revivals of nature reverence.[50] Cicero wholeheartedly accepted the Stoic belief that nature was made to serve man.[51] As the famous Stoic, Zeno, put it, "The end of man may be defined . . . as life in accordance with our own human nature as that of the universe, a life in which we refrain from every action forbidden by the law common to all things, that is to say, the right reason which pervades all things, and is identical with Zeus, lord and ruler of all things."[52] There were dissenters, of course. The poet Lucretius argued against both Aristotle and the Stoics, saying it was absurd that the world, "so foolish the design, contrived so ill," was created by the gods for human use.[53]

Just as Roman religion echoed the animism of the earlier Greek beliefs, so its philosophers often held to the old Greek notion that the world itself was somehow alive. For the Stoics, "the world itself was an intelligent organism; God and matter were synonymous."[54] Lucretius Seneca stated that "the earth's breath nourished both the growths on its surface and the heavenly bodies above by its daily exhalations,"[55] but these concepts vanished in the centuries to come.

Despite their philosophical and cultural veneration of nature, the Romans, like their Greek predecessors, did much to destroy their natural environment.[56] Deforestation took place on a major scale, first in Italy, then in the outer provinces. Wildlife, especially exotic animals, were killed, in part to fuel the famous Roman circuses. In the end the Romans essentially destroyed themselves. Despite knowing the dangers posed by the consumption of water poisoned by the lead used to line aqueducts,[57] the Romans continued to use it in drinking and cooking vessels. Consequently, lead poisoning played a major role in the decline of the Roman Empire.

The Romans not only subscribed to the Greeks' attitudes and behavior toward nature, but also toward technology. The period of Roman domination of the world was largely one of technological stagnation. Only in some areas of civil engineering did innovations emanate from Rome, and some of these have been greatly overrated. The famous Roman roads, for instance, were not much better than those of other nations.[58] As with the Greeks, much of this stagnation was due to a lack of creativity. But, as we have seen, the Romans took a practical attitude toward nature and were extremely practical in the building of their legal and social institutions. Furthermore, they were even more dependent on slaves than the Greeks had been, and shared their mistrust of innovation.

These attitudes toward technology not only had direct social consequence, but they affected the very categories of thought about the world used by classical civilization. The "Greeks and the Romans were not machine-users, save to a very small extent."[59] Therefore they were never tempted to make the analogies between the physical universe and the machine that were common in the later Middle Ages and the Renaissance, when people basically conditioned their ideas and actions toward nature and technology and ushered in the modern world and its new problems.

Nature and Technology in Judaism and Early Christianity

Western civilization, the originator of contemporary technology, is the off-spring of the not always harmonious marriage of the classical civilizations of Greece and Rome and the Judeo-Christian tradition. It is the child of Athens and Jerusalem. If Athens had been ambivalent about nature and technology, so too would be Jerusalem, if for somewhat different reasons.

Christianity came into existence in the womb of Judaism and adopted most of the Judaic ideas about the nature of God, man, and the universe. To understand the vast influence of religion on Western ideas about nature and technology, we must begin with an analysis of the mainstream thought of the Hebrews of the Old Testament era.

The traditions of the ancient Hebrews and the Greeks were very different; indeed they were almost mutually incomprehensible. The amalgam between them created by the early apostles such as St. Paul has always been an unstable one. Where the Greeks were interested in space and in objects, the Hebrews were interested in time and in words.[60] Where the Greek mind yearned for fixity and changelessness, the Jewish mind accepted change. For the Hebrews, God was neither immanent (within the world) nor even simply transcendent (completely outside the world) as some philosophers have alleged.[61] Rather, God worked *through* the physical universe, and history was the chronicle of his dealings with people, above all his Chosen People of Israel.

The Christians, originally a Jewish sect, were intellectually and spiritually the descendants of the Jews. Their beliefs reflected most of this heritage—at least until they undertook to convert the Gentiles after the triumph of the faction led by St. Paul. They then accepted categories of Greek philosophy for the use in Christian theology.

The question of the influence of the Bible—that is, the Old Testament—on later ideas about nature and technology is an inherently difficult one. It involves several linked questions. First, we have to ascertain what the Bible in fact says—a difficult problem that has vexed scholars since it was composed. Second, we have to determine what the Bible meant to those who accepted its authority in pre-Christian and early Christian times—another difficult, largely historical question. We also need to determine how Jewish and early Christian beliefs about the meaning of the Bible affected daily individual and social lives. Finally, we must look at how interpretations and behavior changed over the course of the following centuries.

It is difficult to understand just what the Old Testament taught regarding nature and the place of God and people within it. A good example is the question of wilderness. As we shall see in a later chapter, for centuries Americans accepted—indeed extended—the traditional European distaste for the wilderness and leveled the land whenever they could. Many based this behavior on what they conceived to be a biblical opposition to wilderness. Yet the Old Testament does not even have a generic term for wilderness. What the ancient Israelites feared and hated—and rightly so from their perspective—was a desert in which they could not survive.[62]

The most important single issue relating to the impact of the biblical tradition on the rise of modern industrial civilization and its use of technology to subdue nature involves the meaning of the divine command in Genesis that humanity should multiply and subdue the earth. What in fact does the Bible say about this? Biblical exegesis is a thorny field, with various interpretations dividing theologians and scholars of scripture. It is essential to understand that Genesis is not an unequivocal document; according to biblical scholars, it combines basically two traditions. The oldest of these, the so-called "J" tradition, stems from the tenth century B.C. It gets its name from the fact that it employs the divine name YHWH (for Jehovah) which was transliterated by German biblical scholars as Jahweh. The other tradition used the divine name Elohim and is called the priestly tradition—the "P" tradition—and dates from about 500 B.C. The J tradition appears in Gen. 2:4b–25, the P tradition in Gen. 1:1–2:4a. The accounts of creation found in the two traditions differ in significant respects, and how one interprets them affects how one views the nature of the relationship of mankind and the rest of creation.

The two traditions differ in the order of creation. In the P story, humans and land creatures were all made on the same day, while according to the J tradition the order was man, vegetation, animals, and, last, woman. In the J tradition, all the animals are brought before Adam, who has the privilege of naming them. We have already noted the importance the Hebrew mind attached to words: "Naming was perceived as virtually a creative activity in Hebrew psychology," and this act confirms man's ascendancy.[63]

The issue is not simply whether humans are the highest of the animals and therefore in some sense have "dominion" over them. What is at stake is what that dominion means. Are humans absolute rulers or are they stewards, answerable to God for the creatures—and nonliving nature as well—placed under their control? Do creatures other than human beings—and nonliving nature—have any purposes independent of serving mankind, purposes that must be respected by human beings? It also can and has been argued that both the exploitation and stewardship positions fail to go to the core of the Bible's teachings, and that nature in the Old Testament is really "God's instrument of divine reward and retribution."[64]

To these questions the Old Testament gives varying answers. For one scholar, "The Bible indicates a variety of ways in which nature is subservient to man, but also ways in which man is subservient to nature."[65] The covenant was not simply with the Jews as a people, it was with Israel as a place.[66] The land was a blessing from God, to be accepted and enjoyed as such.[67] But whether the land itself had any rights was another matter. Care of the land, after all, is compatible with the belief that it exists simply to serve man. But this is to address the question improperly. *Both* the land and man exist to serve God. Indeed, the evidence is strong that if man failed to keep his covenant, the land itself would rise up and strike him as God's agent by becoming unfit to support life. But "premeditated decimation of nature is God's prerogative, not man's,"[68] and wanton destroyers of nature will be punished by God.[69]

Similarly, the Jews were enjoined to a humane attitude toward animals and even toward trees. Though the Hebrews rejected anything smacking of animism and pantheism, living creatures were spirits just like people, and had souls in much the same sense.[70] Animals and indeed all nature shared in God's covenant with mankind, could rejoice with God, and nature not only could but would punish humans when they transgressed the covenant.

It seems clear that throughout the Old Testament the Jews were taught that nature had its own ends, independent of mankind. As Glacken writes, "The Lord explains that nature serves man—even limits him—but nature is not for him alone and its significance does not depend on human wants."[71] Elder quotes, with approval, the reminder that God instructed Noah to take on the ark two of every species of animal, without regard to their utility to people.[72] In the book of Job, another commentator notes, "It is clear that God's creatures were meant to exist, whether or not they were useful to humans," and he summarizes, "Religious fundamentalist and secular humanist pretensions to qualitative superiority notwithstanding, biblical tradition typically sees humanity as co-creature with other living things."[73]

The Old Testament view is that there is no unbridgeable gap between humans and their fellow creatures. Nature exists not simply for humans' sake, but for the greater glory of God.[74] Although the Bible is full of statements about nature, "It is difficult—if not impossible—comprehensively to find a theology of nature in the Old Testament: difficult to find expressed a set of theological formulations that adequately fit the mind of ancient Israel."[75] Another commentator notes that "there is no Hebrew word for our modern word 'nature.' The non-human world is historicized, inseparable from human events, and subject to the will of Yahweh in all its particulars, just as man is."[76]

To understand fully the Hebrew attitude toward nature, we must always keep in mind Judaism's desperate struggle against the surrounding pagan societies that worshiped nature, insisting that their God was superior to and outside of nature. But even this perspective can be misleading. For a long time much Jewish worship took place out of doors, and even though natural objects were not worshiped, some locations were viewed as sacred space in a manner not unlike that of the Greeks.[77]

Ultimately, the Jews eschewed nature religion. "Admirable as the creation is, God is not immanent in it; he is a transcendent being" summarizes their thought.[78] There is little of pantheism in the Hebrew outlook. As one scholar puts it, "Ancient Judaism differs from pantheism not in any inability to perceive spirits in animals and plants, but because it does not worship the spirits which it sees."[79]

But God's transcendence did not remove him from earthly concerns. He was both creator and ruler, and humans shared in the glory of his creation. Man was his steward, his viceroy.[80] The Lord "who delivered Israel was the Lord of heaven and earth, not merely the Lord of human history," and "Yahweh is seen as structuring the cosmos not just so he might bless human creatures but also that he might delight in his own works."[81] God's kingdom is statement in itself of the goodness of the earth he has made. Creation itself is his foremost blessing.[82]

Technology in the Old Testament

If the Greeks discovered philosophy and, if disparagingly, at least had something to say about technology, the Hebrews said nothing. Indeed, even in terms of the natural, they had little to contribute. One writer notes that their language "formed no specific expressions for designating the outline or contour of objects and did not even need them."[83] Obviously, at the commonsense practical level, the Hebrews used various tools and techniques, but they took them for granted, making little comment about them. The Old Testament creation stories, replete with accounts, sometimes conflicting, about the beginnings of the universe and the various elements in it, do not deal with the origins of technology at all. The accounts of biblical figures such as Cain and Abel, Nimrod, and Moses show little interest in their tools or agricultural methods. Only in the later Wisdom literature does any concern appear.[84] In any event, the ancient Hebrews "developed no science worth mentioning, and technologically they were inferior to most of the neighboring peoples in the arts both of peace and war."[85] An obvious exception to their lack of concern with technological matters was the tools and techniques used to maintain the various aspects of Mosaic law regarding the preparation of food and the keeping of the Sabbath.

But another—and far deeper—issue arises about what the Bible says about technology. This fundamental question is the other side of the issue of domination of nature discussed earlier. "Technology," it is argued, "can . . . be thought of as a sort of secular fulfillment of a basic outlook about humanity which was already expressed in Genesis."[86] The essential, primal argument for the justification of the human use of technology is that man must help God in the divine plan by subduing the earth, that he is in some sense a co-creator. This argument has been of concern primarily to Christian interpreters of the scriptures, who emphasize the incarnation of God in the person of Jesus and the duty of Christians to further his mission of reconciling man and the physical creation. As one modern Jesuit puts it, "Christian doctrine contains inherently technological predispositions; and . . . technology is structuring a world which is objectively Christian and requires the exercise of Christian virtues."[87] This position, as we shall see, is widely held. But it is also hotly disputed. The contemporary theologian/philosopher Jacques Ellul, an international figure in the discussion of the role of technology in modern society, takes a completely opposite view. For Ellul, the notion of man as co-creator, a kind of Platonic demiurge, is simply blasphemous. "Creation," he writes, "as God made it, as it left his hands, was *perfect* and *finished.* . . . That man as creature is co-creator? To say that is to have a curious idea of the image of God. It is a simple absurdity that is not considered anywhere in the Bible."[88]

How then did technology come into the world? According to Ellul, Adam worked simply because God told him to do so. But then came the Fall, and with it a whole new relationship between man and the rest of Creation. The harmony of Eden was gone, and Ellul contrasts the commands given by God to Adam and to Noah to illustrate his contention that after the Fall the earth was simply delivered into man's hands. Technology came, according to his reading of the Bible, with the sons of Adam's treacherous offspring, Cain. One son made music, another raised flocks, and the third forged iron and bronze tools (Gen. 4:17–22).[89] Thus for Ellul, technology was not part of God's original plan; it is a result of the Fall, and a means necessary for the survival of mankind in a fallen world.

But these speculations were not a concern for the Jews of Old Testament times. For them, technologies simply *were,* and deserved no further attention except as they applied to religious rituals.

The Christian Church and Nature

How did the early Christian church regard nature? We can learn something about its attitude from several sources. The most important, of course, is the New Testament, including the teachings of Paul, who did much to create a new church in the image of his own ideas. Another source is the writings of the fathers of the church, the original great theologians of the early Christian centuries. But the early Christians not only reflected the attitude of the Jewish community from which the new faith had sprung, and the first generation of Christians expected Christ to return in their lifetimes. Thus, as intoxicated with the primacy of history as the Old Testament Jews had been, they paid little attention to the physical world around them, which they expected would soon be transformed by the dawn of the new age.

What did the New Testament have to say about nature and man's relation to it? According to one author, the New Testament views God as an absentee landlord. For him, "The Hebrews achieved this reconciliation by evolving a concept of man's responsibility to God for the management of the earth, a concept which was duly carried over into Christianity, becoming part of the Western heritage."[90] Another disagrees completely, arguing that "the world, for Genesis, was created complete, by a series of divine acts; it was perfect until Adam sinned," and there is in his view no evidence in the New Testament for the idea of Christian stewardship. Rather, man is referred to not as God's steward but as his servant.[91] Certainly Paul was not especially concerned with the topic.[92] As one commentator sums it up: "The Pauline doctrine of redemption embraces all of creation; but neither in Paul nor in the tradition growing from Paul is man's precise relation to the scope of created reality explored in depth."[93] Much turns on the issue of whether all of creation has been affected by the Fall and stands in need of redemption. Says one scholar, "While the Christian doctrine of salvation is widely accepted as offering *man* a way up from the fallen state, it seems to be frequently overlooked that by the same token the rest of the natural world should be saved along with man."[94] Paul spoke of the whole creation "groaning in travail," but as we have seen it is sometimes argued that all of creation has been likewise redeemed. But not all theologians would accept the idea that all of creation had fallen and needed redemption in the first place. The redoubtable Augustine (A.D. 354–430) began a tradition of arguing that the only creature that was groaning was the human creature, the cosmos as a whole had not been disrupted, and the "world" of which Satan was lord was not the physical universe but the "world" of unbelieving and sinful men.[95]

Though the church was still united and the various schisms that would lead to the eventual split between the churches of East and West were far in the future, one could already detect, with the benefit of hindsight, some divergence between the fathers of the Eastern church—the Greek fathers— and those of the West. This difference was also apparent in their views of nature, and became a prelude to the more radical differences in the Middle Ages as well as on their views on technology.

The attitudes of the Greek fathers toward nature were complex. They "displayed deep interest in the natural world" and enjoyed it. But they did not value wilderness, since "the mind trained in the Old Testament scriptures would see no beauty in wilderness, which was always conceived as hostile and dangerous to man."[96] The kind of natural beauty that was often praised was farmland, cultivated land, parkland. A prominent expositor of the beauties of God's creation was Saint Basil of Caesarea (ca. A.D. 331–379).[97] But even he was reproached by his friend St. Gregory of Nyasa when he praised raw land. The Greek fathers largely accepted the Stoic view of natural law, but carefully, since "no first- century Jew, hearing Jesus say that *the earth beareth fruit of itself*, would have been led to think that the origin and source of such natural growth was an impersonal force. The Jewish mind would have attributed the growth to God." However, since in the Hellenistic culture there was a great danger that laws of nature could be taken as the work of an impersonal *natura*, the fathers went out of their way to point out that "though each moment of growth is not directly caused by a separate act of the divine will, natural law is ultimately derived from such an act and is therefore divine law."[98]

But though there was some difference in the intellectual centers of gravity of East and West, it was still not clear-cut. A pillar of the Eastern church, Origen (ca. A.D. 125–254) reaffirmed the Aristotelian view of the hierarchy of nature, and insisted all nature existed for the sake of the highest creature within it, man.[99] On the other hand, one of the major defenders of the intrinsic value of the total creation was Bishop Irenaeus of Lyons (ca. A.D. 130–200). According to Irenaeus, Jesus Christ had come to serve the whole creation, not only fallen man.[100] His theology is theocentric rather than anthropocentric: God created the world not for the sake of man but for his own purposes.[101]

The Greek fathers saw nature not in scientific but in allegorical terms: "For all their interest in the structure and processes of nature and their insistence upon nature as a means by which God reveals his nature," they still hold that "God and nature are not identical, and that the mind must penetrate nature to find God."[102]

The most radical break with Greek thought came in the West in the person of Augustine (A.D. 354–430). In his youth Augustine had been a Manichean, a believer in the radical evil of matter. From this he had moved to Platonism, itself also strongly dualistic. Then he became a Christian, and eventually a Christian bishop. How much of his original aversion to the material world remained in his mature thought? Theologians vary in their opinions. Some, such as the Dominican Matthew Fox, claim that Augustine remained a Platonist, disdainful of the physical works of God. They seek to replace his influence with what they call "creation theology," a reversion to the original Jewish tradition of the goodness of God and his works, which they claim has been obscured since Augustine's time.[103] From a completely opposite point of view, Lutheran theologian H. Paul Santmire argues that Augustine was an "ecological theologian," since "his theology was holistic, rather than narrowly anthropocentric."[104] As we saw earlier, Augustine was one of the first to interpret Paul's words as saying that not the whole of creation but only man fell from grace with Adam and Eve. Things exist for the sake of man, Augustine believed, but they also have ends of their own in their creator's eyes. Man's dominion will be primarily a dominion of understanding.[105] Matter is not simply inert and evil; it is in some sense living since it is caught up in a cosmic process.[106] One critic writes that "despite an almost pessimistic view of human nature, which he considers to be deeply ravaged by sin, Augustine found creation marvelous."[107]

However, despite his appreciation for the progress of civilization, he shared the widespread conviction of his era that history was approaching a climax. It was generally believed that the world itself was becoming senescent. Crops were smaller and less nourishing. For such circumstances, there were clearly too many people. Therefore, Augustine argued, the command of Genesis to increase and multiply had lost its force with time. "There is not the need for procreation that there once was," he held, and "marriage is not expedient except for those who do not have self control."[108]

Could technology help man in his relationship with recalcitrant and failing nature? For Augustine, technology was still art, and the arts, despite their wonders, did not necessarily make life better in the long run. In any case, it was no great problem if there was to be no increase in the number of humans the earth would have to support. That would simply mean that there were already enough souls in heaven, and the world could come to an end.

In the ages after his death, Augustine was generally interpreted as an enemy of the world who emphasized original sin and the consequent need for redemption. His influence on subsequent millennia of Christian thought is almost incalculable, as he helped to set up a radical dichotomy between the spirit and the world of the flesh and created universe that persists to this day.

Chapter Two

RELIGION, NATURE, AND TECHNOLOGY: THE MEDIEVAL EXPLOSION

For the past several centuries the mainstream of historical thought has maintained that modernity began with the Enlightenment and the industrial revolution, and that the period before the Renaissance was one of material and intellectual stagnation, due largely to the influence of religion in the Middle Ages. Even if, as one contemporary historian puts it, "the pervasive assumption that the development of science and technology stands opposed to religious values is a creation of the European Enlightenment and hardly a self-evident truth,"[1] it was accepted as such, and the medieval contribution to technological growth was ignored.

Increasingly, however, it is becoming evident that in at least a technological sense the modern world was on the move long before the Renaissance. There is no longer any question that the medieval period in Europe saw a major change in the way human beings dealt with the world of nature through technology. More and more, we are hearing arguments over how and why this change took place, and what was the general relationship of philosophical and religious ideas to actual social and economic development. As we saw with classical and early Christian civilization (chapter 1), there is often in fact little correlation between professed ideals and actual behavior. But ideas do have consequences, and we shall continue to examine them as if they had causal force, without oversimplifying the issue of how important and direct that force was.

Though the thought of Augustine continued to dominate the medieval world, many of the authentic currents of classical antiquity continued to flow.

Among these were the doctrines of Plato, especially those of so-called Neo-platonism, which emphasized the mystical and mathematical aspects of his teachings as exemplified by his disciple Plotinus. Despite the influence of Plato and Augustine, man's attitudes toward nature and how to treat it varied. As a modern writer sums it up, if somewhat simplistically, in speaking of Western Christendom during the medieval period, "Two distinct points of view could be found. . . . One, the dominant one, claimed that humans had dominion over nature and thus justified humans' technological actions on the environment; contrasted with this was the 'ecological' approach of St. Francis, who felt that all living creatures were equal in God's sight and hence humans had no right to claim mastery over the rest of nature."[2]

The primary medieval view of nature was aesthetic. Nature was viewed as "simply a beautiful mirror of the perfections of its Creator."[3] Usually, following a Platonic motif, nature is presented as a secondary cause, personified in the goddess Natura who is doing the work and bidding of the Almighty who has created her.[4] Not all men of the Middle Ages looked upon Natura as beautiful; she was treated with contempt by many ascetics.[5] But when she was not, the primary meaning of nature was symbolic.[6] According to the revered and vastly influential pseudo-Dionysius,[7] "The symbol was the true and proper expression of reality; nay more, it was through such symbolization that reality expressed itself."[8] Things were not what they seemed; they had a deeper meaning, usually expressed in allegorical form. Even the secular approaches to the beauties of nature reflected this religious attitude.[9] It was not until the rise of medieval science, which looked for the meaning of things in the things themselves, that this way of thought was beginning to be overcome.

The single most important philosopher of the Middle Ages was Thomas Aquinas (1225–1274), although at the time he was not as all-important to Christian thought as he later became. Many of his ideas were condemned by the Catholic church shortly after his death, and it was not until 1323 that he was canonized as a saint.[10] His major significance lay in his use of the ideas of Aristotle. Aristotle was reintroduced into European culture largely through the agency of the Arabs, who were Muslims by this time and regarded as the great enemies of Christendom. But at this time Arabs were culturally as well as technologically further advanced than Christians. The first translations of most of Aristotle's works were from the Arab versions of the original Greek and strongly reflected the interpretations of Muslim philosophers such as Avicenna and Averroes.

Some translations from Arabic into Latin had already been undertaken by Saint Albert the Great, a philosopher of the Dominican order. In his footsteps, Saint Thomas, a fellow Dominican, undertook the revolutionary task of

synthesizing Aristotelian philosophy with Christian theology, which for his time was as bold as modern attempts to synthesize Karl Marx and Christianity. Thomas's major thrust was to insist, against the position of Plato and even against Augustine, that dualism was basically wrong and that the world of nature had its own legitimate existence and rights. Grace did not work against nature, as in the Augustinian tradition, but rather perfected it.[11] The natural order, including the political, had its own rights, though in some areas they were necessarily less than those of the church.[12]

Aquinas had little to say, however, about nature as such. The world was good, being the creation of a good God. "To slight the perfection of created things is to slight the perfection of divine power," he contended.[13] Were it not for the continuing presence of God in creation it could not continue to exist.[14] Yet though the world is good, having been created by a good God, it is still quite distinct, entirely separate from him. Nature is orderly and balanced in its own right,[15] but it is still lower than the spiritual as exemplified by man.

Aquinas retained the Aristotelian concept, fortified by Christian religious beliefs, that the rest of nature existed to serve humanity. Man, being rational and having an immortal soul, was higher than the rest of God's creation, which existed to serve man. Even in the state of innocence before the Fall, man therefore dominated the animals.[16] He should treat nature with some care, since it was a gift of God, but being created *for* man it had no independent rights.[17] The natural world, including animals, could sometimes be possessed by the devil, and then had to be treated accordingly.[18] Above all, Aquinas taught that, since the whole of creation had come into existence to serve human material needs, it would completely come to an end when salvation was at hand. Unlike man, it could not be redeemed.[19]

Despite the charges of heresy against him, Thomas Aquinas was ultimately recognized as a saint, due in large measure to the power of the Dominican order to which he belonged. Saint Francis of Assisi and Saint Benedict, two other important figures in shaping religious attitudes toward nature during the Middle Ages and thereafter, also owe much of their influence to their orders. Indeed, both were founders of religious orders, and Saint Benedict was also the originator of mainstream monasticism in the West.

The monks whom Saint Benedict (ca. 480–550) led were important not only for their dedication to physical work but for their willingness to work with nature and to respect its potentialities and limits. A distinguished sociologist has said that "respect for nature was, for Benedict and his monks, no more than a form of worshiping God." The monks believed, above all, in a life of simplicity; they lived in harmony with nature, but with none of the self-mortification that was involved in the earlier tradition of hermetic holi-

ness. "One does not find anywhere, in the mainstream of Western monasticism, the view that nature . . . is something to be ignored, overwhelmed by,
or exploited."[20]

The work of Saint Francis (1191–1225), unlike that of Saint Benedict, by
whom he was to some degree influenced, was not only controversial in his
own lifetime, but the order he founded had many problems. One wing of it
was eventually condemned by the church, not only because of their views
on nature, but because of the emphasis the "Spiritual Franciscans" put on
poverty in an age when the church was steeped in luxury. Unlike many other
saints, he did not write much, except for his great poem, "Canticle to the
Sun." Still, there is no question as to his significance. Writes one admiring
historian: "The powers of nature which had first been divinized and worshipped, and then in turn rejected by man as he realized the transcendence
of the spiritual, are now brought back into the world of religion . . . thus the
Franciscan attitude to nature and human life marks a turning point in the
religious history of the West."[21] The leading historian of medieval technology,
Lynn White, Jr., proposed that Saint Francis become the patron saint of
ecology,[22] and Pope John Paul II so proclaimed him on Easter Sunday, April
6, 1980.

From where did Saint Francis's new attitudes come? Although Saint Benedict was the founder of monasticism in mainstream Christian Europe, there
had long been a virtually independent church in Ireland. A Celtic monastic
tradition came into existence separate from the mainstream European one,
and it was even more attuned to the importance of nature.[23] Some argue that
Francis was strongly influenced by Irish monasticism. Various oriental ideas
were also being discussed in Europe at the time, but they were closely identified with the dualist heresy of Catharism and, strictly speaking, would have
been eschewed by Francis. But some semipantheist doctrines certainly played
a role, as did the traditions of the Eastern church, which, as we have seen,
were always in consonance with the natural world.[24]

In any case, there is little doubt that Francis was the single most important
Christian nature mystic in history.[25] He was revolutionary in that he called
the dumb animals, down to the smallest creatures, "brother" or "sister,"
thereby recognizing their common origin with man long before Darwin, and
also recognizing his kinship with forces of nature such as "Brother Sun" and
"Sister Moon." Francis, according to White, tried to depose man from his
monarchy over creation, set up a "democracy of all God's creatures," and
"tried to substitute the idea of the equality of all creatures, including man,
for the idea of man's limitless rule of creation."[26] However apocryphal some
of the stories about Saint Francis's preaching to the birds may have been,
they expressed a real meaning.

There is an ambiguity in Francis's literary composition "Canticle to the Sun," however. It is not clear if it is his own hymn in thanksgiving to God for all the works of creation, or one that asks all creatures to praise their creator themselves.[27] This confusion is especially ironic because, according to one legend, he sang it over and over again on his deathbed.[28]

Francis, it should be emphasized, was not unique in his ideas. During the Middle Ages a number of figures arose who proclaimed the holiness of all creation and tried to create spiritual disciplines for themselves and others based on this principle. Hildegarde of Bingen (1098–1179), Mechthild of Magdeberg (1210–1280), and above all Meister Eckhart (1260–1329) labored to this end. Many of them were condemned by the church for various heresies, but their work lived on, and today their ideas and outlook are being revived in some quarters.[29]

Unlike the pagan religions, Christianity desacralized nature. Christianity originated in the cities and spread from there to the countryside—and slowly at that. The very word "pagan" derives from the Latin *paganus*, country dweller. There is reason to believe that as late as the fifth century most of the country folk were still nonbelievers. The peasants still retained many of their long-standing animist beliefs, and, as we shall see, witchcraft owed much to these traditions. But the cult of saints, as one historian argues, "ousted spirits from the material objects of nature and liberated mankind to exploit nature freely. The localized *daemon* of antiquity became the medieval demon, a malevolent fallen angel who shared the saint's abstraction from time and place."[30]

In the Middle Ages many people were loath to speak too positively of nature, for fear of offending God by appearing to diminish their love for him.[31] It would appear, however, that their use and abuse of nature was not based on religious grounds, as White and others have contended. Another historian put it this way: "Readings of early Christian theology that attempt to find a rationale for exploitation are searching for a moral sanction supportive of economic, social, and technological changes that occurred for other reasons."[32]

Medieval Technology

The medieval attitude toward technology was at first cautious. Medieval men continued to take the same positions toward technology and nature as the ancient Greeks, reinforced by the Christian belief that God had made the universe for humans. Especially in the later Middle Ages, the arts, even when their purposes might be "superfluous, perilous and pernicious," could represent the exercise of "an acuteness of intelligence of so high an order that it reveals how richly endowed our human nature is."[33] Along with this belief that the universe is the servant of man and of human intelligence came a lessening of the ancient fear of the natural world. The low point of the Dark Ages was now receding, and medieval man "slowly came to sense the capacities of his powers over the environment."[34]

Ironically, much of early medieval technological development was the result of borrowings from other, non-Christian cultures. But medieval man would soon create a technological explosion that was the seed of modern-day industrial society, that is, the nearly total conquest of nature by technology.[35]

How can one account for this efflorescence of creativity? A major school of thought gives credit to Christianity. Ernst Benz,[36] Lynn White, Jr.,[37] and dozens of other scholars have held that the teachings of the Western Christian church provided the intellectual and moral stimulus for medieval technological growth. What in fact did happen?

Certainly medieval Europeans had much to stimulate their imagination. Religious symbolism and mysticism played a major role in their perception of the world. Ideas derived from the classical heritage and technical knowledge of the Chinese and the Arabs also served to fuel the opening Western mind. But the first beginnings of medieval technology were not based on any concept of progress as we know it today. On the contrary, the early medieval Christians, shaken by the attacks of marauding peoples such as the Vikings and Huns and religious invaders such as the Muslim Arabs, often took an apocalyptic view of the world and came to believe that the Antichrist would use various new devices to confound the faithful. In reaction against this attitude, the famous medieval monk Roger Bacon (ca. 1212–1292) advocated that Christians must also be open to technological progress and new knowledge.[38]

What role did ideas play in this new, more positive attitude toward technology? Foremost was a new approach to the long-standing Christian beliefs about nature, work, and living in the world. As one commentator puts it, "What distinguished the Middle Ages from earlier periods is not only the technologies themselves—metallurgy, optics, mechanics, hydrology—but the development of a religious ideology that encouraged these processes."[39] Basic to this ideology, as White argues, was the fact that while in the Greek tradition sin was ignorance, the Latin tradition had asserted that sin was vice. "The Greek saint was normally a contemplative; the Western saint an activist."[40] For the Greeks, following Plato, to know the good was necessarily to seek it. The West took more seriously Saint Paul's contention that often the will rejected the good presented to it by the intellect.

A major role in the changing attitude toward technology was played by the changing attitude toward physical labor. To work with one's hands was considered degrading by the Greeks and Romans. Did Christianity change that, as many contend? At first blush, this contention sounds plausible. After all, Christ himself had been a carpenter, and the son of a carpenter.[41] But as the medieval historian Jacques Le Goff notes, labor was an issue on which the Bible was ambivalent. Medieval men could, and did, cite it to various results.[42]

The monasteries played a key role in forming—or at least reflecting—medieval society's view of labor. But even that role is controversial. The motto of the Benedictines is usually given as *laborare est orare*, "to work is to pray," but that allegedly was coined in the nineteenth century, as part of "the romantic revival of monasticism."[43] Rule 48 of the order says "idleness is the enemy of the soul, and therefore, at fixed times, the brothers ought to be occupied in manual labor, and again, at fixed times, in sacred reading."[44] The mainstream interpretation of Benedict is that he gave labor a special position of honor, since his monks had to be both laborers and scholars.[45] But a recent researcher stresses the idea that while the first ascetics and monastic theorists favored manual labor, it was "always as a means to a spiritual end," and "work was worship, but it was also a material precondition to prayer and a distraction easily surrendered." Manual labor and technical knowledge always remained subordinate to the needs of the church. In themselves they had little value. This being the case, it is no surprise that in fairly short order manual labor was downgraded by monastic and general Christian thought. By the end of the twelfth century the Benedictine ideal of the "well-rounded" monk had been abandoned.[46]

By the twelfth century most commentators claimed that certain kinds of Christians were best fitted for certain kinds of labor.[47] By the beginning of the thirteenth century, the great Franciscan theologian Saint Bonaventure, defending the right of members of his order not to perform manual labor, maintained that three classes comprise Christian society, all necessary to the proper functioning of the world. The lowest of these was manual labor. To explain these events, Ovitt, for example, suggests that Christian theology had not been the catalyst of technological growth but "reacted to changes in the social and productive order by removing labor and the mechanical arts from their association with the life of the spirit." By the end of the twelfth century, he argues, labor and its products were secularized, and "the theoreticians of medieval culture turned to matters of personal transcendence and left the work of the world to social managers, merchants, and capitalists."[48]

Le Goff essentially agrees and holds that medieval technology developed as a result of economic factors, and that the position of the church was basically reactive. He is especially concerned with the process by which the church, which was once highly suspicious of much of the activity of the merchant class, especially its implicit claim to make time the basis of profit, eventually changed its mind and accepted the claims of capitalism. While at the beginning of the Middle Ages the church had looked down on lucrative professions, considering many trades intrinsically odious, by the end of this period it accepted virtually all of them, including the practices previously condemned as usury.[49]

What made the difference, he argues, were excuses such as "good intentions."[50] The merchant was now justified by his labor and his contributions to the common good. While the mechanical arts were now accepted, the real issue became mental labor, which was good and praiseworthy, and manual labor, which increasingly lost favor. Time, which was once looked upon as a gift of God that could not be sold, was now regarded as valuable. "Nothing is more precious than time," said Saint Bernard. The increasingly widespread use of clocks made it ever easier to measure time, and by the end of the Middle Ages, "The time which used to belong to God alone was thereafter the property of man."[51]

How much, it might be asked, did medieval technology owe to medieval science? Not much, actually. The close relationship between science and technology that we take for granted today is, as we shall see, largely a product of the late nineteenth century. But it can be and has been argued that, even though there were few if any direct connections between science and technology in the medieval period, both owed much to a common conviction: that the world was orderly and created by a rational God. To seek to know the created order was one way of seeking to know God himself.[52] One can also argue that not until Saint Francis had taught the West—or most of it—that other creatures besides man were important could there be an emotional basis for the objective investigation of nature.[53]

In its beginnings, medieval science was a matter of the library rather than the laboratory. A great wave of translations from the Arabic and Greek provided its basic material, which was highly speculative at that. But despite the influence of the Benedictines and their early tradition of working with the hands as well as the brain, the revival of the ancient separation between the two caused late thirteenth-century university physicians to separate themselves from surgeons and apothecaries, who worked with their hands as well as their minds.[54] Medieval science dealt with problems inherited from the ancient Greeks and assimilated ancient science so successfully that, by the time the Renaissance reintroduced Europe to Greek literature and art, there was no comparable need in science. There was as yet no cross-fertilization between science and technology, and in the Middle Ages invention did not depend on science; it was "completely empirical."[55]

Central to medieval science was the belief that a rational God had created an orderly universe. Natural law was both a source of moral norms and a compendium of them. Order in the church and society, however, required the intellectual and practical conquest of time. Thus it was no accident but quite fitting that the archetypal machine of medieval technology and life should be the clock.

Like the notion of work as a proper Christian vocation, the clock entered Western society through the monastic tradition. The monks came to believe that prayers said collectively at regular times were more pleasing to God. To some extent the Divine Office could be regulated by nature directly in the form of sunrise and sunset. But this was less than satisfactory, especially in latitudes where the passage of the seasons changed the length of daylight and dark. The medieval monks wanted something that would measure time independently of ordinary physical nature, and they found it in the clock.[56] Now for the first time in human history time became an abstract entity, and from the monasteries the clock spread throughout society so that most human activities—including ordinary secular work—could be subject to mechanical calculation and regulation. Thus was the real beginning of the machine age and of modern technological civilization. As Mumford put it, "one is not straining the facts when one suggests that the monasteries . . . helped to give human enterprise the regular collective beat and rhythm of the machine."[57]

But the clock did more than rationalize the life of urban Europe and begin the psychological transition to the modern era. The sheer numbers of men needed to produce these new machines that were sweeping the West led to a vast increase in the technological capability of Europe, leading to a clear superiority even over China.[58] The Western world was well on its way to establishing global supremacy sheerly through its technological power.

At the same time, the clock led to a change in the way Western man looked at the world. He became fascinated by machinery. Gimpel notes that "the men of the Middle Ages were so mechanically minded that they could believe that angels were in charge of the machinery of the universe: a fourteenth-century Provencal manuscript depicts two winged angels operating the revolving machine of the sky."[59]

Closely allied to the medieval fascination with mechanism as such was a fascination with power. The men of the Middle Ages were always fantasizing about sources of power other than those provided by nature, such as hydraulic, wind, and tidal energy. Villard de Honnecourt, a noted architect and engineer active between 1225 and 1250, even tried to create a perpetual motion machine.[60] The major practical source of power in the Middle Ages, however, was one that depended on natural energy—the watermill. Such mills were used not only for grinding corn but for a variety of industrial tasks, even for crushing ore, and were a major factor in what Mumford has called the "Eotechnic phase" of development.[61] Second only to the watermill was the windmill.[62] According to White, "One cannot exaggerate the importance of the fact that by the eleventh century in Europe—and, it would seem in Europe alone—every peasant was living daily in the presence of at least one fairly complex, semiautomatic power machine."[63]

Even before the coming of the machine, new technology had radically changed Europe. An agricultural revolution—the basic underpinning of the technological explosion—had been created by the adoption of the new, heavy, iron-shod plow. This new plow not only made it possible to till the more dense soil of northern Europe and move agriculture north of the former heartland of the Roman Empire, but it also caused a radical change in land distribution. The standard size of a farm was no longer determined by the needs of a family but by the ability of a power machine—albeit one based on the cooperation of a plow team—to cultivate the soil. This meant a fundamental change in man's relationship with nature. No longer nature's child, he was now its exploiter.[64] Closely allied with the new plow was a new, rigid, padded, horse collar, probably imported from the East, which made possible the traction necessary for the new plow.[65]

Cultural and Social Developments

An often ignored cultural consequence of the medieval technological explosion was the decline of the acceptance of slavery. The church apparently did not play a major role in slavery's demise, though the enslavement of a Christian was, by the Middle Ages, regarded as a criminal act. The decline of slavery put a premium on labor-saving devices. Unlike the Egyptian pyramids, cathedrals—the pinnacle of European medieval technological achievement in their combination of technical skill and religious devotion—were built by free and well-paid labor.[66]

But there was also a dark side to medieval technology: mining. Slaves were used in the silver mines upon which Athenian "democracy" rested. Serfs were used in the mines of Scotland long after serfdom had been abolished elsewhere in the West.[67] In the later Middle Ages especially, a large-scale mining industry was developed in central Europe. Even though some new devices such as pumps were used, basic techniques remained similar to those used in Roman antiquity. The classic description of the industry, *De Re Metallica* by Georgius Agricola (Latinized version of the name of physician and geologist George Bauer [1494–1555]), remained a practical mining handbook for centuries.[68] It is interesting that Agricola was at great pains to defend mining against detractors who argued that it destroyed nature and that metals corrupted mankind.[69]

Despite the new, more favorable attitude toward nature promulgated by medieval thinkers such as Saint Francis, the technological explosion resulted in a large-scale abuse of natural resources. In fact, it was medieval man, still believing in authentic Christian religion, who initiated the destruction of Europe's natural environment.[70]

With environmental destruction came pollution. The burning of pit coal fouled the air over cities and rural villages alike. Queen Eleanor of England was driven from Nottingham Castle in 1257 by the noxious fumes of coal burned in London.[71] Equally dramatic, and directly related to the demands of mining and coal production, was the destruction of the forests. Europe was still under the spell of the ancient Hebrew disdain for the "wilderness." Furthermore, nature, in the form of wild animals, hostile weather, and rampant disease continually posed a threat to survival. But simple fear of practical dangers alone cannot account for the zest with which trees were cut down and landscapes denuded.[72] Europe behaved toward its forests in "an eminently parasitic and extremely wasteful way."[73] The forests of France, Italy, and above all England were methodically felled for agricultural and industrial purposes. Because wood was the major fuel as well as the major element in the construction of buildings and even machines, this rapaciousness is not surprising. Wood was also used for the construction of ships, which helps account for the special ravages visited upon the woodlands of England.[74]

Technology never exists in a void. Not only does it reflect social systems, it changes them. What kind of new social institutions were emerging in medieval Europe as a result of technological advance?

First, there was the growth of new cities. During the "Dark Ages," after the fall of the Roman Empire, urban life had declined as people found more protection in smaller areas, and the economic infrastructure that had made the continued prosperity of the large cities possible slowly crumpled. But with the gradual repulsion of alien invaders—Vikings, Huns, and Muslim Arabs— a new peace settled on the land, and new techniques increased agricultural productivity, making the support of cities possible again. New technologies also gave birth to new economic opportunities in the cities,[75] which were perceived as an opportunity to rise out of feudal society. "City air made men free."

These new cities were often crowded and dirty, and had traffic congestion that rivals our own.[76] But the city was a vibrant place, where opportunity lay. It was in the cities that the new social institutions affected by the new technologies had their greatest strength.

Primary among these was the guild, a group of workers in a single field who controlled the conditions of work and the level of workmanship. Master workmen normally owned their own shops. Thus the guilds, which had legal monopolies, in theory at least ran much of the economy and were a filter through which technological changes passed.[77] But one must not be misled to think of the Middle Ages as a period of harmony in which happy worker-owners dominated. Increasingly there was a division between capital and labor, and strikes were not uncommon.[78]

One of the most important social developments in the Middle Ages was a drastic change in the status of women. Although they played a major role in the early Christian communities, masculine prerogatives had reasserted themselves in society within a few centuries. The status of women in Christianity was in some basic respects higher than it had been in pagan society; women, like men, now had immortal souls to be saved, and, in Christian teaching, men could no longer divorce them if they were unable to bear children.[79] But basic Christian beliefs about the vanity of the flesh soon combined with lingering Hellenistic philosophical dualism to create a situation in which women were increasingly downgraded. By the third century, women were forbidden to preach or hold any church office except that of the deaconate,[80] a minor form of holy orders from which they were also excluded soon enough.

A major enemy of women was Tertullian, the Western father of the church, who, among other things, is noted for taking literally Christ's injunction about becoming a eunuch for the sake of God (Matthew 19:12) and castrating himself. In his *On the Apparel of Women*, he spoke of the female sex as "copartners with the fallen angels in revealing the secrets of technology . . . and [sharing] with devils a blasphemous curiosity, which probes the hidden secrets of nature for the purpose of modifying what is intrinsically good," a charge later subsumed into the literature on magic and witchcraft.[81]

The church was always uncomfortable with sex. As time went on, the traditional practice of priestly marriage was gradually abolished. In the latter part of the eleventh century, Pope Gregory VII issued a flat prohibition against clerical marriage.[82] This ban was part of a general campaign of reforms within the church that involved issues other than sex, and was not unrelated to the problem of defending the lands and properties of the church against the claims of relatives of monks, priests, and bishops. It is telling that of all the canonized saints, the women almost without exception are either virgins or, in a few cases, widows.

This is a matter of significance not simply in itself but in relation to the question of nature and technology. Women were looked down upon not simply because they supposedly were more sexual than men but because they were thus closer to nature. Sex itself was evil because it was natural and spontaneous, of the flesh rather than the spirit. Though some feminists trace the exclusion of women from Western society to the Renaissance, historian Ovitt holds that it was sometime in the eleventh and twelfth centuries in which "the status of women was altered for the worse." He suspects that "one reason for the increased pressure to eliminate women from participation in both sacred and secular public life must be traced to the broader project of subduing the natural world through technology." The reason, he claims, is that "Women . . . are part of generative forces that lie outside of masculine control and serve as a constant reminder of the limitations of masculine power."[83]

As for basic living conditions, most people lived in a chronic state of undernourishment. Plagues were frequent. Between 1348 and 1351, the famous Black Death, a form of bubonic plague, killed 25 million out of a total of 80 million Europeans. But such public health problems did have a silver lining: the drop in population made more resources available to the living. This turn of events may have had a great deal to do with the Western world's dynamism in this period, giving it a head start toward modernity.[84]

Though there were some attempts by rulers and guilds to control technology in the workplace, the major and essentially failing attempts took place on the battlefield. War increasingly became the major activity of princes in the medieval period. The reasons why are a matter of conjecture. It has been argued that Western Europe was different from virtually all the rest of the world in having a merchant class that was aggressive and no stranger to arms.[85] Though one recent historian has argued that "technological changes introduced into field combat during the two millennia from about 500 B.C. to 1500 A.D. were frequently minimal,"[86] Europeans not only embraced new military techniques but did so avidly. The Franks were the last horse-riding people of the Eurasian continent to use the stirrup, but the first to realize and embrace its uses in warfare.[87] The introduction of the gun and gunpowder—from China, originally, but soon developed by Europeans beyond the Chinese originals—revolutionized warfare, not only ending the supremacy of the mounted knight and the essence of feudalism but also leading to new arts in fortification.[88] This interest in warfare was reflected in the economic patterns of the era. Though the still nascent states of the Middle Ages were quite weak by modern standards, most of their energy and economic resources went to the service of war, within Europe and soon outside of it.[89]

What was accomplished by Europe during the Middle Ages? In the beginning, Europe had been underdeveloped relative to the rest of the world; by the end, it was far ahead of other regions. During the period, the peoples of western Europe, at first under siege by hostile neighbors—some of whom, such as the Moslems, were technologically superior—had gained initiative and were poised for global expansion. The seeds of later science and technological progress were planted. Above all, the stage was set for the beginning of a completely new and unique relationship between humans and nature in the world, one which was to result in an almost complete conquest of nature through technology. The medieval technological explosion thus ushered in the beginnings of the modern world.

Chapter Three

TECHNOLOGY FLOURISHES: THE RENAISSANCE AND THE REFORMATION

The Renaissance was crucial to the development of the contemporary world, but not for the reasons given in most conventional histories. Though progress continued to be made in all areas, it was not a period of many great technological advances and innovations. But it did herald the beginnings of a radically new attitude toward the relationship between humanity, nature, and technology that still dominates our thinking today. However, this new outlook was not part of the heritage of revived classical literature and ideas that was central to the Renaissance. Indeed, it emerged in opposition to them.

The core of the early Renaissance was based on natural magic, the belief that the ancients had found ways of mastering the world through spells and incantations and that this knowledge could now be revived so that it could be shared. Because even a magician must carefully observe the world so he can call on the appropriate spells, such observations became instrumental in laying the groundwork for the later scientific revolution. A prototype of this transition from magic to science was alchemy, which became the foundation for the later science of chemistry. The alchemists sought to change natural substances by use of the occult, and their final goal was to change lead into gold. At this time the line between magicians and scientists was not yet clear-cut. Even Sir Isaac Newton believed in and practiced alchemy.

The men of the early Renaissance wanted many things from the magical control of nature, but above all they wanted personal wealth and gain. The natural magicians believed, like many of the ancient Greeks, that metals and jewels actually grew in the earth, and one could find them by means of magic. But unlike the new natural scientists who would finally triumph over them, the natural magicians had a respect for nature. They, and all popular writers, thought of nature as a nurturing mother.[1]

This manipulative but respectful "natural magic" was finally defeated by two rival intellectual forces: the Catholic church and the rise of the mechanistic philosophy that basically prevails to this day. The whole structure of natural magic was condemned as heretical by the church in the sixteenth century, and the mechanistic philosophy of Francis Bacon (1561–1626) and René Descartes (1596–1650) replaced it.

The triumph of the mechanistic philosophy was preceded and accompanied by intellectually earth-shaking developments in the natural sciences. Man had always believed that the earth was the center of the universe. For Plato and Aristotle, Augustine and Aquinas and their contemporaries, this belief was a fundamental tenet of their thought about the world. Jew, Greek, Roman, Muslim, and Christian embraced this conviction. Then Nicolaus Copernicus (1473–1543), a Polish astronomer, successfully challenged this concept with his theory that the earth revolves around the sun.

Central to the new role of astronomy, both practically and symbolically, was Galileo (1564–1643). His distinguished career as a mathematician and astronomer was cut short when he was tried and convicted in 1633 by the Inquisition for teaching the Copernican system, which was then considered heretical. But, as with Copernicus, his real importance lay not in what he said directly about the planets and the sun but what he implied about the whole nature of the universe. The physical universe was now denatured, that is, reduced to something radically different from the organic living being that even the Neoplatonists had envisioned. There was now a big chasm between the contemplating human mind and that which it contemplated. One could no longer observe the real essence of things—what they really were in themselves—but only those things that could be measured.[2] In time one could no longer say that the midday sky was blue, only that certain wavelengths of light were apparent. Mind could not know quality, but only quantity. The modern world of science was emerging.[3]

A crucial role in changing men's thinking was played by Descartes.[4] He wanted to create an airtight philosophical system subject only to discussion in logical terms and unassailable as such. Hence his famous assumption in *Discourse on Method*: "*Cogito, ergo sum*" ("I think, therefore I am"). Central to his system was an absolute separation between the material and the spiritual. The mind was spiritual, the body material. Given the radical separation between the two, Cartesians thought of all things in the universe, including plants and animals, as mere objects, rigidly subject to the laws of mathematics and physics. Since God had created man in his own spiritual image, man had the right to use this nonliving matter as he saw fit.

Though Descartes had a tremendous impact in philosophy for several centuries, especially in his native France, it was Francis Bacon, an Englishman, who was most influential, both practically and symbolically, in shaping our modern attitudes toward nature and technology. Bacon was not only a philosopher and dabbler in science, but also a statesman; he was Lord Chancellor until he was impeached for bribery and ended his life in disgrace.[5]

Bacon's ideas were expressed primarily in the *Novum Organum* and *The New Atlantis*. Bacon's discussions of science dealt specifically with what today we would call technology, a word that had not yet gained currency in his day. He was not interested in the old idea of science as contemplation, which he regarded as futile and impotent; he was interested in knowledge as power. Once man knew the secrets of nature, he could bend nature to his will. This manipulative view of nature was much like that of the natural magicians. "Bacon, in fact," one critic writes, "suggested that the term 'natural magician' should be revived and applied to the side of natural philosophy dealing with the 'production of effects'—what we would today call technology."[6]

But now this magic would be public domain and used not for the benefit of the magician but for all humanity.[7] Bacon felt men should seek knowledge not "for pleasure of mind, or for contention, or for superiority of others, or for profit, or fame, or power, or any of these inferior things; but for the benefit and use of life."[8] He was motivated above all by charity in the biblical sense, and it marked the first time in history that this theological virtue was used as a defense of technology.[9]

Though he is concerned with technology, Bacon talks much about scientific method. One must dissect nature, he writes, using an anatomical metaphor. Several critics have held that his use of language, in which the masculine investigator wrests secrets from a feminine nature, is redolent of the witch trials of the period.[10] One notes that Bacon proclaimed "his virile intention to inaugurate the 'truly masculine birth of time,' the new era in which mankind would increasingly gain the power to 'control and subdue [nature], to shake her to her foundations.'"[11]

The New Atlantis was reprinted eleven times from 1627 to 1676; it has gone through over a hundred editions since, becoming a recognized classic. It takes the form of a typical utopia of the time. In the story, voyagers from England come upon Bensalem, an unknown island in the Atlantic, and return to describe it to their countrymen, in a manner reminiscent of the work of Saint Thomas More. But on this island, scientists rule. The scientific academy Bacon first suggested in the *Novum Organum*, which advocated a new scientific endeavor, has now become the dominant institution of Bensalem society, in the form of Salomon's House. Here a scientific elite not only carries on research but actually controls society, despite the presence of a titular though essentially powerless monarch. The imaginary regime is a paradigm for what would later be called technocracy.[12]

Following Bacon's death, his idea of setting up a group of men devoted to the growth and progress of science became a reality. In 1644 the Royal Society was founded in England for the purpose of bringing together "natural philosophers" to examine ideas, exchange data, and promote Baconian ideals.

So far we have chronicled a movement toward a scientific-technological conquest of nature, organized by men who believed that nature was subject to rational laws that men could understand and then use to control natural forces. But the era also witnessed a great revival of witchcraft—real and alleged—throughout Western Europe, much of it in England. Most Europeans—and some accused witches—believed that, through contact with Satan, witches could gain powers to put aside the ordinary laws of nature. Why did the revival of witchcraft take place, and how was it related to the rise of science? One would assume that a sharp rise in witchcraft is contradictory to a growing belief in the scientific conquest of nature.[13]

Besides the belief in natural magic, other factors were involved in the rise and fall of witchcraft. Among these was the continued existence of religious traditions that included strong elements of superstition. Another was the great arbitrariness of life for the overwhelming majority of the population; even if the poor suffered most from the vicissitudes of fortune, the well-off were equally subject to fire, plague, and earthquake. It is most important to note, however, that the natural magicians were commonly literate or semiliterate males, working with ancient texts, while virtually all witches were illiterate women, operating on the basis of the traditions of the countryside.[14] One could argue, as many feminists have done, that the defeat of witchcraft represented the final suppression of women and nature by male technology.[15] In a similar vein, one can say that the invention of forceps and the emergence of the male-dominated medical profession took control of the process of childbirth from midwives and gave it to men; the forceps serve as a Baconian symbol for wresting power from a female nature.

The belief in witchcraft came to an end not so much because the beliefs of Bacon and his followers triumphed, but because those who had been concerned with the machinations of the devil became more concerned with the social unrest of the poor. Arson began to replace witchcraft as a manifestation of dissent. It is noteworthy that in late seventeenth-century England, prosecutions for witchcraft declined while those for malicious damage to property multiplied.[16]

The Reformation

The Renaissance primarily affected the Catholic nations of Europe and Anglican England, and it was here that the rising tide of scientific rationalization rose highest. But the future world of the twentieth century was being shaped elsewhere as well. The sixteenth century was also the era of the Reformation, a time that greatly affected men's views of nature and technology.

What were the views of Reformation leaders on the subject of nature and technology? Neither Luther nor Calvin felt any need to treat nature in a systematic fashion.[17] The relation that concerned them was not between man and nature, but between man and God. Nature was primarily the stage on which the drama of salvation was to be played. By relegating nature to a secondary place at best in the dynamics of man's all-important relationship with God, they emphasized how insignificant nature was compared to the human will, on which they focused their attention. When they did discuss nature specifically, they tended to echo earlier Christian themes, especially those of Saint Augustine, who was a major influence in the thought of Luther.

Martin Luther (1483–1546) was the central figure in the Reformation. Luther believed, as had most Christians before him, that man was the center of creation and that the natural world had been created simply to serve man. Genesis "plainly teaches," he argued, "that God created all these things to prepare a house and an inn, as it were, for the future man." He personally shared the common view of his time that the world was dangerous and alien and that its end was near. Since the Fall, nature existed under the curse of a God of wrath. He could not share the view of Saint Francis that nature was a path to God, and he often failed even to share Saint Augustine's view of it as a great cosmic harmony.[18]

Yet, despite his basically anthropocentric view of the natural world, Luther never denied man's solidarity with nature. God was not wholly separate from nature: he was immanent in it, though usually hidden.[19] He agreed with Augustine that nature was full of wonders, indeed miracles, and that, come the consummation of the world, it would be transformed. Not only would humans be resurrected bodily, but "Then there will be also a new heaven and earth, the light of the moon will be as the light of the sun, and the light of the sun will be sevenfold. . . . That will be a broad and beautiful heaven and a joyful earth, much more beautiful than Paradise was." Although Luther shared many of the motifs of medieval Catholic theology with regard to nature, albeit in somewhat attenuated form, his emphasis was elsewhere: "Rising above nature in order to enter into communion with God became a hallmark of Protestant thought."[20]

John Calvin (1509–1564) shared many of Luther's ideas, but strongly stressed the notion that man's domination of the earth was not simply a fact of God's goodness, but in a sense a vocation. One had a duty to God to take the words of Genesis quite literally and extend one's control over nature. God is lord over creation, but also lord over men. His will is the most important moral agent in the universe, and what he wills for man is to subdue the earth in his name. The conquest of the world outside was, of course, a corollary to the conquest of the world inside. As a modern-day observer puts it, "When one contemplates the conquest of nature by technology one must remember that the conquest had to include our own bodies. Calvinism provided the determined and organized men and women who could rule the mastered world."[21]

The ultimate result of the teachings of Luther and especially of Calvin was a turning away from nature as a source of value in the Protestant world, or what one contemporary Lutheran theologian styles "the secularization of nature in the Reformation tradition in the nineteenth and twentieth centuries." As he says, "Nature came to be viewed . . . as a mere thing, a world of objects, closed in upon itself, moved only by its own laws, not open to any other dimensions of reality, and therefore a world which humans must transcend if they are to be rightly related to God."[22]

Renaissance Ideas about Technology

We have explored how Western ideas about nature were changed during the Renaissance by a new interest in science and by the thinking of the Reformation theologians and their emphasis on the relations of man to God rather than man to nature. To what extent and how were the ideas of modern man vis-à-vis technology also changing?

The Middle Ages were a time of great interest in technology and of great technological progress. But the men of the era were always concerned exclusively with its immediate practical aspects and had little interest in its history or theoretical implications.[23] With the advent of the Renaissance many things changed. The revival of ancient philosophical and literary texts with which the Renaissance is usually associated was accompanied by a spate of translations of classical texts in the various fields of technology,[24] and of a revival of the work of the Roman architect Vitruvius (late 1st century B.C.–early 1st century A.D.).[25] Perhaps even more important was the general humanist spirit of the Renaissance itself, with its interest in the world and its belief in the power of man to control nature through human agency.[26] During the fifteenth century the intellectuals of the West were becoming more interested in the mechanical arts through the stimulus of the Renaissance.[27]

This new attitude toward the mechanical arts bespoke a new attitude toward not only technology as such but toward social relations as well, presaging the construction of new hierarchies in which technology and work were king. Works as disparate as Bacon's *The New Atlantis*, which reflected on technocratic rulers, and Campanella's *City of the Sun*, which "exalted [labor] as a central and decisive element in the formation of man,"[28] illustrate this point. Given this change in human attitudes, it is little wonder that new inventions and new practices began to sweep through the advanced nations of Europe. Yet during this era the medieval split between technology and science continued in full force, despite the growing prestige of technology. Leonardo da Vinci (1452–1519), the great multivalent Renaissance artistic and technological genius, insisted he was a "man without letters." Physicians continued to maintain their distance from surgeons, since the latter worked with their hands.[29] Until the eighteenth century the contributions of science to technology were minor.

The predominant technology during the Renaissance in terms of its economic importance was mining. Mining had always been associated with the holy, in part because the earth was considered sacred to certain deities. Until the end of the Middle Ages, religious ceremonies accompanied new mine openings even in Christian Europe.[30] The ancient Greeks and Romans had had mines, but were not obsessed with taking minerals from the earth. As a noted economic historian has put it, they "showed little disposition to *ransack* the subsoil for these underground riches."[31] Mining began to decline in the Roman Empire in the third century, and was not revived until the tenth. Medieval miners had relatively high status and up until the thirteenth century were free and largely self-governing.[32] This idyllic social system did not last, however. After the Peasants' War of 1523, the miners' guilds were abolished and a period of labor struggle between technically free miners and capitalist owners—including joint stock companies—began.[33]

As we saw earlier, the great complaint against the mines was their defilement of nature. The mines required fuel for various operations involved in processing ore. The only available fuel was wood, and much of the deforestation of early modern Europe was the result of increased mining activity, spurred on by the needs of war. Agricola was at great pains to defend the industry against these charges. Some of his arguments were specific and resemble contemporary arguments about the effects of mining, for example, that land used for mining is later good for crops. On the philosophical level, the argument is that metals are necessary for a prosperous and humane civilization.[34]

Although German miners may originally have been influenced by the quasi-mystical natural philosophy of the early Renaissance, as the mines evolved and became larger and deeper, mining became well suited to the new mechanical philosophy and its later extrapolations. According to Mumford, *"The mine is nothing less than the concrete model of the conceptual world which was built up by the physicists of the seventeenth century."*[35]

The Triumph of the Experimental Philosophy

We have seen how the philosophy of Descartes and his followers distinguished sharply between man and nature, and laid the groundwork for what he called the need for men to be "masters and possessors" of nature. But this philosophy would, in short order, be superseded by that of Bacon, whose aspiration to create a scientific outlook based on experiment would fulfill Descartes' hopes in the practical realm. As one scholar has commented, "In seventeenth-century philosophy the concept of mastery over nature has achieved its definitive modern form, the one which has remained authoritative and substantially unaltered down to the present day. The rough path marked out by Bacon quickly became a well-travelled road."[36]

The first major scientist to take this road was Sir Isaac Newton (1642–1727), a Cambridge mathematician and physicist (or natural philosopher, as physicists were still called in his day). Newton, among other things, discovered the law of gravitation and began the development of the calculus. He is considered by many to have been the greatest scientist of all time, and his ideas revolutionized the worldview of the West. Newton was a Christian, but his scheme, by making nature a mere thing (unlike the numinous universe of many Renaissance philosophers), in time led to the rejection of Christianity by many. A monarchical God ruled this dead world from outside, but for Newton the "physical phenomena themselves were not thought to be divine in any sense, and when science made it more and more difficult to believe in such a god, the divine disappeared completely from the scientific world view."[37]

As noted earlier, Newton was a serious student of alchemy. He devoted much research to the alchemical tradition, so much so that the famous modern economist John Maynard Keynes, after an extended study of Newton's work in this area, could say that "Cambridge's greatest son" was "not the first of the age of reason," but "the last of the magicians." Though Newton was a one-time believer in witchcraft, by life's end he completely denied the possibility of Satanic intervention in the affairs of this world.[38]

Perhaps more than anyone, Newton was responsible for the new and generally accepted image of the world as a great machine, created by God and then left to go about its business according to his divine laws. God was viewed by many as in effect a great clockmaker, who had wound up the world and was now sitting back to observe his handiwork.

For Newton's followers the emphasis was always on the predictability of nature and its consequent susceptibility to human control. Thus a fellow member of the Royal Society, chemist Robert Boyle, spoke of nature as God's "great... pregnant automaton."[39] Such a nature need not be given any special consideration. In his *Inquiry into the Vulgarly Received Notion of Nature*, Boyle holds that "this veneration, wherewith men are imbued for what they call nature, has been a discouraging impediment to the empire of man over the inferior creatures of God," and that as long as men "look upon her as such a venerable thing, some make a kind of scruple of conscience to endeavor so to emulate any of her works, as to excel them."[40]

When members of the Royal Society spoke of the power over nature the new experimental philosophy would provide men, they meant *men* literally and not women, whom they regarded as inferior and incapable of science. Society secretary Henry Oldenberg, writing in his preface to a book by Boyle, said that the society's members were men whose business it was "to raise a Masculine Philosophy." Joseph Glenvill, a fellow of the Society, held that although it was Eve who first gave in to the wiles of the Serpent, "Our *masculine powers* are deeply sharers of the consequential mischief; and for that reason the *Woman* in us still prosecutes a deceit."[41]

Major dissent against the desiccated worldview of experimental philosophy came from theologically oriented scientists and thinkers who fought back with an ancient argument for the existence of God—the argument from design—in new scientific guise. Not only was common sense appealed to as such, but a whole new school of thought called physico-theology arose. It described the wonders of the natural order in detail, daring anyone to believe that something so wonderful could have come about through mindless, accidental causes, as was implied or expressed by most of the natural philosophers. A major figure in this school was John Ray (1627–1705), possibly the greatest naturalist of the seventeenth century and a pioneer in systematic biology. His major polemical work, *The Wisdom of God Manifested in the Works of Creation* (1691), asserts that God created the earth and everything on it all at once, and calls on us to marvel at how wonderfully interrelated it all is. Ray's friend, the Reverend William Derham, published a book called *Physico-Theology* (1713), which is in many ways more elaborate and is almost modern in its discussion of such topics as the food chain, the interdependence of creatures, and the relation of the various physical features of the earth to its living inhabitants. Regarding physico-theology, a contemporary scholar goes so far as to say that "I am convinced that modern ecological theory . . . owes its origin to the design argument. The wisdom of the Creator is self-evident, everything in the creation is interrelated, no living thing is useless, and all are related to the other."[42]

But no longer did people agree on the premise of physico-theology that God had created the universe for man. This idea itself became a matter of debate among the leading philosophers of the seventeenth century, most notably Benedict Spinoza (1632–1677) and Gottfried Wilhelm Leibniz (1646–1716).[43] Spinoza is of special interest to us because he has been hailed as an early patron saint by contemporary thinkers who deplore the split between man and nature that the modern world has inherited from Descartes, Bacon, and their followers.[44]

Unlike Descartes and Bacon, Spinoza did not think of God as outside the universe in the act of creation. God is an *immanent* cause. In that sense all nature is divine, albeit possibly "material" rather than "spiritual." For a Spinozan, since dualism is rejected at the outset, the distinction between the material and the spiritual does not, indeed cannot, exist. Since man is in nature and is part of it, this means that man shares in the divine substance, and it follows that "the human mind is part of the infinite intellect of God."[45] Since we are part of nature we can directly apprehend it. "Clear knowledge of total nature comes to be in us, 'not through conviction of reason but through a feeling, and enjoying of the thing itself.' "[46] Thus despite Spinoza's rationalism, intuition is vindicated. His views are decidedly nonanthropocentric[47] and also serve to create a bridge to the thought of the Orient, as we shall see.

What of Leibniz? Lovejoy notes that he argeed with Spinoza in maintaining that " 'we find in the world things that are not pleasing to us,' since 'we know that it was not made for us alone.' "[48] But for Leibniz, unlike Spinoza, final causes do exist, although we may not always be able to discern them, and belief in them is important in scientific work. The earth is not degenerating, as is widely believed, but improving, showing God's providential goodness.[49] What appears to be disharmony in nature, such as volcanic eruptions, is really an illusion, stemming from the inability of our senses to perceive the harmony that really exists.[50]

A supreme optimist, Leibniz was caricatured as such in Voltaire's *Candide* for his belief that all things natural and societal were constantly improving. Because of God's preestablished harmony, nature responded to human activity as man cultivated it, refined it, and cared for it. The progress of mankind was inevitable, and therefore, he argued, so was the progress of the physical universe under human tutelage.[51]

It was not until the next century that philosophy finally made the crucial break with *all* attempts to reconcile God and nature. The work of Immanuel

Kant (1724–1804) marks a great turning point in Western thought. Until that time there were constant quarrels about exactly how God and the natural world were related, and how one might learn something from nature about God and his laws. But it was assumed that they *were* related. The great English skeptic David Hume (1711–1776) had done his best to undermine such attempts by his arguments against causality, and hence against all concepts of natural law. Although Hume had some followers, the mainstream of Western thought balked at the implications of his views, which denied all philosophical validity to claims about morals. A middle ground was needed and, in effect, supplied by the basic teaching of Kant that there was a radical disjunction between two worlds. The first is the world of "phenomena," the empirical world that can be apprehended by our senses, and is subject to the laws of science. The second is the world of "noumena," of ideas that are directly apprehended by the mind, such as extension, space, duration, and other basic and necessary categories of thought. It is in this latter world that our ideas of duty, good, and other moral categories exist.[52] This world is subject to the laws of reason. In Kant's system, which became dominant not only in his native Germany but in most of the Western world and still has its adherents, there can be no connection between these two worlds. Thus both science and moral philosophy are saved, but at the cost of denying the latter any of the "objective" certainty that was soon felt to be the exclusive quality of scientific activity. Since individuals could differ about what one could conclude through reason, they eventually came to believe that while one could agree, on intellectually compelling grounds, about certain observations and theories of science, questions of moral judgment were simply matters of individual opinion. Science was objective, philosophy was not.

But nature was not Kant's main interest; he was primarily interested in human conduct. In the field of ethics, he is known in large measure for his principle of universality, which declares that we should never do anything that we would object to anyone else doing. Thus, even though he divorces humanity from nature, modern environmentalists can heartily accept at least one element in his system, as it applies to pollution. He also raised another issue that is key to modern environmental thinking. According to one critic, in Kant's philosophy "the idea of a duty to posterity assumes, perhaps, for the first time, a central place."[53] This is perhaps an overstatement, since it ignores certain nuances of the Christian doctrine of stewardship which, while stressing the primary duty to God the Creator, implies that we also have a duty to future generations to preserve our world. But Kant certainly introduced this idea to the mainstream of secular moral philosophy.

During the early modern era, while some philosophers were radically changing how Western man thought about himself and his place in the world, others were effecting political change in human culture. The most important of these was the Englishman Thomas Hobbes (1586–1679), who, along with Bacon, was originator of the modern synthesis.[54] Hobbes claimed to be the true founder of political science, giving a scientific basis to what had simply been loose speculation. But what was Hobbes's model of science? Well into his life, he was introduced to Euclid and underwent a conversion experience. Geometry, he said, was the first science God had given to man. This was so because it was completely self-contained. Once the axioms were defined and understood, there was no question concerning the conclusions; truth was guaranteed. In his major work *Leviathan* (1651), as in his other works, Hobbes begins by defining his terms carefully. Once you accept his definitions, as in the case of Euclid's *Elements*, you cannot escape from his system.

We must note carefully what is being said by Hobbes about science and all human knowledge. All we can really know is that which we have made ourselves.[55] This truth applies not only to government but to the world as a whole. Before this time, the mainstream of human thought had always assumed that there was a world "out there," independent of human knowledge, and the role of the thinker was to understand it. After Hobbes, this was no longer so. For Bacon, knowledge was the basis of power; for Hobbes, knowledge and power are still the same thing, but now power is knowledge. The world no longer exists "out there," to be discovered; it now exists "in here," in human ideas, to be made. In such a universe, nature can have no rights against technological manipulation because it does not exist except as we know it through technological control.

Hobbes has no teachings about what we today would look upon as environmental or ecological issues. Like most men of his era, he foresaw no basic problems in feeding humanity and was not concerned about overpopulation or even about pollution. At the same time, however, his gloomy view of human nature and politics—which some have compared to that of Augustine—assumes conflict, and conflict implies scarcity of some mutually desired end. As a result, some present-day ecological/political theorists have looked upon him as their predecessor. They argue that only through a strong sovereign power like the one Hobbes envisaged, which can as easily be a congress as a king, can human beings be made to deal collectively with what these theorists see as the overarching social conflicts created by overpopulation, resource depletion, and pollution.[56]

Among Hobbes's contemporaries were members of the extreme Puritan party, who wished to abolish not only the monarchy, but also hallowed social institutions such as privately held land. These groups, the Levellers and the even more radical Diggers, were utopians who relied mostly on the Bible for guidance.[57]

Preeminent among the Puritans was Gerrard Winstanley (1609–1676). He and his followers advocated the abolition of organized religion and favored complete political and economic equality among all men. Especially interesting are his ideas on nature and technology. His primary concern was with justice, and his views reflect that concern. He believed that in the beginning, man had lived according to reason, but somehow he fell; original sin as usually understood played no real role in his system. This fall affected all of nature, corrupting the earth and causing it to "bring forth poysonous Vipers, Todes and Serpents, and Thornes and Bryars." But nature too will be reborn in the coming Golden Age, the millenarian hope is for all of creation, and nature, too, is "groaning under bondage, waiting for a restauration."[58] But even before the complete millennium, when true government would finally be restored and the Commonwealth comes into being, nature will serve all men. But this service would not come about in the way Bacon imagined it, that is, under coercion. According to one commentator, in Winstanley's ideas, "Nowhere is found the desire that mankind will take possession of the earth as a means of displaying man's power and greatness. . . . Winstanley seeks to realize a beautiful world in which great emphasis is placed on tenderness and gentleness."[59] Winstanley, says another, "proposes an ethic about our relationship with nature which should not be violated by technological progress."[60]

The Diggers flourished for only a few years, primarily during 1649 and 1650, and were soon scattered, becoming a footnote to history. But the period of the British Civil War, which ultimately ended with the Glorious Revolution of 1688, became the backdrop for the work of John Locke (1632–1716), the great theorist of parliamentary supremacy as well as of modern capitalist individualism.[61]

Philosophers generally agree that Locke differed from Hobbes about the "state of nature," the hypothetical condition of human beings before the existence of organized society or government. In this condition humans were free; they left it by creating a sovereign. Locke is usually interpreted as holding that this prepolitical condition was tolerable if inconvenient.[62] Thus the obligation to obey the sovereign created to escape it was not as binding as it was for Hobbes, to whom returning to a condition that was horrible, in his view, was the greatest possible social evil. One modern political theorist who is especially concerned with ecological issues holds that Hobbes was a believer in strong government because he was above all impressed by the existence of scarcity. Locke, on the other hand, was not. "It was precisely this that allowed John Locke, whose political argument is essentially the same as that of Hobbes in every particular except scarcity in the state of nature, to be basically libertarian where Hobbes is basically authoritarian." Locke, he maintains, was one of the philosophers of the "Great Frontier";[63] his "justification of original property and the natural right of man to appropriate it from nature thus rests on cornucopian assumptions: there is always more left; society can therefore be libertarian."[64]

Locke's economic views are as important as his political teachings, and they are closely related. Above all, he assumed that man is an acquisitive animal; he is an "infinite consumer." But, as one critic points out, "If man's desires are infinite, the purpose of man must be an endless attempt to overcome scarcity."[65] Even if one assumes, as Locke did, that there is plenty to go around, there still can never be enough. For Hobbes, human happiness consisted in the ceaseless quest for power after power. For Locke it was the ceaseless quest for wealth after wealth. In Locke's time, wealth meant land, and it was the duty of the new parliamentary government, indeed of any government, to aid the landowner. "The increase of lands and of the right employing of them is the greatest art of government," he tells us.[66] Note the word "increase": it is not static but dynamic property that Locke wants the state to champion.[67] Locke is above all the political and social philosopher of the modern liberal capitalist state and its aim to create ever more goods of consumption through the conquest of nature: "The central theme of Locke's whole political teaching . . . is . . . increase."[68]

Anyone who increased his share of wealth by acquiring more land performed a service for all. According to Locke,

> He who appropriates land to himself by his labor does not lessen but increase the common support of mankind. For the provisions serving to the support of human life produced by one acre of enclosed land are . . . ten times more than those which are yielded by an acre of land of an equal richness lying waste in common. And therefore he that encloses land, and has a greater plenty of the conveniences of life from ten acres than he could have from a hundred acres left to nature, may truly be said to give ninety acres to mankind.[69]

So much does this "enclosure" of nature increase wealth that even the natural law prohibiting waste is no longer binding.[70] Note Locke's imperative not to leave things to nature: *that* is the true waste, as he explicitly states, for "land that is left wholly to nature . . . is called, as indeed it is, waste."[71] Man is now no longer part of nature, he is "effectively emancipated from the bonds of nature" and he no longer has to follow it. Indeed, "the negation of nature is the way to happiness."[72]

It is of course through technology that one achieves this domination of nature. It is the technology of the agricultural proprietor that makes it possible for an English day laborer to live so much better than an Indian chieftain in America, in Locke's view, and makes land in human hands so much more valuable than land left to the waste of untouched nature.[73]

But the Baconian model, the Enlightenment, and their creed of progress through scientific dominion over nature were not exclusive, even in the eighteenth century. A backlash was already emerging in the form that largely triumphed, at least in nonmaterial culture, in the nineteenth century as the Romantic movement. As we shall see, the great paradox of this century was that, at the very time that industrial capitalism, now fully armed by science and technology, was taking over the material world through the industrial revolution, Europeans and Americans were becoming more aware of the beauties and importance of the very nature they were conquering and destroying. They were now also attempting to restore nature to its old privileged status as an ideal and as a physical reality as well.

Chapter Four

TECHNOLOGY TRIUMPHS: THE INDUSTRIAL REVOLUTION

Are the physical ways in which human beings live, including their use of technology, the products of their cultures and ideas? Or are their cultures and ideas the products of their physical life? Whether technologies or ideas are ultimately more powerful, the influence they have on each other is reciprocal and occurs not all at once but over a period of time. We have been looking at some aspects of the culture of early modernity. Now we will look at the material changes that were taking place while this culture was developing, and we must remember all the while that there is a dialectic between technology and ideas.

The Western world had long been poised for the leap into modernity. This was especially true in England, which had become the undisputed leader in industrialization.[1] A new culture was arising—an industrial culture that "stresses sobriety, paternal wisdom, caution, thrift, and the postponement of sensuous gratification in order to achieve long-term financial prosperity—all amounting to the familiar work ethic."[2]

A leading role in industrial culture was played by the inventor. Inventors had always existed, but now they had a special new social role. The inventor, if he was successful, owed his fame and fortune to innovations in technology. Yet he was not moved by fame and fortune alone. Many of these men were motivated by a genuine desire to make life easier and more prosperous, as well as by an interest in technology for its own sake.[3] They received the active backing of the state, not only in Protestant countries such as England, but in Catholic nations as well.[4] As Bacon had foreseen, in the early years of industrialization an important role was played by large extended families that pooled their capital, acting much like modern corporations.[5] On the continent

47

as well as in England, there was a great interest in technology among the educated long before the industrial revolution. Diderot's *Encyclopedie* (1751– 1780) contained eleven volumes of plates depicting the various technologies already in use in Europe on the eve of what is conventionally thought of as the industrial revolution of the nineteenth century.[6]

One can say that the industrial revolution did not really begin in the early nineteenth century, but had its origins much earlier. The distinguished economic historian John U. Nef has traced the origins of the industrial revolution to the sixteenth century.[7] It depended, in turn, on technological achievements going back to the Renaissance: "The printing press; the blast furnace; the furnace for separating silver from . . . copper . . . ; boring rods for discovering the nature of underground strata, etc."[8] Much was also owed to the contributions of immigrants; skilled craftsmen were a major form of technological transfer among nations. Countries wishing to progress technologically courted the skilled, while the lands of their birth often forbade them to leave. Skilled people thus carried with them the intrinsic value held by patents today.

What other factors were at play in the earliest industrial revolution? Nef believes that science played a scant role, though the early scientific and technological revolutions "each touched the other."[9] What of government? During the century after the Reformation, the English crown made numerous attempts to control manufacturing. In 1559 Queen Elizabeth I refused to grant a patent for a knitting machine that would have caused unemployment among her subjects. Much later Charles II gave a charter to the hoosiers' guild, protecting them against mechanized competition until the nineteenth century.[10] There were attempts to regulate mining as well, but ultimately most attempts by the state to check the complete freedom of rising industrial capitalists were rendered ineffectual.

The role of the steam engine in the industrial revolution is debatable, though conventional wisdom has made it central. Such an engine was already conceived and probably built by Hero of Alexandria in Hellenic Egypt.[11] Translations of his works, published in Europe in the late sixteenth century, ignited new interest in the idea. But whereas in ancient civilization his engine was only a toy, given the lack of interest in labor-saving devices in a slave society, eighteenth-century England was ready for more. Mining was central to the later industrial revolution, and deeper and deeper shafts meant greater and greater problems with water removal.[12] Thomas Newcomen invented a new kind of pump in 1712, but it remained to James Watt (1736–1819) to patent the prototype of the modern steam engine in 1769. From a new machine developed initially for the mines arose a technology that would soon transform all kinds of operations, not only in the manufacturing industry but in transportation as well.

Leaving the socioeconomic effects of the steam engine aside, a controversy still rages about how much Watt's new engine owed to technological tinkering and how much to theoretical science. The noted twentieth-century American chemist James B. Conant once remarked that "science owes more to the Steam-engine than the Steam-engine owes to science."[13] The whole matter is part of the larger question of the relation of science to technology in the industrial revolution. Landes has argued that "in spite of some efforts to tie the Industrial Revolution to the Scientific Revolution of the sixteenth and seventeenth centuries, the link seems to have been an extremely diffuse one."[14]

Why did the industrial revolution of the modern era come into being? Why first in England? Certainly there was new interest in technology, as attested by its adoption as a new concept. In ancient times and throughout the Middle Ages, *techne* had meant any skill. But not until the late seventeenth and early eighteenth centuries was the word "technology" used in the modern sense. Christian Wolff (1679–1754) defined it as "the science of the arts and the works of art," with art meaning the mechanical arts. Jacob Bigelow, in his *Elements of Technology* (1829), identifies technology with the practical application of science.[15] But this new interest needed an area of activity in which to be expressed. It was above all in Britain where the conditions— primarily economic, but social, cultural, and political as well—were ripe for a modern-day industrial revolution.[16]

A leading contemporary authority accepts the conventional concept of the nature and scope of the industrial revolution and says that the words, when capitalized, refer to the "first historical instance of the breakthrough from an agrarian, handicraft economy to one dominated by industry and machine manufacture."[17] Many have held that the industrial revolution began in England because the Commercial Revolution had already given it an economic head start with profits from overseas trade, which could be diverted into domestic economic activities.[18] But British leadership in the industrial revolution was not due primarily to economic or even technological factors. One economic historian takes this position: "[That] the Industrial Revolution was essentially and primarily a sociocultural phenomenon and not purely a technological one becomes patently obvious when one notices that the first countries to industrialize were those which had the greatest cultural and social similarities with England."[19] What characteristics of British society might have been salient for industrialization?

Above all, English society was fluid,[20] making it possible for knowledge to flow from the realm of technology into that of science. Britain's nobility was unlike that on the continent; British income and wealth were much more evenly distributed. Britain was unlike either France, where commercial enterprise meant the end of noble status, or Germany, where strict laws kept social groups apart.[21]

Religion may have played a role. The historian David Landes takes Max Weber's view that there was a definite positive link between Protestantism and business success, and he attributes much of England's priority in modern industrialism to this factor. John U. Nef, on the other hand, suggests a somewhat more complex relationship, recalling that the post-Reformation confiscation of church lands went furthest in England. Many of these lands contained iron and coal, and the new owners, mostly landed gentry who were close to the new Protestant monarchs, were only too eager to exploit them.[22]

This development was significant. The exploitation of iron and coal was the key to Britain's early preeminence in the industrial revolution, and in turn led to a whole new approach to industry, that is, the increasingly widespread use of mass production. The basis of mass production was the creation of tools and machines to make other machines. A major impetus came from America, where the relative scarcity of skilled labor and a larger market based on even greater economic equality made mass production more economically necessary and culturally palatable than in Britain. The "American System," as it came to be called in the mid-nineteenth century, was described as "a method of manufacture in which complex mechanical devices were produced in a series of sequential machine operations" involving the creation and use of interchangeable parts. It soon spread to Britain.[23]

But it was in England that the manufacturing system was most welcomed and lauded. Dr. Andrew Ure (1779–1857) wrote *Philosophy of Manufactures*, which became widely popular because it extolled the division of labor first idealized by Adam Smith (1713–1790) in *The Wealth of Nations* (1776).[24] Ure lavished praise on the assiduous skill that workers needed to tend the new machines, but he failed to call attention to the fact that a cardinal principle of the new system was to build "skill" into machines, so that workers needed little themselves.[25]

Now that workers needed less skill, the supply of labor was greatly enlarged, which was of course one of the industrial revolution's aims. By the 1830s, for example, between one-third to one-half of the workers in English cotton mills were under age twenty-one, and much more than half were women. Wages were abominably low, forcing women and children to work to help support their families, and the hours were dreadful by any standards. Working days for children in the 1830s could last as long as eighteen hours. The results of such a system were inevitable. Not only were terrible problems created for the health of the workers, especially the women and children, but the family itself was shattered.[26] Equally important, a class cleavage of monumental proportions was instituted.

At the same time that ordinary workers were being forced into even deeper poverty and loss of social status, a new class was being created. The bourgeoisie had long existed, but the technical requirements of the industrial system were now bringing a new group to the fore. In 1812 a British editor noted that about fifty years earlier, "the general situation of things gave rise to a new profession and order of men, called Civil Engineers." John Smeaton (1724–1792) is said to have been the first person to call himself a civil engineer, and the Society of Civil Engineers was founded in 1751.[27] Centuries earlier, Bacon had envisioned a world run by technocrats. The raw material for this new breed was now appearing.

In time, the industrial revolution spread from Britain to the continent. The Germans moved especially fast to catch up, going Britain one further by pioneering the systematic merger of science and technology—the hallmark of contemporary industrial society. Industrial society, the prime expression of the use of technology to control nature, was conquering the world.

The industrial revolution began in England for another important reason. England had embraced the basic premises of capitalism, that (1) the search for private economic gain is the major motivating factor in man; (2) this is a good thing in itself; and (3) economic freedom to seek private profit is the basis of all freedom. The corollaries are that the search for profit must be allowed free reign regardless of the possible consequences to nature, and the primary function of technology is to serve the private profit of those who can create and control it.

The role of religion in this transition is a matter of dispute.[28] Ironically, capitalist ideology triumphed in large measure because people came to believe that "laissez-faire" was in accord with nature, or at least the nature of man and society. The groundwork for this belief was laid by the physiocrats, who also invented the term.[29] Largely French, this first school of economists taught that all human wealth came from land, and that a natural order would prevail if only government would leave the economy alone and replace the existing mazes of taxes with a single tax on the produce of land.

Adam Smith (1723–1790), founder of capitalist ideology, largely agreed. But he could not accept the basic idea that all wealth came from land and business was simply parasitic on agriculture. For Smith, business too was productive and should be left alone by government. His real interest was in the wonders of the market system itself, this natural order which, if left alone, would bring prosperity to humanity.

Though the proponents of capitalism argued, like Smith, that it was the only system in accord with the nature of society, society did not take to it easily. Paradoxically, a powerful state was necessary throughout Europe to institute the new system. As one scholar has noted, "There was nothing natural about *laissez- faire.* . . . The road to the free market was opened and kept open by an enormous increase in continuous, centrally organized and controlled interventionism."[30] Free-market liberal capitalism was ultimately, then, the creation of politics.

Foremost among those objecting to the new order were the workers. Labor warned that it would become simply another factor in production, just another commodity. But labor, unlike corn and steel, was defined primarily by human concerns and abilities. Unlike other commodities, it could protest and organize and seek government intervention in the market system on its behalf. Protests against laissez-faire became increasingly prevalent as industrialism progressed.

If England was home to the industrial revolution, it was also home to the modern proletariat. The condition of these workers was rooted in the Enclosures which, beginning in the fourteenth century, drove entire families off the land, reducing them to wanderers and beggars and eventually laborers in the new factories.

During the period 1811 to 1816, workers in the English textile industry, primarily weavers, fought back. In 1809, Parliament had repealed all protective legislation in the woolen industry. As a result, in 1811 masked men, operating at night for several months, smashed newly introduced machines throughout the textile region. They sent a letter to Parliament, claiming that their actions were legal under their charter. The declaration was signed "Ned Ludd, Sherwood Forest." Ludd was a mythical character, but the aliases "King Ludd" or "General Ludd" were widely used by labor leaders protesting new machinery. From this beginning the Luddite movement was born.[31] Much debate ensued, and even Lord Byron defended them in the House of Lords. In 1816 the Luddites reappeared in riots in Nottingham and unrest soon spread throughout the textile centers of England. But these outbursts were effectively put down by Parliament, which sent twelve thousand soldiers to suppress them. The movement's ringleaders were executed or transported to Australia, and the movement died out. Similar riots and violence took place in France and Germany in the next decades but achieved few results; indeed, the word "sabotage" stems from this period, referring to the French workers who used their wooden shoes (*sabots*) to kick in machines. Still, regardless of the rigor of worker resistance, capitalism dictated the use of whatever machines were necessary to save labor and money.

Industrialized War

Less contested were the benefits of technology as they applied to the pursuit of war. Some critics, hostile to both war and the modern industrial state, have argued that the military is the paradigm of modern industrial capitalism. Not only has war "been perhaps the chief propagator of the machine," according to Lewis Mumford, but *"the army is in fact the ideal form toward which a purely mechanical system of industry must lead."*[32] Certainly every effort has been made in the modern era to turn the military services into an organized, rationalized, disciplined body capable of being directed toward a predetermined end. The institution of drill to effect discipline is a familiar part of this process. Drill was originally used by the Romans, but it was reintroduced in the sixteenth century by Prince Maurice of Orange and Nassau.[33] Uniforms came soon after, and in time had the effect of necessarily stimulating mass production and thus large- scale industry. Communications helped to make armies a coordinated machine as well. Human and animal messengers and crude semaphore devices were the major means of communicating until the 1840s, when the electromagnetic telegraph made it possible for a central command to keep in touch with and control a large army in the field.[34]

Firearms took some time to become established as a major weapon of war. During the Hundred Years' War (1337–1453), arrows remained the weapon of choice. Firearms were of no major use until the last quarter of the fifteenth century.[35] But the cannon began gradually to transform the battlefield. It was first used, ineffectively, by the French in 1494–1495. Sir Walter Raleigh (1554?–1618) felt that cannons were no great improvement over the weapons of the classical era. In defending Syracuse against the Romans, Archimedes, according to Raleigh, "had framed such engines of war as . . . did more mischief than could have been wrought by the cannon or any instruments of gunpowder, had they in that age been known."[36] But by the eighteenth century it was a different story. Every army of Europe acknowledged the superiority of artillery on the battlefield. When firearms first came into use, it was a common complaint that even a commoner, a relative coward, could strike down the bravest, most gallant knight. Artillery, which killed impersonally at a distance, was even worse. But even the likes of Frederick the Great of Prussia, who had downplayed artillery as being contrary to the real military spirit, learned through his decisive defeat by Napoleon at Jena that the old ways had met their match in the new industrialized science of war.[37]

By the late nineteenth century the revolution in warfare was complete. Technological change followed social imperatives. Anglo-German naval rivalry in the decades prior to World War I illustrates this pattern: "Within limits, tactical and strategic planning began to shape warships rather than the other way around."[38]

Throughout modern history, war and industrial civilization have evolved together, and the industrialization of war has had profound effects on war and society alike. The same technological forces that made it possible for people in the advanced industrial nations to become materially prosperous compared to those in simpler societies also made it possible—often mandatory—for everyone to participate in war. The same mass production that was to generate a cornucopia of goods for the civilian sector was in fact pioneered by the military. Eli Whitney was the first to inaugurate this system through the wholesale manufacture of rifles, made possible by a certain military market.

The first fully industrialized war was the American Civil War. The Civil War, for example, was the first war in which the railroad played a major military role. The Europeans, however, though they sent military observers galore, learned little from this war, calling the American armies "unprofessional."

But if modern war and industrialization came into existence in tandem, which had the greater effect on the other? Lewis Mumford, an opponent of war and a skeptic about the benefits of industrialism, gives priority to war, finding in war and preparations for war prologues and analogues of all that he dislikes about modern society. Historian John Nef abhors war and much about modern industrial society, yet he denies the thesis of German sociologist Werner Sombart (1863–1941) that the demands of war were a major stimulus to the growth of modern capitalism.[39] Nef argues that the first large factories were not built to serve military needs, nor did war inevitably lead to industrial concentration,[40] and that "the role of war in promoting industrial capitalism has been small compared with the role of industrial progress in bringing on war."[41]

The Revival of Interest in Nature

Just when nature seemed to be falling completely under man's dominion, Westerners began to appreciate it more. Along with the technological conquests of nature exemplified in industrialism came a new willingness to accept nature's autonomy and downplay the anthropocentric emphasis on human superiority that underlay Western man's assaults. Having more or less succeeded in taming the natural world, people now wanted to reexamine it.

The Enlightenment glorified man and human reason, and in the course of doing so, reinforced the dignity of all humanity. But it also extended aspects of this egalitarianism toward the natural world. The famous French essayist, the Baron Michel de Montaigne (1533–1592), wrote that "there is a certain consideration, and a general duty to humanity, that binds us not only to the animals, which have life and feeling, but even to the trees and plants. We owe justice to man, and kindness and benevolence to all other creatures who may be susceptible to it. There is some intercourse between them and us, and some mutual obligation."[42]

It would be some time before these sentiments of human sympathy and obligation toward the natural world became widely accepted. But as early as the mid-sixteenth century, John Bradford, a Protestant martyr in the reign of Queen Mary, explicitly rejected the traditional medieval scholastic teaching that animals existed solely for the use of man. In later decades it became common to say that the world had been created for the glory of God and that he cared as much for the humblest of creatures as for man. Cambridge Platonist Henry More argued that creatures were made not simply to serve man but to enjoy themselves. It was "haughty presumption" to think that this was not the case.[43] Increasingly, various scientists—astronomers, zoologists, botanists—began to suggest that man was not the center of creation as was once believed. Anthropocentrism—a worldview that places man at the center of the created universe—became less widespread among the educated. Nature was now considered interesting— and valuable—for its own sake.

It took a long while, however, even among the English, for animals to be respected. In 1612 a preacher explicitly rejected the "too, too pitiful" philosophy of Protagoras, who had objected to the mistreatment of animals. Those who killed animals simply for pleasure, as was the case in bear baiting and cock fighting, could claim that "Christianity gives us a placard to use these sports."[44] The passages in the Old Testament that emphasized stewardship toward God's creatures were rationalized away by virtually all Christian theologians. But people were beginning to keep animals as pets, and in time began to see in their animal pets traits they spoke of in human terms, such as wisdom and fidelity.

Animals were not the only subjects of concern. Europeans were becoming increasingly worried about the future of the woodlands, fearing the loss of this valuable natural resource. The modern conservation movement, with its internal divisions between those who are apprehensive about the potential loss of economic resources and those who value wilderness for its own sake, was not yet born. Although there was as yet no concern with wilderness, forest preservation was a pressing issue. Jean Baptiste Colbert, the economic adviser to Louis XIV of France, convinced him to proclaim the French Forest Ordinance of 1669, regarded as a landmark in the history of European forestry. It organized and collated a vast amount of existing legislation and introduced new principles as well.[45] Now, for the first time, government was systematically engaged in protecting a nation's natural resources.

The most influential eighteenth-century naturalist thinker was the famed Frenchman, Count George Buffon (1707–1788). Buffon was interested in the interaction between man and his physical environment, and how each had affected and altered the other.[46] Buffon believed that the most important event in human history had been the domestication of the dog. He said that nature altered by man was superior to an unaltered nature, and he introduced the all-important concept that "the state in which we see nature today is as much our work as it is hers. We have learned to temper her, to modify her, to fit her to our needs and our desires. We have made, cultivated, fertilized the earth; its appearance, as we see it today, is thus quite different than it was in times prior to the invention of the arts."[47] If this idea is valid, nature no longer exists independently of humans. When we look at nature it seems to say we are looking at a mirror of our past selves.

Buffon was a conservationist who believed that it had been necessary to cut down much of the earth's forests because of utilitarian reasons, but that there was a vital need to use care and foresight with what remained. The British had long had a major deforestation problem. Even before the Norman conquest (1066), England had lost all but 20 percent of its woodlands. By the beginning of the twentieth century, the percentage of woodland (4 percent) in the United Kingdom would be the lowest in Europe. Until then, however, the English still widely believed that cutting down forests was a path to civilization. Prime Minister William Gladstone went on famous tree-cutting exhibitions. The Germans did not share his enthusiasm. When Gladstone visited Bismarck in 1895, the German chancellor made a point of giving him a gift of a young oak to plant when he returned home.[48]

The conservation movement grew slowly in Britain.[49] By the seventeenth century, theologians were disputing the lawfulness of forcing any species of God's creatures into extinction. The scientific concept of the balance of nature can be said to have arisen from the theologians' concept of the orderliness of creation. The widespread interest in natural history in the nineteenth century led to the same conclusions, albeit in a secular way, and made for a strengthened interest in nature for its own sake.

The last element in the new reverence for nature was an appreciation for wilderness, defined as land untouched by man.[50] If Buffon is right, there may not be such a thing, especially on an island so long and densely inhabited as England, but search for it the English did. Through changing tastes in landscapes—of which the informal "English garden" is simply one manifestation—a type of nature mysticism evolved. By the end of the eighteenth century, the "appreciation of nature, and particularly wild nature, had been converted into a sort of religion."[51] This was a general European phenomenon closely allied to romanticism, but the English went further than any other nation in their devotion to it. When people no longer need land for farming, it becomes easier for them to appreciate it in a wild state. An interest in wilderness is necessarily concomitant with a prosperous civilization. It is no accident that Britain, home and leader of the industrial revolution, had the most pronounced interest in wild nature. The wilderness's greatest charm is seen most in contrast to industrial civilization. Solitude, not simply silence, appealed to English people as diverse as Queen Victoria and John Stuart Mill. Mill based his opposition to population growth partly on the fact that fewer people enhance the possibilities for solitude.[52] The wheel had come full circle. Nature had been conquered by technology, but many of the conquerors were having second thoughts about vanquishing the losers.

Reactions to Industrialism

A major reaction against the industrial revolution was the Romantic movement, which arose in the world of art, music, and literature as a rebellion against classicism, the belief that everything has an ideal form toward which all artists should aspire. But romanticism had significance beyond literature and the arts. It was also a repudiation of much of the industrial revolution. Says one critic, "Romanticism was a protest against the exploitative and alienating aspects of capitalist society."[53]

Romanticism is generally thought of as a preference for emotion over reason in matters of culture, of self-expression over discipline. Plato, Aristotle, and Thomas Aquinas had believed that in nature, in what was "out there" independent of human cognition, there was an objective reality that could provide guidance to both knowledge and personal conduct. Nature was primary, but only as apprehended and grasped through reason. The Romantics made nature not simply something to be seen through reason, but a way of seeing in and of itself. Instead of Descartes' "I think, therefore I am," the Romantics said, "I feel, therefore I am." The classic approach permitted a dispute about the merits of two rival paintings or sonatas to be settled by reference to a standard capable of being discussed. The romantic approach effectively denied any standard. What was good was whatever individuals or groups found appealed to their inner vision of reality.

The conservatism of Edmund Burke (1729–1787)[54] and his followers served to maintain the Romantic denial of the sovereignty of reason. Central to the triumph of conservatism in Britain and elsewhere during the nineteenth century was pessimism about nature's ability to support humanity. A major contributor to this pessimism was the English clergyman and amateur economist, Thomas Malthus (1766–1834). His *Essay on the Principle of Population* (1798) had great influence on the dean of nineteenth-century British economists, David Ricardo, as well as on Charles Darwin.[55] Malthus is well known for his thesis that while the means of human subsistence increase only arithmetically, human populations increase geometrically, a contention still vigorously debated today in the argument between neo-Malthusians and their opponents. Basic to his thesis were his ideas about soils; like many of his contemporaries, he believed that mankind always cultivates the best soils first. He also held that human beings should follow the injunction of Genesis to increase and multiply, yet they are subject to the laws of nature as well, with populations kept down by war, famine, and pestilence. The means of checking growth voluntarily were by "vice" (by which he meant primarily birth control, which he deplored) and "moral restraint." He considered it immoral to marry and beget children that one could not support economically.

This Malthusian pessimism was in direct opposition to the basic optimism expressed by leading British proponents of industrial growth. Some compromise seemed necessary. William Morris (1834–1896) was a major figure in the reconciliation of optimism about human nature with pessimism about the consequences of the industrial revolution.[56] Strongly influenced by the art critic John Ruskin, Morris believed that England had been happier in the Middle Ages when workers were able to take craftsmanship seriously. In his famous utopian novel, *News from Nowhere* (1890), he portrayed a world of decentralization, communal ownership, and high art. He wrote in 1894 that, aside from his desire to produce beautiful things, "the leading passion in my life has been and is hatred of industrial civilization."[57] This hatred extended to civilization's capitalist masters. In the latter part of his career he was a leading socialist agitator. Along with Ruskin, Morris was a major inspiration for the Arts and Crafts movement that was important not only in his native England but also in the United States around 1890 to 1910.[58]

Although thinkers such as Burke, Malthus, and Morris were important in shaping Western man's ideas about nature during the period of the industrial revolution, one man towers above them all. The life and work of Charles Darwin (1809–1882) are central to modern thought, especially in the English-speaking world, not only because of the ideas he created but those he reflected. Since the beginnings of Western history, philosophy and theology reflected a search for unchanging essences behind the world of appearances, a quest that united virtually all thinkers from Plato through the early modern era. Immanuel Kant (1724–1804) argued, convincingly to many, that there was no way to get from the observable world of phenomena to the real world of eternal, necessary ideas.[59] In the early nineteenth century, this search itself would be abandoned. Efforts to comprehend the process of change replaced the quest for unchanging truth.

Darwin's importance was in taking the concept of process—pioneered in metaphysics by Georg Wilhelm Friedrich Hegel (1770–1831)—and extending it to biology. Change, not essence, was primary here. Darwin denied that species were fixed, concentrating rather on how they changed under the influence of their environments and each other.[60] His first real inspiration came from a reading of Malthus, which convinced him that competition for scarce resources was the key to biological change.[61] Darwin fundamentally transformed ideas about man and nature.[62] Even though some biologists also disagreed with his ideas, Darwin's most outspoken enemies were religious people who felt that he had not only banished man from his worldly throne but, more basically, dethroned God himself by presenting the universe as being without conscious purpose. Darwin thus completed the revolution in human thought that began when Copernicus displaced the earth from the center of the universe: he displaced man from the center of creation, indeed destroying the notion of creation itself.

One cannot understate the significance of Darwin's concepts for man's image of nature and of himself. Even after the triumph of the heliocentric theory, the world remained a place ordered for man and was ultimately benign. In Darwinism the world was still ordered, but no longer benign; rather nature was "red in tooth and claw." Nature was akin to Hobbes's presocial state of nature. Bitter mortal combat united all creatures in a complex relationship whose motif was "kill or be killed." The universe was not only cruel, but soulless, possessing its only "meaning" in the laws of science. "Darwin's discovery," as one commentator has stated, "rounded out the cosmological design first charted by Newton; it furnished the basis for the final reduction of the organic and human world to the physical laws governing the inorganic universe."[63]

Not all Darwinians took a pessimistic view of human nature, however. Though Darwin had specifically denied that evolution could have any moral direction, his theories were (and are) continually misapplied to support moral or ideological crusades. The theory of evolution was proof, for some, that nature had been straining, since the first single-celled amoeba, to produce its highest product, the Victorian Englishman. Evolution was thought of as a progressive movement. Thomas Henry Huxley (1825–1897), a distinguished biologist who was one of Darwin's major defenders, pressed for human control of future evolution and endorsed the idea that there could be an evolutionary ethics which dictated that whatever contributed to the progress of the species was good.

The Beginnings of Technocracy

The French Revolution and its era had produced its share of utopian thinkers who looked to the creation of a new world freed from the restrictions of the ages; thus the climate in France was receptive to the creation of a new order.

It now remained to create an order that kept not only with the ideals of the past but reflected the new social realities spawned by rising industrialism.

Count Claude Henri de Rouvroy de Saint-Simon (1760–1825) was one of the most influential thinkers of the nineteenth—and indeed the twentieth—century. He fought in the American Revolution, then returned to France and supported the French Revolution but wound up in jail for his efforts. There, he claimed, Charlemagne appeared to him in a dream and told him to become a great philosopher. Following this command he became prodigious in producing books and ideas. He spent his fortune in spreading his beliefs, made an unsuccessful suicide attempt, and finally died in poverty.

Saint-Simon was an important precursor of Karl Marx; he was "the first social thinker to recognize the important role which industrialism was to play in human affairs."[64] His intent was to provide a blueprint for a new way of life in a new society that would be consonant with the triumphant new industrialism. His disciples, "who flourished in the first half of the nineteenth century . . . were the most enthusiastic propagators of an enlarged, updated Baconian vision."[65] In the area of vocabulary alone, between 1800 and 1832, he and his disciples coined the words—and concepts—"individualism," "positivism," "industrialism," and "socialism."[66] But his major contribution was his vision of a society in which all other claimants to political and social power are swept aside in favor of what he called the "industrials." His thesis was that in each era of history, society would and should be dominated by those who contributed most to it. In Saint-Simon's France, this group was made up of those who had the technical and managerial skills necessary to keep industrial society going. They were not the old landed aristocracy of his ancestors, but neither were they the new merchant class that had taken over after the revolution. They were in effect the technocrats. There is a clear line of descent from Saint-Simon's ideas to those of the technical elite that has run modern French society since World War II—indeed to the whole concept of "postindustrial" society.

What was needed, according to Saint-Simon and his followers, was not yet another ideology but an organized social order. According to Langdon Winner, they wanted "the building of a technological society ruled by a technically competent aristocracy." There would be no room for democracy or anything like it in this new order. Members of Parliament would be chosen on the basis of professional competence alone, without any elections. This would not be a question of personal rule, since decisions would necessarily be made according to the impersonal dictates of science.[67] Here we find a preview of the argument, made during the late nineteenth century and early twentieth centuries, for a government run by experts and for the replacement of politics with public administration.

They disdained not only democratic politics, but all politics which would be replaced by objective science. They aimed to put an end to not only politics as we usually think of it—conflict among leaders or potential leaders—but government itself. The Saint-Simonian motto was that society should go "from the government of men to the administration of things."[68] Later this call became the battle cry of the European anarchists and the cornerstone of what little Marx and Engels tell us about political society after the inevitable world Revolution.

All of this presupposed an era of plenty resulting from the new industrialism, and the followers of Saint-Simon did all they could to bring this change about. They were concentrated among the alumni of the Ecole Polytechnique, founded in 1794, and have been credited as the main force in the economy of the Second Empire in France, especially in the banks and railways.[69] But even more important than the visible works of the Saint-Simonians has been Saint-Simon's intellectual legacy of antipolitics.

Nineteenth-century France was marked by other important social thinkers who challenged the established order in favor of utopian designs of their own vision. Noteworthy was the utopian socialist Charles Fourier (1772–1837), who was much touted by Marx and Engels. Fourier was quite probably insane. But in his elaborately detailed fantasies, he touched upon many key problems of social organization in new ways. Unlike the mainstream of the whole utopian tradition since Plato, he rejected austerity because he recognized the new powers technology had given to man to rise above the niggardliness of nature. Indeed, Fourier "cleansed luxury of its Christian theological stigma and demoted poverty from its ideal position as a crowning virtue."[70] He also broke with the utopian tradition of self-control by accepting the importance of passions—including sexual attraction—in human life. Details of his ideal society of "phalanxes" need not concern us here. What is important is how widely admired his ideas were in his own time, not only in France but also in America. From 1841 to 1847 the American utopian experiment of Brook Farm was conducted according to Fourierist principles—without the bureaucratized free love he supported—largely due to the influence of Albert Brisbane, a former student of Fourier and an important American journalist.[71] Indeed, some forty Fourierist phalanxes were founded in America around this time. Fourier was significant not only for his belief that technology could make a utopia of luxury and nonrepression possible, but he was also, as Engels noted, "the first to declare that in any given society the degree of woman's emancipation is the natural measure of the general emancipation."[72]

Marx and Engels looked explicitly to men such as the British industrialist Robert Owen, Saint-Simon, and Fourier as vanguards for their ideas. In them they saw socialists—utopians—who were basically correct in their diagnoses of the evils of capitalist industrialism and in their suggestions about what kind of society should replace it; yet they failed to show how existing society could be changed into socialist society.

Nineteenth-century political life was not simple, whether on the Left or elsewhere. Marxists would have to struggle to gain control over the socialist movement intellectually as well as organizationally. One of their rivals was the anarchists. Perhaps anarchism's most important intellectual leader was Prince Peter Kropotkin (1842–1921). Imprisoned in his native Russia for working with the anarchist NAROD group, he escaped to western Europe. But his political activity also led him to jail in France in 1883. He was released in 1886 and went to England, where, like Marx before him, he devoted himself to writing. He returned to Russia after the Revolution in 1917, but he opposed the Bolsheviks and soon retired from active politics.

Kropotkin's greatest contribution was to challenge the basic premise of the Darwinian system, arguing that cooperation, not competition, was the way by which nature operated. Science did not lead to a picture of nature red in tooth and claw; such clashes did happen, but the progress of nature—and evolution did have a direction—was not based on them. A noted geographer and naturalist, Kropotkin argued in his masterwork, *Mutual Aid: A Factor in Evolution* (1902), that it was cooperation between species that had led to evolutionary development. This was just as true within humanity as in the world of nature. Just as the Darwinian postulate of the centrality of competition led to the concept of the necessary centrality of competition in social and political life, so for Kropotkin the centrality of cooperation in nature led to the centrality of cooperation among people, which was basic to the tenets of anarchism. His major work, devoted directly to social questions and entitled *Fields, Factories, and Workshops* (1898), took up the cudgels for a decentralized economy and political system in which technology would be introduced when all those concerned, including the workers, were convinced it would be useful. In his ideal system work in the fields would continue with work in the factories.[73] A balance between agriculture and industry was essential.

A major conflict thus exists between the anarchist tradition and the mainstream of Baconian-inspired social thought that advocates the complete, unbridled conquest of nature by man through science and technology under the aegis of a centralized state dominated by technical experts.[74] The Baconian view certainly seems to have carried the field, but an undercurrent of dissent remains. Writes one dissenter: "It is well to remember . . . that the ecological impulse toward community, the impulse toward the idea of nature and the ideals of simplicity, mutual assistance, and cooperation, is a seemingly ineradicable impulse in Western thought."[75]

We have seen how the industrial revolution and the philosophy behind it triumphed in the modern world in both civil and military life, despite dissent from workers, Romantics, nature lovers, and others. But the modern world we have examined has been essentially the world of Europe. But European civilization had an offspring in the United States, where modern industrial civilization flourished above all. We will now turn to that story.

Chapter Five

AMERICANS CONQUER THEIR CONTINENT, 1492–1865

The New World reflected the varied and conflicting dreams of the peoples of Europe. Their dreams combined the desire for material plenty with the wish for spiritual fulfillment—dreams that have divided North America, particularly the United States, ever since.

Whether or not Viking adventurers such as Leif Eriksson[1] were the first to visit the New World, America as an area for European colonization still dates from the voyages of Christopher Columbus. Columbus combined the dual thrusts of Old World utopian longings: he was "saturated with medieval legends about the location of the earthly paradise," but he was also greatly influenced by contemporary millennial movements, as "the economic objectives of spices, incense, jewels, and gold is perfectly comprehended by the image of the material paradise." Columbus, a religious visionary, was also interested in the "paradise of innocence and beatitude. He wished to open the world to the Gospel in fulfillment of the Biblical prophecy of the Second Coming."[2]

These utopian aims, combining material paradise and religious duty, were important among the English who colonized America as well.[3] The Puritans especially had this sense of destiny:[4] their colonies were extensions of this special realm of God. Their utopian vision has remained an important facet of America's national sense of itself.[5] The Puritans and Quakers "dreamed long ago of establishing God's New Israel in the wilderness. In some form, their idea of America as a chosen land has been with us ever since."[6]

America also offered another kind of utopia, the long-standing ideal of a life lived in simple harmony with abundant nature, an Arcadian paradise. As we shall see, this pastoral myth has long haunted America and conditioned much of its attitudes toward technology. From the outset the Arcadian myth was a counterpoint to the main direction of American history, the ruthless conquest of the continent.[7] Since this New World was deemed virgin land (its native inhabitants were not using it properly by European standards), it had to be broken in for proper use. This would demand technology. The particularly American attitudes toward nature and technology were thus not adopted over time—as they were in Europe—they were coeval with the very founding of America.

New attitudes developed swiftly, although the colonial period retained many "medieval" characteristics in its social structure and culture.[8] Its society was going in the direction of strict order and hierarchy,[9] and it had a manipulative, exploitative attitude toward nature and soil, embracing a capitalist mentality toward the land even before its technologies were able to bend the land fully to human will. One observer argues that American history from the time of Columbus to the present "could be 'explained' by the scramble for real estate."[10]

Just how fearful and intimidating was America's vast and untamed wilderness? According to one scholar, "The New World forest . . . was not so very unlike European forests. It could even be hospitable, and certainly lacked any mystical power to defeat inevitably all the utopian schemes of European dreamers."[11] But another takes a diametrically opposed and more convincing view, holding that "the heart of the bias against wilderness was the ancient association between security and sight. The American pioneer reexperienced the situation and the anxieties of early man. Neither felt at home in the wilderness."[12]

To this natural aversion was added a religious belief that all wilderness—and here the Puritans echoed the tradition of the Old Testament—was still unredeemed and in the power of the devil. Conquering it was therefore a religious imperative.[13] One major exception to the Puritan distaste for wilderness was the noted divine, Jonathan Edwards, who had a deep poetic appreciation of nature, at least as God's instrument. Edwards wrote that "almost all men . . . love life, because they cannot bear to lose sight of such a beautiful and lovely world."[14]

Besides the Puritans of New England, others had come to America for religious reasons. Most important of these were the Quakers, who founded the commonwealth of Pennsylvania as a religious refuge. Quakers differed from Puritans in their attitude toward nature. They may not have been admirers of wilderness as such, but, like William Penn, their founder, they wished to preserve natural beauty.[15] In clearing the land, the settlers left one acre of trees for every five acres they cut down. Penn even appointed a forester for the colony.[16]

American Indians, Nature, and Technology

In no other respect did the colonists vent their attitude toward nature more clearly than in their treatment of the Native American population. Since Columbus, indigenous Americans were called Indians, in the mistaken belief that the newly discovered lands were part of Asia; this misnomer remains in use to this day—in films, in popular culture, and in children's games, though "Native American" has gained acceptance. In the colonies of Spain and Portugal, it took a papal edict to finally formalize the principle that Indians are indeed human beings, and, like slaves, are still fellow children of God. It was not easy for Europeans to grasp how large the world really was and that all its people were somehow descendants of Adam and Eve. Because the Indians had souls to be saved, there were some limits—in theory at least—as to how badly they could be treated.

In North America, the French, even before the papal action, had looked upon and treated the Indians as human beings; their missionaries sought to convert them, but at great peril. The English settlers, however, were Protestants, unbound by any papal pronouncements, and often had little regard for souls outside their own particular sects. They felt a great sense of racial superiority, not only vis-à-vis blacks but toward many of the ethnic groups of Europe. The settlers came to regard the Indians as creatures of Satan, a living part of the "hellish" wilderness they feared and hated.

The Indians had an attitude toward nature that was radically different from that of the settlers. The Europeans instinctively objected to the Indians' desire not to conquer nature but to live in harmony with it. The "noble savage" might in the abstract be looked upon as benign by *philosophes* in distant Europe, but these American Indians had something the colonists desperately wanted and needed: land.[17] Unable to exploit natural resources the way the colonists could, the American Indians found themselves ruthlessly subjugated and ignored.

Indicative of colonial sentiment toward the Indians is a statement made in 1756 by the relatively humane Benjamin Franklin: "I do not believe that we shall ever have a firm peace with the Indians, till we have well drubbed them."[18] After the American Revolution, a torrent of settlers moving westward brutally swept the Indians aside and onto miserable reservations, where their modern descendants still languish in poverty and squalor. Ironically, with the birth of the modern environmental movement in the latter half of the twentieth century, some white Americans have come to recognize that Native Americans were far wiser and successful than the European settlers in their relationship with nature, and to regret how they had been treated.

What exactly were American Indian attitudes toward nature? One expert, expressing conventional wisdom, writes that "the Indian view of the human role in nature is almost the reverse of the Western European understanding of 'domination.' Greater power resides in natural forces and spiritual beings than in the hands of mankind." The "unmistakable Indian attitude toward nature is appreciation, varying from calm enjoyment to awestruck wonder. Indian poems, songs and descriptions are full of natural images that reflect a pure interest in environmental beauty."[19] Much of this sentiment attributes to the Indians attitudes that are more those of modern American environmentalists than traditional Native Americans. There is no question that the American Indian did far much less to modify his natural environment than did the white settlers who seized it from him.[20] There is, however, little reason to believe that the failure of the Indians to alter their environment was the result of choice rather than lack of opportunity. According to one scholar, the Indian "shared with the early white settlers . . . a feeling that there were limitless horizons toward which he could expand." The Indian was "quite willing to take advantage of advances in technology to further exploit his environment."[21] The prime evidence is the way in which the Indian tamed the horses that had escaped from the early Spaniards and become wild. The economy and social structure of most Western tribes came eventually to be based on the use of the horse.

The Emerging Economy of Waste

Though the settlers acted to remove hindrances to their use of the land such as the Indians, they did not use the land well. From the outset, agricultural practices were poor. The English colonists were notable for their terrible husbandry. Because labor was far more scarce than land, waste was not considered a problem. A tradition of "resourceful wastefulness" arose. One historian gives as an example a surveying party led by William Byrd in 1733 on which the men, rather than take the trouble to climb some trees to get the chestnuts, simply cut the trees down.[22] For the colonists, the greatest labor-saving device "happened to be the wasteful use of land."[23]

The English settlers were not attached to the land as were agricultural peoples throughout history. William Penn began the practice of selling it in Pennsylvania, and such sales soon spread throughout the colonies. While New England towns usually forebade the selling of land to outsiders without the permission of the community,[24] land soon became simply another commodity. Pressures on the land were great, with settlers wanting to get as much out of it as fast as they could. Early in Massachusetts history, for example, it even became necessary to pass a law against overgrazing on the meadows of Cambridge.[25]

Settlers "assumed an attitude toward nature which was very different from the one which had prevailed in the economically stable, socially static resource picture of the old world."[26] This was not simply the case only in the northern colonies but in the South as well. In short, the focus of the American dream had shifted from a search for a political utopia to the conquest of an economic one. While the "vast fertile lands of America might serve as a lure for the oppressed and discontented of the Old World," soon "they acquired value only as they promised to furnish profitable staples for commerce."[27]

We have noted earlier that the British crown did not want to see industries arise in the colonies. Yet they did. The American colonists were, in general, inept and unskilled—indeed, this is one reason why nineteenth-century Americans used machines so widely.[28] The noted historian Henry Adams testifies to this long-term lack of skill on the part of his countrymen: "[Americans] struggling with the untamed continent in 1800 seemed hardly more competent to their task than the beavers and buffalo which had for countless generations made bridges and roads of their own."[29]

This general lack of skill was not due to any disdain for technology as such. John Smith had successfully requested that the London Company send on their second ship to Virginia not only gentlemen but skilled workers of various kinds. But there was little technological innovation. The technology of the New England villages was essentially the technology of the old England simply transferred. Benjamin Franklin learned printing on a press little improved since Gutenberg.[30] Early American technology was nonscientific in origin and specific to the tasks involved.[31] On the whole, it was quite traditional.

The colonists were by no means content with their lack of technology. As early as 1643 the Plymouth colony and Massachusetts Bay had granted monopolies to private individuals to set up ironworks, and other colonies soon followed suit.[32] Generally in the colonial period, the economy was very weak and completely dependent on Britain for growth capital and for markets. Despite this British domination, the colonists lived well, and wealth was distributed relatively equitably.[33]

During the colonial period, the colonists' exploitation of nature continued unabated. The invention of a new axe for chopping wood made it easier to cut down trees, and this was done with gusto. A visiting Frenchman reported that "an American has no idea that anyone can admire trees or wooded ground. To him a country well cleared, that is where every stick is cut down, seems the only one that is beautiful or worthy of admiration."[34] For the settlers, the clearing of forests was a symbol of progress. One idea all colonists were imbued with was the idea of progress.

From the outset the South was not so concerned with creating a religiously ordered society like that of New England. Theirs was an order of class—and soon of race distinction—and they had even less reason to restrict their depredations against nature. British policy minimized colonial expansion westward, and thereby inhibited the colonists' attack on the natural resources of the whole continent. At the same time, the British mercantilist policy attempted also to put a brake on colonial industrial expansion and hence on technological change. With the coming of the Revolution, both these controls would be lifted and Americans would be able to get down to the business of using technology to conquer nature.

The Revolutionary Era

The American Revolution was fought to provide the colonists political freedom from the British crown. But the ideals that inspired the leaders of the Revolution also involved concepts of nature and the relationship of humanity to nature. The revolt against British economic domination was implicitly also a fight to have an independent technological destiny.

It is quite true that many of the Revolutionary leaders were inspired by the desire to better themselves materially. The merchants especially chafed against British restraints on trade with the West Indies, and often resorted to smuggling to evade them. But not all were moved simply by crass economic motives. Many of the colonial leaders were seeking to establish a republic based largely on the model of classical republicanism which they felt had been destroyed in Britain itself by increasing monarchical presumption.[35] The earlier age had been one of virtue, simplicity, patriotism, a love of justice and of liberty. The present was venal, cynical, and oppressive.[36] American revolutionaries sought not only "liberty" but "the public good." It was assumed that the community was an organic unit whose interests did not conflict with those of the individuals who comprised it. Republicanism, like Puritanism, was essentially anticapitalistic and more concerned with individualistic drives that threatened the life of the community.[37]

From the outset, the Revolutionaries were interested in the relationships among luxury, liberty, and science. The mercantilist policies of Britain that limited colonial industry had the effect of providing "an incentive for Americans to revive the spartan virtues of their forebears and to emphasize the public good over private gain. . . . Plain living . . . became a symbolic measure of one's patriotism."[38] Many viewed the Revolution as necessary to keep

Britain from corrupting American society and destroying "our frugality, industry and simplicity of manners."[39] Samuel Adams was worried that material goods were diverting New Englanders from concern with their spiritual growth and political freedom,[40] and hoped that the Revolution could create a "Christian Sparta."[41]

Others were skeptical. Carter Braxon, a Virginia signer of the Declaration of Independence, argued that classical republicanism had been based on frugality and hence virtue, reinforced by sumptuary laws, but "such schemes were inapplicable for America. However sensible they may have been in naturally sterile countries which had only a scant supply of necessities, 'they can never meet with a favorable reception from people who inhabit a country to which Providence has been more bountiful.' "[42]

Braxon was right. Samuel Adams's dream of a Christian Sparta was short-lived. Within a few years after independence he abandoned his hopes for a more refined republic and turned to convincing Americans they were basically just like other nations.[43] Thomas Jefferson could not understand how Americans could so easily discard the simplicity and rusticity of the Revolutionary years,[44] but discard them they did. By the late 1780s more and more people argued that it was unrealistic to expect virtue in the public, and they denied that republicanism depended on virtue.[45]

The American republic was created by the drafters of the Constitution as an experiment that would—for the first time in human history—reconcile freedom with a central power. But it failed under the pressures of an underpopulated land and endless bounty, the very things against which the Constitutional Convention had been warned by Gouverneur Morris: "Growth, expansion, progress, acquisition, and material prosperity."[46]

The classical republican tradition makes it difficult to assess the post-Revolutionary constitutional settlement to provide a framework for the American treatment of land and nature. The vision was clearly at odds with the actual conditions of American life—the need to people and subdue a vast unspoiled continent. Furthermore, nature did not loom large in the classical republican tradition. As one modern theorist has put it, "The West has not only failed to cultivate *virtu*, but... nature and wilderness... have... no important role in classical political thought." Classical political thought in general "involved man's relationship not to the natural environment but to fellow man."[47] Yet there is no question that there existed among the men of the Revolutionary era a de facto ideology that was quite developed. The Americans conceived of themselves as an agricultural people, and hence virtuous. Farming "constituted a source of cultural value and a sign of virtue, a moral as well as an economic condition." Not only was agriculture important, so was machine technology. Native manufactures "provided an essential defense of freedom."[48]

Benjamin Franklin was a city man, devoted to his beloved Philadelphia, but he did not defend urbanism as such.[49] Despite being a city dweller, he looked on agriculture as the true source of virtue.[50] He distrusted industrialism, arguing that it "sprang from the national poverty and is nourished by it."[51] Yet by 1784 he had also felt the same post-Revolutionary disillusionment with republican virtue as Carter Braxon and Samuel Adams: "Is not the Hope of one day being able to purchase and enjoy Luxuries a great spur to Labour and Industry? May not Luxury, therefore, produce more than it consumes, if without such a Spur people would be, as they are naturally enough inclined to be, lazy and indolent?"[52]

Thomas Jefferson was equally ambivalent. He was torn throughout his life by contradictory impulses, being at once a physiocrat in his basic beliefs and a cultivated man who admired the cosmopolitanism of urban life. His basic allegiance to farming as a way of life, based on what he considered its greater conduciveness to democracy, received classic expression in his often-quoted statement in *Notes on Virginia* (1784):

> For the general operations of manufacture, let our workshops remain in Europe. It is better to carry provisions and materials to workmen there, than bring them to the provisions and materials, and with them their manners and principles. The loss by the transportation of commodities across the Atlantic will be made up in happiness and permanence of government. The mobs of great cities add just so much to the support of pure government, as sores do to the human body. It is the manners and spirit of a people which preserve a republic in vigor.[53]

Jefferson's devotion to the pastoral ideal was based on the concept of America as "Nature's nation." "America derived her virtue and vitality . . . from independent and direct contact with nature."[54] But as president he found himself forced to accept manufacturing as vital to the nation, and sought explicitly to distance himself from his previous opinions. In 1805 he complained that his former views had been misunderstood; he had meant to apply them only to the great cities of Europe, not to his own contemporary America. Through all this, however, he still believed that it would be possible to avoid the worst horrors of European industrialism. The factories in America would be decentralized, within the context of agriculture, and not concentrated in large cities as in Europe. They would not weaken but strengthen republican institutions.[55]

The final factor in Jefferson's ultimate acceptance of manufacturing was the War of 1812. It was a reluctant conversion, based on the belief that the nation needed manufactures to preserve its security in a hostile world, but a conversion nonetheless. Jefferson wrote in 1816, in reference to his *Notes on Virginia*, that "experience has taught me that manufactures are now as necessary to our independence as to our comfort."[56]

Jefferson's views did not go unchallenged. Foremost among his opponents was Alexander Hamilton. Hamilton, as secretary of the treasury in Washington's cabinet, urged America along the combined paths of industrialism, a national debt, and a strong central government in his classic *Report on Manufactures* (1791).[57] A major supporter of Hamilton's views was his assistant at the Treasury, the wealthy Philadelphia manufacturer Tench Coxe, who advocated an economy balanced between agriculture and industry, but whose real heart lay in the new proposition that industrialism and republicanism could be made synonymous: "I consider it possible to convert men into republican machines."[58] Central to Coxe's ideas was the necessity to industrialize America, though it was to be a new kind of industrialism, fit for republican America. As a member of the Pennsylvania Society for the Encouragement of Manufactures and the Useful Arts, Coxe argued that the scarcity of labor held out by opponents as a barrier to an American manufacturing industry could be easily overcome by utilizing "water mills, wind mills, fires, horse, and machines ingeniously contrived."[59] Thus in effect the machine itself would take the place of labor.

Despite such aspirations, progress was slow at first. There was no immediate surge of economic and industrial independence after the Revolution, mostly because Americans had no general technological capacity. As one historian explains it: "When the republic began there were *even fewer* active inventors then there had been when the states were colonies. A major reason for this was that as soon as the independence of the United States was achieved, everyone who could moved west to settle the wild land and become a farmer. Among the first to go were the mechanics."[60] Successful projects such as the Erie Canal, built from 1817 to 1825 under the supervision of the largely self-taught John B. Jarvis, were the exception rather than the rule.[61] The United States was behind England in virtually every branch of technology,[62] and England was determined to keep it that way. Immigrants from England to America were required, prior to leaving the home country, to sign a certificate that they were not "manufacturers" or "artificers."[63]

But this relative lack of technological skills did not prevent the westward-moving colonists from exploiting the land and its resources as much as their abilities allowed. Land, above all, was to be the key to realizing the nation's dreams, however they were interpreted, no matter that the lands were already occupied by Indians.[64] Faced with their presence, the American government expected the natural course of settler pressure to induce the Indians to leave as their hunting grounds disappeared. When this did not happen, Congress took a more active role by passing the Indian Removal Act in 1830. Designed to get virtually all of the tribes east of the Mississippi out of the area, the act led to forced evacuations such as that of the Georgian Cherokees to Oklahoma, in what came to be known as "the trail of tears."[65]

Once the land was occupied, Americans were flagrantly careless in their use of it, disrupting natural ecosystems and mining the soil to produce the largest crops possible in the shortest period of time. With no incentive for careful husbandry, little took place. Erosion and depleted soil meant little to them: they could always move further west.

Nature and Technology in America between the Revolution and the Civil War

To comprehend fully what was taking place in nineteenth-century America, we must realize that for most Americans there was no conflict between nature and civilization. Nature was only nature after it had been exploited and civilized. The wilderness was to be conquered for the full potential of nature to be realized. Technology was thus not opposed to nature; it was the only means by which nature could become its real self.

Most early nineteenth-century Americans were not enamored of nature in its raw state and were eager to alter it. This attitude is manifested in both the mainstream of popular writers as well as in such figures as Cooper, Emerson, and Thoreau, who are frequently quoted to the contrary. A few may have argued that man needed unspoiled nature, as did the famous Philadelphia physician Benjamin Rush, who wrote in 1800 that "man is naturally a wild animal, and . . . taken from the woods, he is never happy . . . till he returns to them again."[66] But, according to one historian of American attitudes toward nature, "in spite of such sentiments Romantic enthusiasm for wilderness never seriously challenged the aversion in the pioneer mind."[67]

This general attitude was, despite some surface appearances to the contrary, shared by the leading literary and intellectual figures of the day. James Fenimore Cooper (1789–1851), America's most important novelist of the period—in both popular and critical terms—is a case in point. One school of thought holds that Cooper took refuge from the vulgarities of Jacksonianism and capitalism in a romanticized eighteenth century, loving the pristine wilderness and hating the squalid frontier.[68] Some claim him as a pioneer conservationist, concerned with the wanton destruction of American natural resources in the process of settlement.[69] But Cooper, although he often rhapsodized over the lost wilderness, recognized, like the pastoral Jefferson before him, that wilderness had to go for the sake of a civilization that was, in the last analysis, preferable. For Cooper, the final destiny and fulfillment of nature was for it to *become* civilization.[70] Though Cooper portrayed the Indian and the wilderness in idyllic tones, both he and his readers knew such a portrait was overdrawn, and that both were doomed, rightly so, for the sake of the necessary conversion of wilderness into civilization. Even his hero Natty Bumpo shared the basic assumptions of the advancing civilization,[71] and Bumpo's feelings of closeness to the wilderness were only temporary and did not deny the superior importance of the new order to come. Cooper was telling a morality tale, but the moral was the inevitability and desirability of the transformation of wilderness into civilization. Though he wrote some important didactic works, such as *The American Democrat* (1838), Cooper's fame and influence were based primarily on his novels about the frontier.

Ralph Waldo Emerson (1803–1882) is a different case. Though he wrote some poetry, he was primarily an essayist and lecturer who sought to support himself through his writing and lecturing after a crisis of faith forced him to leave the Unitarian ministry in which he had been ordained.[72] As a matter of principle, Emerson believed that inconsistency could be a virtue, and indeed he was virtuous in his ambivalent attitude toward nature and technology. He was a leader of the transcendentalists, who invoked nature as a norm by which to judge society, and identified nature with "the divine indwelling in man."[73] Nature validated the insights of the individual who took the natural world seriously.[74]

Emerson believed that people's view of nature was all important, and that they had a determining effect on societal institutions. He was basically a Romantic himself, in his methods as well as in his conclusions. Reason alone could lead only to science; it was only through feeling that one could truly understand the real world: "Nature can teach man more than Newton's materialism."[75] When in the woods, Emerson became what he called a " 'transparent eye-ball,' who *is* nothing and sees all, who is part and parcel of God."[76]

There is no question that Emerson was a lover of nature. "The world is so beautiful," he said, "that I can hardly believe that it exists," and "In the woods is perpetual youth."[77] He wrote in his *Journal* in 1834 that "I do not cross the common without a wild poetic delight. . . . Thank God I live in the country."[78] But what was "the country" for Emerson? Friendly critics have claimed that he replaced the Puritan picture of nature as a howling, unredeemed wilderness with one of nature as a harmonious expression of divinity.[79] There is actually no evidence to suggest that Emerson, any more than his transcendentalist colleague Thoreau, had a taste for truly untamed nature.

Emerson's views on technology were directly related to his views on democracy, that is, on the need for every man to have some role in determining his destiny, and led to his inconsistent views about technology. From 1834 to 1844 Emerson was especially enthusiastic about technology's ability to help man control the world.[80] In 1837 he welcomed the news that New England was destined to become the manufacturing center of America. He was especially enthusiastic about railroads, the nineteenth century's prime exemplar of technology.[81] In "Nature's Nation" he argued that technology "would not abrogate America's claim to be 'Nature's nation.' " "By transforming presently poor and uncultivated earth, the railroad would guarantee the country's agrarian basis and fulfill America's promise as Eden." Technology and transcendentalism both "represented explosive new forces directed against outworn conventions of thought and behavior."[82] For Emerson America would escape the "dark, satanic mills" found in industrializing England.[83] His ideal was a technology without cities. Like Jefferson, he believed that in America science and technology could be made to serve a rural ideal.

But these views were not to last his lifetime. In 1847 he traveled to England and was shocked to see the results of a more mature industrialism. He was appalled—as Engels had been before him—by the squalor of Manchester and became concerned about the consequences of unchecked technology. Grateful for the benefits of American republicanism, he wrote to his daughters after his visit, instructing them to "thank God that they were born in New England."[84] He was most impressed by the way industrialism had transformed English civilization: "The machinery has proved, like the balloon, to be unmanageable, and flies away with the astronaut."[85] By 1851 he asserted in Concord, in a famous phrase, that "things are in the saddle/ And ride mankind."[86]

If Emerson's views on nature were convoluted, those of his personal friend and fellow trancendentalist, Henry David Thoreau (1817–1862), were even more so. Thoreau has been described as "the most controversial of American writers."[87] In large measure this controversy relates to his ideas about nature, since "even what he thought about Nature is something of a puzzle."[88] Thoreau has enjoyed a great modern reputation as a lover of nature and a patron of wilderness. This esteem is largely undeserved, though Thoreau was known for his aversion to what technology was doing to America.[89] He wrote that "no really important work [can] be made easier by . . . machinery,"[90] and "Our life is frittered away by detail. . . . We do not ride upon the railroad; it rides upon us."[91] Yet the picture of Thoreau as a lover of wilderness and raw nature is a false one. He fully believed, as had Cooper before him, that it was part of American destiny to conquer nature and subdue the continent. For Thoreau, the American makes himself "stronger and in some respects more natural" because he "redeems the meadow." For him, "The weapons with which we have gained our most important victories are not the sword and the lance, but the bushwack, the turf-cutter, the spade and the bog-hoe."[92] But his acceptance of this conquest was ambivalent. His basically Romantic outlook inevitably involved a rejection of the humanist tradition upon which this conquest was based. "There is no place for man-worship," he argued. As a citizen of Concord, however, he was concerned that the vast primeval forest which had once covered New England was being destroyed, and bringing at least some of it back was one of his major wishes.[93]

Paradoxically, Thoreau was not only a worshiper of nature but, as a transcendentalist, he was also trying to overcome nature in the name of the spiritual. For him, as for men such as his friend and mentor Emerson, "Spiritual progress . . . requires the steady extermination of the primeval and the expansion of a more civilized terrain."[94] In *Walden* he wrote, "Nature is hard to be overcome," and "We are conscious of an animal in us, which awakens in proportion as our higher nature slumbers."[95]

In actual practice, Thoreau was no lover of nature untamed. His most famous foray into nature, his two-year stay at a cabin owned by Emerson, recounted in *Walden*, was an exercise in hypocrisy. One critic notes that Thoreau, "as we know, was no wild man, no hermit of the woods despite his brief sojourn in a cabin within easy walking distance of his home, toward which he sometimes walked for dinner."[96] Every day or two, Thoreau went into Concord for the provisions necessary to sustain his life in the woods.[97] His cabin at Walden Pond was made largely of the artifacts of a machine civilization.[98]

Indeed, Thoreau often found nature wanting. "Nature is not made after such a fashion as we would have her. We piously exaggerate her wonders, as the scenery around our home," he wrote in *A Week on the Connecticut and Merrimack Rivers*.[99] Wilderness actually terrified him. Of an excursion to Mount Katahdin in Maine he wrote, "Vast, titanic, inhuman Nature got him at disadvantage, caught him alone, and pilfers him of his divine faculty," and spoke of the "savage and dreary scenery" and of "the grim, untrodden wilderness . . . even more grim and wild than you had anticipated."[100] For Thoreau such "primeval, untamed, and forever untamable *Nature*" is suitable only for men who are more akin to rocks and "wild animals."[101] In short, he was not a lover of nature per se, but a pastoralist in a traditional American sense. For modern-day Americans, he has little to contribute in solving twentieth-century technological and ecological problems.

Writers such as Cooper, Emerson, and Thoreau were not the only ones to reflect and influence American thinking about nature. The historian and politician George Bancroft (1800–1891) was a major popular and intellectual figure in this sphere. His ten-volume *History of the United States* (1834–1874), though virtually unread today, exerted a major influence upon nineteenth-century thought.[102] The premise of Bancroft's *History* was simple: "Material changes on the American continent only verify spiritual progress. . . . In 'little more than two centuries' America has been changed from an 'unproductive waste' "[103] into something far more natural in American usage, a garden.[104] The wilderness as such was not to be despised, however; it was a "kind of temple" in which was forged the manhood of Washington, the other founding fathers, and heroes such as Daniel Boone.[105]

But the wilderness, however salutary, is not to be valued in itself. Bancroft contrasts the Hudson Valley of his own time with the primitive scene he imagines Henry Hudson must have seen,[106] which he paints in highly charged, nonlaudatory terms. Above all, "His sole criterion for evaluating the Hudson Valley is utility. By that standard the landscape holds no aesthetic appeal, no appeal of any kind." Most Americans shared the impulses behind Bancroft's "proud vision of the Hudson Valley civilized from its loathsome wilderness."[107]

America's ambivalence toward nature manifested itself not only through literature but in the visual arts as well. The work of Thomas Cole (1801–1848), a prominent member of the Hudson River School of Romantic painters, and George Catlin (1796–1872), one of the first American artists to specialize in Indians and the frontier, embodies this ambivalence. Cole "constantly idealized a juxtaposition of the wild and the civilized";[108] his *The Oxbow* (1836) is a "schizoid canvas [which] exactly conveys Thoreau's belief that man's best environment combined wildness and civilization."[109] Although his paintings glorified nature, Cole also found the natural world menacing. He is perhaps best known for a sequence, *Course of Empire* (1836). Described by one writer as a "dramatic illustration" of the survival of the ideal of classical republican simplicity,[110] it nevertheless "ends with vegetation threatening to engulf a destroyed civilization."[111] Cole clearly sensed something was wrong with America's assault on nature, and that the country might someday have to pay a terrible price for its headlong pride.

George Catlin specialized in portraying Indians, and was so widely recognized in his own time that Daniel Webster tried to get the federal government to buy his paintings for an "Indian Gallery." The Southern majority in the Senate voted it down, because they feared it would create so much sympathy for the Indians that the expansion of slavery westward through Indian territories might be inhibited.[112] Though he celebrated wild scenery, Catlin "had the vision to see that the primeval glories of nature in this country could not last forever."[113] Unlike Cole, he expressed his views on man and nature not only through painting but also in words. As early as 1835 he argued that Indians, buffalo, and the wilderness would not have to be destroyed completely by civilization if the government would only protect them in "*a magnificent park* . . . what a beautiful and thrilling specimen for America to preserve and hold up to her refined citizens and the world, in future ages! A *nation's Park*, containing man and beast, in all the wildness and freshness of nature's beauty."[114]

Politics and Nature

While all this soul-searching about nature—primarily by intellectuals—was going on beneath the mask of a general lack of concern, the American political process continued to reflect and condition America's attitude toward nature. Andrew Jackson (1767–1845), the seventh president, expressed a common sentiment when he asked "what good man would prefer a country covered with forests and ranged by a few thousand savages to our extensive Republic, studded with cities, towns, and prosperous farms, embellished with all the improvements which art can devise or industry execute?"[115] "Philanthropy could not wish to see the continent restored to the condition in which it was found by our forefathers." This belief that America's great glory was attributable to her conquest of nature was shared by Jackson's opponents, the Whigs, as well.[116]

Not all agreed. Horace Greeley, a distinguished American editor and eventually a presidential candidate in the post-Civil War period, urged Americans in 1851 "to spare, preserve, and cherish some portion of your primitive forests."[117] But Greeley wanted to show Europeans the natural basis of American political virtue, and he was not really concerned about the exhaustion of natural resources. Not until the Civil War era did some American leaders realize that there is a limit to nature's bounty.

Between the Revolution and the Civil War, the movement of settlers across the continent in search of new land exacted a tremendous toll on the American landscape. European settlers always believed they had a God-given right to spread across the American land, a belief formalized as "manifest destiny." This term, coined by John L. O'Sullivan in the New York *Morning News* in December 1845,[118] meant different things to different people, but basically it put forth the idea that Americans had the right, even the duty, to control the whole continent.

The government played a major role in fostering this belief. Whatever their differences, Jefferson and Hamilton and the presidents who followed their era were in basic agreement on manifest destiny. It was the duty of the federal government, by means of land sales and grants, to enable farmers and agriculture to prosper. Canals and railroads were built even before settlers appeared on the scene. Putting a rudimentary infrastructure into places, "government . . . typically played the role of pioneer."[119] Despite fluctuations in the terms of sale that often yielded great profits to speculators, public land policies were designed to put as many farmers on the land as possible.[120]

Farmers were not the free yeomanry Jefferson had idealized, so a central government was important to them. From the outset they were capitalists when presented with the opportunity. "Subsistence farming was practiced more because of the impractability of commercial agriculture than because of a desire for complete self-sufficiency." In the Jacksonian era, the American yeoman was moving rapidly toward land speculation and commercial farming. To do so necessitated being near transportation, and to make this possible, the government gave public land to build railroads and it "removed" Indians. As a delegate to the 1850 Ohio Constitutional Convention put it, "The earth must be subdued."[121] To do so required not only getting rid of Indians and encouraging railroads but making new banking arrangements as well. The farmer was above all a merchant of agricultural products.[122]

Science and Technology

Not only were Americans taming the continent through agriculture during the first half of the nineteenth century, they were asserting their newly discovered powers over nature through technology as well. Specifically, three inventions had a powerful effect. Whitney's cotton gin unified the South around cotton; Slater's reproduction of English textile machinery bound New England to the South; and the steamboat made it possible to settle the West while binding the West to the East.[123]

The American willingness to embrace technology as a democratic imperative stemmed in some measure from the early bias of American science—compared with much of traditional European science—toward the practical. American science during the colonial period was weak and derivative.[124] But despite the emphasis on the practical, early America was no scientific desert, and the Founding Fathers themselves were personally interested in science. Thomas Jefferson was something of a universal genius, and science and technology—including gadgetry—were among his interests. Benjamin Franklin is also known for his passion for science, and in 1743 he produced a "Proposal for Promoting Useful Knowledge among the British Plantations in America." Much later he founded the still-extant American Philosophical Society, an attempt at a democratic counterpart to the Royal Society of England. George Washington had an interest in scientific agriculture. James Madison, in his second inaugural, pushed the idea of a national university, a "temple of science."[125] [125] Early universities did more for scientific education than at any other time until the late twentieth century. They emphasized science's potential benefit for mankind, especially American mankind.

"Here, then let us found an Empire—An Empire of Science. . . . Let us erect an edifice which in all coming time will prove the glory of the American name; thus shall we serve our country, the country of our fathers and the country of our children," said Samuel L. Stoddard in his address to the Columbian Institute in 1827.[126] His sentiments, combining enthusiasm for science with nationalistic fervor, were typical of the period.

Technology soon became identified with democracy, though neither term was in widespread use in the early nineteenth century. The term "technology," in the American context, was first used publicly by Jacob Bigelow, the first professor of "the application of science to the art of living" at Harvard, in his *Elements of Technology*, published in 1829.[127] He observed in an address at the newly founded Massachusetts Institute of Technology in 1865 that when he first used the term it "was not in use nor was it widely understood." Events were soon to change that. By the time of his MIT speech, technology was well on its way to achieving the goals set forth by Bigelow, that is, "control over the natural environment and the creation, as a partial replacement, of a man-made environment."[128]

But the most important proponent of technological advance in America was the German-born immigrant John Adolphus Etzler of Pittsburgh, who sent a public message to Congress describing what technology could accomplish if given free reign.[129] Etzler in 1833 addressed "To All Intelligent Men" a detailed prospectus entitled *The Paradise within the Reach of All Men, without Labor, by Powers of Nature and Machinery*. In it he outlined his plans to harness tidal action, subdue the winds, and capture the heat of the sun to serve people's needs.[130] Reprinted several times in England, it was rather scornfully reviewed by Thoreau in the lead article of the *Democratic Review* of November 1843 entitled "Paradise (To Be) Regained." Thoreau chides Etzler for what he calls "a transcendentalism in mechanics" rather than "a reformation of the self." Thoreau seems genuinely convinced by Etzler of the possibilities of new technologies, but disagrees with him about human motivation, saying that it must stem from the "crank within" and that no "really important work [can] be made easier by machinery."[131] But despite the philosopher Thoreau's scorn, the engineer Etzler remained an important prophet and an influence on many, including the German designer of the Brooklyn Bridge, John A. Roebling.[132]

American technological activity was viewed as part and parcel of the triumph of republican institutions, still conceived of in the Jeffersonian usage in which "republican" meant what we today would call democratic, or close to the people. Steamboat inventor Robert Fulton said that "every order of thing which has a tendency to remove oppression and meliorate the condition of man directing his attention to useful industry, is, in effect, republican."[133]

When seeking government support for work on his "torpedo," Fulton argued that "republicans are those who labor for the public good."[134]

Early in the process of technological growth in America, the nation's leading statesmen clambered aboard the bandwagon. In 1817 Jefferson, John Adams, and James Madison all officially accepted honorary membership in the American Society for the Encouragement of Domestic Manufactures.[135] Most Americans were fascinated by machinery and embraced new technology. Foreign travelers during this period testified to the nation's obsessive interest in power machinery. French observer Michel Chevalier said the typical American "has a perfect passion for railroads; he loves them . . . as a lover loves his mistress."[136]

Economic historians agree that, at the time of the Revolution, the United States did not possess a skilled working class such as that of England or the continent. The young nation therefore turned to the machine. Writes one scholar: "The very 'backwardness' of the American economy and technology made introduction of the new factory ways easier."[137] It is not clear precisely where the factory system actually began, but for our purposes we can date it to a factory in Waltham, Massachusetts, in 1814, where all the processes used in producing a common item—in this case, cotton cloth—were found under one roof using a single power source. Since most of the new factories used steam instead of water power, they could be located anywhere. European travelers, such as the already-mentioned French economist Chevalier, were pleasantly surprised to see factories located not in tawdry cities as in England and on the continent, but in bucolic settings. The New England factory "sprouted in an unspoiled rural landscape."[138]

The real pioneer of the American factory system is conventionally considered to be Eli Whitney, who was apparently the first to develop one basis of the system—the use of interchangeable parts—at his musket factory near New Haven, Connecticut, in the years after 1801.[139] Equal credit should probably go to Oliver Evans, who, thirteen years before Whitney's gun factory, had built a flour mill in Newcastle County, Delaware, in which he installed belt conveyors, screw conveyors, endless chain-bucket elevators, and other elements of the modern assembly line.[140] So far ahead of his time was Evans, however, that when he predicted steam carriages in a petition to the Pennsylvania legislature for a patent in 1786 he was deemed insane.[141]

Several things should be noted about Whitney's project, which has been called the "first contract for mass production in the American manner."[142] The idea of interchangeable parts was not original to Whitney. Both the French and the British were already familiar with it, but both these nations preferred to use time- honored skills instead. Whitney had no other choice since, as he put it in explaining and defending his system, its purpose was "to substitute correct and effective operations of machinery for that skill of the artist which is acquired only by long practice and experience; a species of skill which is not possessed in this country to any considerable extent."[143] As a result, machinery rather than people became specialized.[144] In addition, it was a government contract that gave rise to this "first" experiment in mass production. Moreover, it was in the interest of a military project, freeing America from dependence on European sources of weaponry, that it took place. Still, Whitney and his partners, after ten years of toil, barely made a financial profit.[145]

The overwhelming desire to create industry without also creating an industrial working class was illustrated by the Lowell experiment. The leading producer of cotton goods, the largest American industry before the Civil War, was in Lowell, Massachusetts. American industrialists were conscious of the Jeffersonian strictures about the dangers of following in Europe's path; they sought to create a new kind of industry that would fit in with the nation's presumed destiny as a primarily rural and class-free country. Francis Cabot Lowell, chief founder of the industrial city that was to bear his name, had visited England in 1810–1812 and admired the textile machinery of Lancashire; but he was horrified at the condition of the English working class. He and his associates were determined that New England would not pay the same price for modernization.[146] Unlike European and other American factories, those to be built by Lowell and his co-organizers were to hire only young women from the surrounding area, treat them in a special way by creating a "total institution" that would meet all their needs as genteel young women, and then make it possible for them to return to the farms and small towns from which they had come. By having a rotating rather than a permanent population, Lowell and his associates, who "opposed the idea of a long-term residential force that might lead to an entrenched proletariat,"[147] satisfied their need for labor while preserving the Jeffersonian republican ideal of a classless society.

Every effort was made to make life for the young ladies of Lowell as comfortable as possible. The women lived in strictly regulated boardinghouses, their morals were supervised, education and religious practices were encouraged. Several magazines were even published by them.[148]

But all this harmony would not last. The first strike took place in Lowell as early as 1834, not simply because of a reduction in wages but as a protest against the paternalism of the system. The death of the Lowell experiment came with the flood of Irish immigrants to the city who found work replacing the girls who left in droves, unhappy with the system and unwilling to work alongside the immigrants.[149] The noble experiment of Lowell and his associates had failed the test of time. A new permanent proletariat of immigrants had undermined the attempts to combine factory production and pristine republican social institutions.

During this era, mechanical progress was proceeding at an ever faster pace. Farmers increasingly took advantage of improvements in agricultural technology, though this only increased their capitalistic orientation, bringing them under the control of outside financial forces from which they had to borrow money to pay for new equipment.[150] Great strides were made in transportation. Emphasis was placed above all on speed, getting there first so as to be able to settle the land and get its products to market. America developed a passion for steamboats, built more for speed than safety.[151]

The steamboat was soon replaced by the railroad, and water transportation by land travel.[152] Railroads were constructed in the quickest way possible, and new locomotives were designed for the special conditions of American usage.[153] Mining also made great progress during this era, with coal and iron found in the East and the gold and silver in the West.[154]

But these new developments in technology were accompanied by, and to some extent occasioned by, radical changes in the American social system. In a truly republican society, after it had abolished the aristocratic standards inherited from the British Empire, there was, at least for white men, only one real possibility for differentiation: wealth. Burgeoning capitalism opened the way for dividing Americans into the rich and the poor.[155] This division was strongly conditioned by the new technologies. The new ruling class arose spontaneously in the person of the "businessman," a type that arose in the course of westward expansion[156] but soon became a national figure. Some contend that the new group was not really a fixed class, since in the highly speculative age that was the early nineteenth century, fortunes rose and fell so rapidly and readily that the sums necessary to enter business "were within the capacities of numerous men."[157]

Be that as it may, the period between the Revolution and the Civil War was far from being one of general economic equality, even among free white men. The position of those increasingly perceived as the working class was worsening.[158] This is quite contrary to the impression one receives from reading Tocqueville and other foreign travelers, but we must remember that they tended to speak mainly about how meager American fortunes were compared with those of wealthy Europeans, and how many of the rich Americans had formerly been poor. Despite the firm belief of many Americans in the possibility of social and economic mobility, by the Civil War the rich "in the cities at least . . . constituted an elite that became increasingly segregated by exclusive clubs, high social life, intermarriage, foreign travel, and business alliances."[159] Not only was there extreme social inequality among Americans in the period prior to the Civil War, there was an increasing spread of poverty, marked by slums in the large Eastern cities, and aggravated by a complete lack of public agencies devoted to social welfare. Nor was there genuine social mobility. Some moved slightly up the ladder, but rarely.[160]

In retrospect, it is striking how long it took most American workers to become class conscious. Even though there was labor organization and agitation in the decades before the Civil War, the labor leaders of the era were quite likely to be small proprietors rather than workers as such.[161] The strikes that took place, and there were many, were conducted by skilled workers. Owners, buoyed by the courts and fresh supplies of immigrant labor, had little trouble in breaking them: "Both the labor movement and the reform movements . . . represented struggles to return to a past that had gone. Capitalism they regarded as the radical force, ruthlessly destroying the little liberties and amenities of another day, a new and alien power rising within the republican framework created by an earlier revolution."[162]

Many argued that workers could escape proletarianism by becoming farmers. In the 1850s Horace Greeley popularized the slogan of the new Republican party, "Vote yourself a farm." This was part of a whole mystique, later embraced by historians such as Frederick Jackson Turner, about the frontier as a safety valve for eastern labor. But according to a modern historian, "The theory . . . has been thoroughly demolished."[163] Virtually no eastern workers had either the skills or the money necessary to travel to the frontier to start a farm, and virtually none of them made it as farmers.[164]

Despite all this social upheaval, new technology was generally accepted as good in itself. Most Americans welcomed innovations—whatever their impact—without cavil. After a stay in America in the 1820s, the German economist Friedrich List wrote that "everything new is quickly introduced here, and all the latest inventions. There is no clinging to old ways, the moment an American hears the word 'invention' he pricks up his ears."[165] By European standards, American technology was far from polished, but it enabled Americans to do what they wanted: produce as much as possible for as many people as possible as quickly as possible.[166] Americans were satisfied that technology was dealing with nature as they wished nature to be dealt with, that is, it was conquering the wilderness and helping to realize the American dream. Their virtually complete lack of interest in controlling technology and their acceptance of its role in changing the social structure was itself a statement about politics, expressing the still robust belief that technology and its material abundance could solve all of the important problems of the republic.

Chapter Six

AMERICANS EXPLOIT THEIR CONTINENT, 1865–1940

The decades after the Civil War saw major changes in American attitudes toward nature and technology. "Nature" became a more multivalent concept. Increasing skepticism was being expressed about the relationship of technology and republicanism. At the level of philosophy and social thought, these years saw an increasing acceptance of the ideas of Darwin, the belief in nature as constituted by an amoral process of evolutionary chance in which competition leads to higher forms of animals and societies.[1] Though Darwin and other scientists did not overtly draw the conclusion that later forms of life were superior to earlier ones—except in survival capability—so ingrained was the belief in progress that it was assumed that they must be. Nature was no longer primarily a substantive presence in the form of forests, animals, plants, or the weather; it increasingly became a basic force in human affairs, taking the place of God as the ultimate touchstone of reality and the source of values.

However, it was nature in its substantive sense that became problematic in the period of the Civil War. George Perkins Marsh (1801–1882), who was originally trained as a lawyer and later became foreign minister to Turkey and Italy, was a self-taught geologist and naturalist. In 1864 he published a book called *Man and Nature* (later subtitled *Physical Geography as Modified by Human Action*), which went through ten editions. Marsh did not want to go back to primitive nature and recognized the need for change, but he believed that such change must be taken with "informed care."[2] Though he found dangers in its use, Marsh was a supporter of irrigation in the West.[3] Despite his concern for the natural environment, he remained quite Baconian in his basic attitudes, writing of nature that we must "learn . . . how to emancipate ourselves from her power."[4] Marsh's book, though widely read both in America and in Europe, had a limited immediate practical impact, but it deeply influenced a number of important people.[5] William Cullen Bryant, for example, recognized him as an authority and quoted him in an editorial in the *New York Post*. In a tangible way, Marsh also helped make Yellowstone Park possible. Though later editions of *Man and Nature* advocated great parks in the natural state,[6] Marsh helped convince officials that Yellowstone should be created not for aesthetic or spiritual reasons but for practical ones, such as the health of human visitors. "Marsh's arguments became a staple for preservationists."[7] As for Yellowstone itself, formally created by President Ulysses S. Grant in 1872, the park was established primarily for such things as its recreational value; neither the text of the bill nor the debates in Congress indicate otherwise. "It is clear that no *intentional* preservation of wild country occurred."[8] The motives of its creators were narrowly anthropocentric; the land was being preserved for human utility in the same way that trees were cut down to serve human ends.

The Advancing Frontier

The Americans of the post-Civil War period, freed from the distraction of battle, resumed the westward march the war itself had stymied. Not only a geographical phenomenon, "The Frontier" was also a state of mind. The essence of the frontier outlook was summed up in the mystique of the West, a promised land in which the redemptive powers of nature could be completely exercised. In America's romantic view of its continental rendezvous with destiny, "the West was a limitless national reservoir of spiritual strength."[9] As Bancroft had argued early in the century in his *History of the United States*, Washington alone had not redeemed us because we were still clustered on the Atlantic seaboard, close to corrupt Europe. But once the great interior was settled, under the leadership of Andrew Jackson, a complete product of the West, "Americans would achieve an organic relationship with the virgin land. There every man would become a child of nature."[10]

One noted American historian argues that the West was doomed from the start. Because it was based on irrigation, it was what the German social theorist Karl Wittfogel would have described as a hydraulic society, one that "demonstrates once more how the domination of nature can lead to the domination of some people over others."[11] Despite its promise of freedom, the West soon became "a land of authority and restraint, of class and exploitation, and ultimately of imperial power."[12]

Settlement of the frontier had been deliberately encouraged by the federal government. The Homestead Act of 1863, passed in large measure as a redemption of the campaign promises of the new Republican party, made it easy, in theory, for anyone to claim land for himself simply by settling on it and working it for a few years. According to conventional wisdom, the act was central to the opening of the West to the average man. Reality was quite otherwise. The act was actually aborted by the land speculator and the railroad monopolist, because it was "incongruous with the industrial revolution."[13] Why then has the myth surrounding the Homestead Act been so widely and so consistently accepted over the years? Historian Henry Nash Smith ties the act's role to the larger myth of America as the garden of the world. If homesteading did not work, increased population would be a burden rather than an asset.[14] More fundamentally, however, Americans, given their faith in the democratic promise of their continental experiment, believed that it *must* be possible, against all the experiences of other civilizations, somehow to combine an agriculture based on irrigation with a system of free small farm holdings.

A major propagator of the frontier myth was historian Frederick Jackson Turner. He began to propound the uniqueness of the frontier and its importance for American democracy in a major paper given in Chicago in 1893 and in his classic *The Frontier in American History*, published in 1920. According to Turner, the frontier "had been responsible for forcing Americans to discard 'complex European life' in favor of the 'simplicity of primitive conditions.' "[15]

Turner's ideas enjoyed great vogue for several generations. But there is reason to believe that his thesis is as egregiously overstated as was the earlier faith in the efficacy of the Homestead Act. One later historian faults him for discounting too easily the influence of European ideas and technologies, Eastern development capital, and especially the demands of Eastern and European markets.[16]

A factor often ignored in discussing the democratic potential of the West is the question of who would do the work. At first slaves were suggested for the job. Jefferson Davis, arguing against the admission of California to the Union as a free state in the Compromise of 1850, held that slavery was better suited than free labor to "an agriculture which is based on irrigation."[17] Though slavery came to an end with the Civil War, there was cold logic in his view. The irrigated economy of the West needed large numbers of workers for short periods of time. In order to survive, it could not provide long-term support for them, since their contribution was brief. The logical solution was servile labor, migratory in nature. Chinese and East Indians were imported for the task, then Japanese. Finally the West settled on migratory labor from Latin America, above all Mexico.

The movement toward the frontier was not as helter-skelter as some would believe. By the late nineteenth century, the federal government took a very direct role, creating the U.S. Geological Survey in 1879, which was to explore and map all of Western America, just as the U.S. Bureau of Ethnology was to collect anthropological data on its Indian inhabitants. Both were headed for a long time by a most remarkable man, John Wesley Powell (1834–1902).[18] One of the most widely known and respected scientists of his time, Powell was popular and famous as an intrepid explorer. But he was not only an explorer but also a philosopher, if an antiphilosophical one. He especially disliked the influence of the still revered Emerson, arguing against transcendentalist faith that "the history of science is the discovery of the simple and true, in its progress illusions are dispelled and certitudes remain."[19] Yet Powell retained faith in America as a special place. "The merit of Columbus," Powell asserted, "was his faith in science. . . . The new world was the trophy of science. . . . The new world became the home of republics."[20]

As a result of his explorations and mappings, Powell advanced great plans for the West that called for larger expenditures of federal money and a much greater degree of federal control. Land speculators were of course bitterly opposed to such measures.[21] Powell advocated the use of technology to control nature. After the disastrous Johnstown flood in Pennsylvania, caused by the breaking of a dam, he still argued against letting one mishap dampen popular enthusiasm for using modern science to control rivers that might otherwise "run to waste."[22] As a modern historian has put it, "The West was to be, in Powell's mind ought to be, a technological civilization, militantly modern, bent on the complete domination of nature."[23]

Though the federal government continued to be a major presence in the West after Powell's time (and still owns much of the land in most Western states), in the twentieth century the small farmers have increasingly been replaced by larger ones. Corporations, in the form of agribusinesses, have triumphed, usually with government assistance. Ironically, their only real competition comes from the demands of growing Western cities for water. With his faith in reason, Powell might well have found the urbanization of the West an even greater aberration than the commandeering of the land by a few.

Powell's advice was ignored, and Congress rejected his General Plan for the arid lands of the West. All that remained of Powell's vision was his drive to harness streams to the service of man, climaxed perhaps in the era of the New Deal with its vast hydroelectric projects. Liberal senator George Norris of Nebraska, father of the Tennessee Valley Authority, echoed Powell when he praised the TVA as a step toward "the dawning of that day when every rippling stream that flows down the mountainside and winds its way through the meadows to the sea shall be harnessed and made to work for the welfare and comfort of man."[24]

The Conservation Movement

In the last decades of the nineteenth century, the modern conservation movement was born. Conservation, since its first stirrings in the Middle Ages, had meant "protection."[25] But protection of what, for what purpose? From the very beginning, the American conservation movement represented an uneasy practical alliance between two radically different philosophies about nature. Some conservationists, originally the mainstream, were primarily concerned with the preservation of natural resources for practical use by future generations; for lack of a better term, we shall call them utilitarian conservationists.[26] The others, the preservationists, believed that nature itself was a source of value and should be preserved—in as primitive a state as possible—for its own sake. They believed that whatever benefits humanity received from nature's existence were basically incidental.[27] Utilitarian Gifford Pinchot and preservationist John Muir battled for the support of their respective positions from Theodore Roosevelt, America's first real conservationist president. Pinchot won that battle, but the ideological war continues to this day.

Gifford Pinchot (1865–1946) had studied forestry in Europe and worked on the Vanderbilt estate in North Carolina before entering government service. He thought of himself as a "scientific forester," and when in 1898 he took office as chief of the Division of Forestry it was with the expressed desire "to get forestry out of the dark and into the woods."[28] His major concern was with the utilitarian principle of "sustained yield." To his mind there were only two relevant, guiding interests to be considered: "humans and natural resources."[29] Pinchot "vigorously opposed efforts to allocate lands in the national forests for parks and other noncommercial purposes," writes a historian. "Hostile to what he called 'preservationists' he played a major role in establishing the dominant mood of conservation as efficient, scientific, economic development."[30]

Chief among the preservationists was John Muir (1838–1914). It is hard to imagine two men with like goals who were more different in specific ideology and life-style than Pinchot and Muir. Born in Scotland, Muir came to the United States and studied at the University of Wisconsin, where he overcame his Calvinistic background in the study of science and transcendentalism.[31] Soon after graduation, he walked a thousand miles from Indiana to the Gulf of Mexico in 1867. He settled in California and, for the next ten years, wandered through the High Sierra, studying botany, geology, and natural history. Appreciating nature in a fashion that owed much to Emerson and Thoreau, his primary interest, however, was wild nature in its raw state.[32]

An avid wilderness hiker, Muir was also an active writer and publicist for his causes, and a founder in 1892 of the Sierra Club. Originally an elite regional hikers' organization, it always had a political orientation. Only in the mid-twentieth century, however, did it become the influential national organization that it is today.[33]

By the turn of the century Muir had become the primary spokesman for wilderness preservation. As one historian has noted, "Muir promoted his cause with the passion of a zealot, for to him nature was a sacred reservoir that must be preserved for future generations. . . . He believed with Emerson and Thoreau that by shedding the artificialities of civilized society and penetrating the *wild* one could experience the rapture of Divine presence."[34] Muir found in nature the key to a higher spirituality. Inviting Emerson to visit his beloved mountains and see the truth for himself, he wrote, "I invite you to join me in a month's worship with Nature in the high temples of the great Sierra Crown beyond our holy Yosemite. It will cost you nothing but the time and very little of that for you will be mostly in Eternity." Emerson, despite his sixty-seven years, accepted and had a great time.[35]

Muir was not simply concerned with future generations as such. He was the exact opposite of the anthropocentric conservationist: nature had rights of its own independent of humanity. He believed that one could discover through science that man's dominion was not a necessary part of the universal plan.[36]

"Why," he asked on another occasion, "should man value himself as other than a small part of the one great unit of creation?"[37] Muir's challenge clearly was not simply to Pinchot and scientific forestry, but also to the whole American attempt to tame the continent, and indeed to the entire belief system about the relationship of man and nature that had prevailed in the Western world since the time of the ancient Greeks.

The clash between the conservationist positions of Pinchot and Muir was not simply philosophical; it was practical and even personal, as dramatized by their fight over the Hetch Hetchy Dam. In 1902 the reform mayor of San Francisco, James D. Phalen, began buying up water rights in the Hetch Hetchy Valley of Yosemite National Park in his own name, to avoid alerting water interests, eventually transferring the rights to the city. In 1912 the city renewed its rights in an attempt to dam the Tuolumne River and flood the valley to provide power. Muir and the Sierra Club were up in arms. The fight was a long and complex one,[38] with Pinchot pressing for the dam for utilitarian reasons. The city, he held, needed the water and power. Pinchot's side won, and Congress in 1913 passed the Baker Act, granting the city its rights to the water and the dam. It is worth noting in passing that, as in so many environmental clashes throughout American history, the votes against the Hetch Hetchy Dam came primarily from those who lived at some distance and did not stand to benefit from it. Those actually on the frontier have always tended to see nature less romantically than those who find it a rare refreshment to their spirits.

At first blush, Theodore Roosevelt, the nation's first great conservationist president, appears unlikely to have been a great friend of the environment. A firm believer in Manifest Destiny, he spoke in his *The Winning of the West*, an epic history written in his thirties, of "the spread of the English-speaking peoples over the world's waste space" as "the most striking feature of the world's history." Only "a warped, perverse, and silly morality" would, he felt, condemn the American conquest of the West.[39] He had no difficulty in relegating Indians to the periphery of the political and economic landscape. "I don't go so far as to think that the only good Indians are the dead Indians, but I believe nine out of every ten are. . . . The most vicious cowboy has more moral principle than the average Indian," he said in 1886.[40]

What then made Theodore Roosevelt such an ardent conservationist? In part it was due to his love for the West, an area where he had gone to recuperate, with notable success, after a childhood illness. But, above all, it was his interest in the "strenuous life." If Americans were a superior race it was because of their ability to meet hardship and danger. That ability must be maintained if they were to survive and triumph in the Darwinian struggle. War could, to some extent, be a means of keeping nationally fit, as Roosevelt believed and stated; but contest against the rugged outdoors was also important.

Roosevelt was a man of action as well as words. Regarding the presidency as a "bully pulpit," he used its powers to advance conservationist aims. As president he successfully pushed for water conservation and forest reserves, working with Congress, when possible, to take land out of the control of private speculators and reserve it for public use, acting unilaterally when necessary and constitutional.[41] His last great effort as president was to call a governors' conference at the White House in 1908, which was attended by virtually every important leader in America and resulted in the creation of a National Conservation Commission to study and make reports on the nation's conservation needs.[42]

Though it would be a full generation before another president with anything like Theodore Roosevelt's conservationist enthusiasm occupied the White House (in the person of his distant cousin Franklin Delano Roosevelt), the mainstream of conservationist thought did not dry up with the passing of John Muir. His place was soon taken by Aldo Leopold (1887–1948), a longtime worker for the U.S. Forest Service and one of the founders of the Wilderness Society.[43] He was, ironically, one of the first graduates of the Yale Forestry School, set up with an endowment from the Pinchot family, and started his career as a conventional representative of the dominant school of wildlife management. Over time, however, he evolved into a major spokesman for a radically different kind of conservation, one which put man back into the center of the biotic community.

Leopold has been referred to as "an authentic patron saint of the modern environmental movement," and *A Sand County Almanac* as "one of its new testament gospels."[44] Leopold did not spring forth overnight as the prophet of a new dispensation. Throughout his career, from his earlier experience as a conventional wilderness "manager" for the federal government in the West until its culmination as a professor at the University of Wisconsin, he was moving toward a basic break from the dominant attitude in which he had been trained.[45]

Leopold's creed has been represented as a major departure from previous ideas, he himself as someone whose "land ethic" presented "a community-oriented ecocentric alternative to the homocentric ethics of ecosystem management."[46] But despite its origins in the pseudo-mysticism of the mainstream of traditional ecology, Leopold's self-expressed intentions appear more modest.

Indeed, in *A Sand County Almanac* he sounds an almost anthropocentric note, saying that those who prefer wild geese to television are a minority,[47] and goes on to write that "when we see land as a community to which we belong, we may begin to use it with love and respect. There is no other way for land to survive the impact of mechanized man, nor for us to reap from it the esthetic harvest it is capable, under science, of contributing to culture."[48]

The land ethic is about precisely that—*ethics*. It is an attempt to extend to nonliving substance—if the land can be said to be really "nonliving"— the same rights we extend to living human groups. "That land is a community is the basic concept of ecology," he goes on, "but that land is to be loved and respected is an extension of ethics."[49]

"An ethic," Leopold writes, "ecologically, is a limitation on freedom of action in the struggle for existence." The first ethics dealt with the relations between individuals, the second with the relations between the individual and society. "There is as yet no ethic dealing with man's relation to land and the animals and plants which grow upon it," but "the extension of ethics to this third element in human environment is, if I read the evidence correctly, an evolutionary possibility and an ecological necessity."[50]

Central to Leopold's land ethic are his views on conservation. "Conservation," he writes, "is a state of harmony between man and land." For it to take place, the center of gravity of land-use ethics must be altered, since "land-use ethics are still governed by economic self-interest, just as social ethics were a century ago."[51]

How Leopold would have reacted to the interest in ecological ethics by various philosophers and theologians in the decades after his death is a fascinating speculation. There is no question that in the period immediately prior to his death he was generally pessimistic about the future of any land ethic. He admits, interestingly, that a taste for wilderness for its own sake is a paradoxical product of industrialization and modernization, that "wild things . . . had little human value until mechanization assured us of a good breakfast, and science disclosed the drama of where they come from and how they live." Yet he goes on to say that the "whole conflict thus boils down to a question of degree. We of the minority see a law of diminishing returns in progress, our opponents do not."[52]

During the latter part of the nineteenth century, many Americans, especially the urban upper class, paid increasing attention to nature. There was a big boom in spas, and New Yorkers especially began to go to the mountains. The Adirondacks, once scarcely penetrated, became "almost crowded" by the 1870s.[53] In 1872 the State of New York established a 715,000 acre "forest preserve" in the Adirondacks with the stipulation that it "shall be kept forever as wild forest lands"; the decisive argument for the preserve was a utilitarian one—that forest land is needed for an adequate water supply.[54]

Many of the city-bred refused to give up their familiar comforts, and for this they were not only disliked by the local inhabitants but satirized by the press. Various intellectuals complained that there was something undemocratic in a process whereby rural areas were taken over and spoiled for the locals by those of greater means.[55] "Gentrification" has a long history, but the phenomenon was first a rural rather than an urban one. Though more and more people began to seek the outdoors and the healthy exercise found therein, the fact remained that "Americans were incredibly homocentric in their approach to their surroundings."[56] After Theodore Roosevelt left the White House, his successors, William Howard Taft and Woodrow Wilson, continued many of his conservation projects, but gave increasing attention to Pinchot's emphasis on using the continent's resources for people and for the growing American economy rather than "locking them up"—to use the pejorative phrase of conservation's opponents—for future generations.[57]

American Attitudes toward Technology

Americans were becoming increasingly ambivalent about nature. Still willing to exploit it, they at the same time sought increasingly to preserve certain aspects of it, usually for human use but sometimes, in the case of a few such as John Muir, for its own sake. This same ambivalence was present in their attitude toward technology. For the most part, they continued to embrace it without qualification, but many were having second thoughts about its beneficence as well.

Through most of the nineteenth century most Americans were enamored of technology not simply because it was useful but because, they felt, it was beautiful. Much of their aesthetic appreciation was of course based on a utilitarian underpinning. Technology was beautiful in large measure precisely *because* it was useful, because, furthermore, it served republican ends.[58] So great did this aesthetic love of machinery become that many argued that "machinery might supplant the function of the fine arts in a republic altogether."[59] The old equation between nature and civilization still reigned. For the bulk of Americans, to "be both Nature's nation and a rapidly developing industrial power was . . . not a contradiction, but the fulfillment of America's destiny."[60]

The post-Civil War period marked the final triumph of technology on the American farm. During the war itself, farmers had been offered credit on generous terms to buy imported agricultural technology, which was especially useful because their sons were off to battle.[61] America pioneered the mechanization of agriculture. During this period technology increasingly dominated the activities of the American farmer. New developments took place not only in production technology but in land laws, harvesting, and marketing.[62]

Industrial growth proceeded rapidly. The railroads continued to lead the way, literally as well as figuratively. The first joining together of the nation by rail occurred on May 10, 1869, at Promontory Point, Utah, when the Union Pacific Railroad joined the Central Pacific. But the geographic scope of the railroads was not all that made them powerful forces for change. The railroads made the existence of large cities possible, but also enabled large industrial corporations, which were on the rise, to receive their raw materials and send out their finished products to their customers.[63]

Vast new industrial combines emerged, dwarfing the individual both economically and socially. Monopolies proliferated in such industries as oil, steel, and meat packing. Central to the post-Civil War technological expansion was the vast amount of new energy—from other than human and animal sources—now available. One scholar sums it up by noting that "the total horsepower for doing basic industrial work from all sources—steam, electricity, wind, water, and work animals—was approximately 2,535,000 in 1860 and 46,215,000 in 1900. This was nearly an 18-fold increase in energy applicable to production in one lifetime."[64]

Business and government worked hand in hand in increasing American productivity. Despite the attacks of some reformers on monopolies and the occasional attempts to curb them, what was good for large enterprise was considered, by and large, to be good for the country as a whole.[65] Vital to the industrial growth of the period was a new attitude on the part of workers. A "work ethic" geared to industrial society had to be created by industry, and it was. From a nation of farmers, unaccustomed to regular, disciplined activity, America had evolved into a nation of primarily industrial workers, accepting the new constraints of large-scale factory life.[66]

The Movement of Ideas

During the period following the Civil War, the intellectual and literary mind of America was slowly coming to grips with the new realities of an industrialized civilization. Prominent among those who reacted with mixed fascination and dismay to the new world were the brothers Henry and Brooks Adams, direct descendants of presidents John and John Quincy Adams. Lost aristocrats in a democratic era, they were still trying to hold fast to the old republican ideals in a world in which they seemed hopelessly anachronistic. Henry Adams (1838–1918) found in the new discoveries of science a cause for despair, especially in the second law of thermodynamics, entropy, which postulates that as time goes on there will be more and more disorder in the universe, ending in a final "heat death" in which all differentials of temperature and motion would cease.[67] In his works, above all in his classic *The Education of Henry Adams*[68] and *Mont-Saint-Michel and Chartres*,[69] he contrasted the medieval world with its (alleged) order and peace to the modern world with its disorder and chaos. For Adams, the "Virgin and the Dynamo" became the symbols of two opposing civilizations.

Adams was above all impressed—and depressed—by the influence of technology. Writes one critic:

A sense of the transformation of life by technology dominates *The Education* as it does no other book. . . . He regards industrial power as an objective "cause" of change in American society.

Throughout *The Education* we hear the voice of the historian calling attention to technology as an impersonal and largely uncontrolled force acting upon human events.[70]

As early as 1862 Henry Adams had told his brother Charles that "man has mounted science and is now run away with. . . . Some day science may have the existence of mankind in its power, and the human race commit suicide by blowing up the world."[71] Henry Adams's prediction of nuclear weapons is not startling. He propounded something he called the "law of acceleration" similar to what many contemporary ecologists dub the "law of exponential growth," according to which changes in science and human society were taking place at a faster and faster rate.[72]

Brooks Adams (1848–1927) has been described as a "congenital rebel."[73] Trained at Harvard Law School, he soon abandoned the practice of law to write mordant histories, influenced strongly by Darwin and Marx, in which he propounded deterministic theories akin to those of his brother; but he focused more sharply on skepticism about the ability of the American capitalist class to cope with the forces unleashed by technology throughout the world. His first and most popular book, published after much soul searching and discussion with his brother, was *The Law of Civilization and Decay* (1896).[74] The capitalists, he believed, controlled everything, not only the economy but the courts and government as well. Unfortunately, they were, he felt, incompetent as administrators. Such high-level ineptitude, in Adams's opinion, would have disastrous social consequences. A later work, *The Theory of Social Revolutions* (1913), concluded that "the extreme complexity of the administrative problems presented by modern industrial civilization is beyond the compass of the capitalist mind."[75] In other words, technology had created a system too complex to be dealt with by minds formed in an earlier era.

But Henry and Brooks Adams were in a distinct minority. The mainstream of American social thinkers and philosophers hailed what technology was doing for the nation. But American literature was telling a different story, increasingly questioning the effects of technology upon the American landscape and, above all, on republican institutions. A major figure in the turn-of-the-century rejection of technological civilization was the literary giant Samuel Clemens, better known as Mark Twain (1835–1910).[76] Brought up on the frontier, "Child of the Southwest in its early boom days . . . cradled in profitless schemes and nourished on dreams of vast potential wealth,"[77] Twain evolved during his career into a complete pessimist about the world. "Everything about the human condition is pathetic," he wrote. "The secret source of Humor is not joy but sorrow. There is no humor in heaven."[78] Ultimately for Twain, as for Henry Adams, the universe was a soulless machine that had scant regard for man.

But even within that universe, institutions do make a difference. The institutions being created by technology were dooming the old virtues, and with them the republic. In describing steamboats on the Mississippi, Twain noted that while the passengers could look at the beauties of the river, the pilot, required to concentrate on his technological task, could not.[79] In his novel, *A Connecticut Yankee in King Arthur's Court*, Twain's hero, Morgan, a former manager in the Colt arms factory in Hartford, is miraculously transported back to medieval England. He uses his knowledge to make himself a powerful figure and ultimately seeks to overthrow the monarchy and the Catholic church and institute a republic, with himself as president. He initially defeats his enemies, but eventually is overthrown by the population led by his arch foe Merlin, who represents age-old superstition and magic. Having lost most of his followers and drenched the land in blood with modern weaponry, he sighs, "My dream of a republic to *be* a dream, and so remain."[80] Twain's pessimism was not simply about the past, but also about the America he saw around him. Technology could not save republican government; quite the contrary.[81]

Most contemporary critics failed to grasp the hidden moral of Twain's cautionary tale and took the position that Morgan—and by extension their America—was correct in trying to use technology to create a democratic society. Though Twain was a fan of the "socialist" Edward Bellamy,[82] he was generally viewed as an apologist for the existing order and no utopian himself. Such a mistake could not be made about his contemporary, the prolific and popular novelist William Dean Howells (1837–1920), who produced two openly utopian novels, *The Traveller from Altruria* (1894) and, thirteen years later, *Through the Eye of the Needle*. Both were influenced by the author's peaceful, nonrevolutionary Marxism and especially by the English utopian-socialist William Morris.[83] Throughout his life Howells was ambivalent toward the city, glorifying rural virtues while analyzing the urban life he in fact led.[84]

"Technological nightmares recur in important novels of this period," as a critic has noted.[85] One extreme reaction against what technology seemed to be doing to republican ideals is found in the works of Jack London (1876–1916), an immensely popular novelist and short-story writer who is perhaps best known today for his story, *The Call of the Wild* (1903). London combined in his ideas and work a curious blend of devotion to Nietzsche and to socialism. He was expelled from the Socialist party in due course because of his devotion to Nietzsche's elitism. As part of his protest against an overly technologized America, he produced a novel, *The Iron Heel* (1907), which portrayed an America in which the majority of the population were in effect slaves to a new ruling class that now dominated the productive process. The moral was clear: new centralized machinery negated the hopes of democratic control.

If American literature was expressing a growing disenchantment with technology and its products, there was no question that the American political climate in the late nineteenth and early twentieth centuries contained the elements of a major protest against capitalist industrial civilization. This protest was exemplified in utopian communities such as the Shakers and the Oneida community;[86] social thinkers such as Henry George, Edward Bellamy, and Thorstein Veblen; and movements such as populism and progressivism. Henry George (1839–1897) was one of the most important American social thinkers of this era. At the time, his major book, *Progress and Poverty* (1879), outsold everything published in America, save the Bible.[87] George was a reformer in the old American tradition who basically believed that science would improve the human condition. His problem lay in explaining a situation ignored by the dominant social thought of his day: how was it that progress—in whose existence he firmly believed—had led to increased poverty rather than a higher standard of living for the population at large? In typical American fashion, George rejected the idea of political revolution and even of class conflict between workers and industrialists; he located the villain in the land speculator, who reaped where he had not sown, and profited automatically by the increased value of land caused by an increasing population. The wealth that accrued to the land speculator was the cause of the misery of the farmer and worker. Thus progress—in the form of the increasing growth of cities—led automatically and inexorably to poverty.

The solution to the situation caused by this paradox was as simple as the paradox itself: tax away the unearned increment of the landowner through a "single tax" on the increased value of land. George's ideas gained adherents throughout the nation. Many men ran for office—usually without success—on the platform of the single tax. But Georgeism never became a long-lasting political movement and was soon lost in the maelstrom of social discontent, leaving only the legacy of a small group of economists and ideologues devoted to his ideas. The banner of opposition to the existing order was carried primarily by agricultural interests unhappy with the fruits of increasing industrialization and centralization.

The main thrust of farmer revolt came with the movement known as populism, which took its name from the People's party, formed in Omaha in 1892.[88] The Populist movement has been the subject of much controversy. Some consider it simply an uprising of a displaced rabble against modern society. Others regard it as an important part of a long American tradition of democratic revolt against the status quo.[89] The mainstream of modern scholarship seems to hold that the Populists were not opponents of industrialism and technology as such, but sought mainly to control and democratize these forces through the political process.[90] Their major demand was for an inflationary policy that would make it easier for them, by raising the market value of what they produced, to pay off the debts that threatened them with loss of their farms. Basic to the Populist appeal was a desire to return to the virtues of preindustrial America. One can readily understand these wishes given the hard-money policy of the government, which "in effect supported a stepping-up of the pace of industrial concentration—the revolution sweeping away the remnants of Jeffersonian America."[91]

Eventually the Populist impulse subsided as its greatest champion, William Jennings Bryan, who used it to capture the national Democratic party nomination in 1896, failed in several attempts to win the presidency. In part his failure was due to the unwillingness of urban labor to rally to the cause, distracted by the Republican symbol of the "full dinner pail," the claim that Republicans could serve economic needs better than the Populists.[92] Unlike farmers, city dwellers were still largely content with the new industrial dispensation, or at least they were wary of the Populists' remedy for it. In time the farmers themselves became distracted as the discovery of gold allowed farm prices to rise. Formerly Populist farmers, now more economically content, became dependent and docile suppliers to a centrally controlled complex of food production and distribution, often becoming junior partners in agribusiness.

The Progressive Movement

But the centralization and hierarchy created by rising industrialism and its technologies did not threaten only the farmer. A major segment of the old established middle class was also imperiled, above all smaller businessmen and members of the hitherto "free" professions, such as law, medicine, and pharmacy. Yet, while they condemned the political and economic consequences of technology, the spokesmen for this new revolt against the system accepted its central premise—the need for technology to flourish unchecked and to dominate the economy and the life of the nation. This ambivalence was the essence of the movement that came to be known as progressivism.

The adherents of progressivism differed on many issues, but all were united in the belief that it was too late for society to reject industrialism.[93] The only solution was a close relationship between business and government,[94] and the hope was that a government which accepted rather than fought big business would be able to control it. Progressivism flourished in an era in which mergers were rampant not, as was generally believed, because they were called for by the exigencies of technological efficiency, but because they were profitable financially, and in tune with the new ideology of concentration. Many decisions seemed on the surface to be required by the demands of efficient production, but most were really designed to aid the owners in their struggle with organized labor.[95] Long before modern automation, the worker was becoming a cog in a machine.

Even when the motivations for change were more forthright, the problem remained. Republicanism might have been compatible with particular new technologies as such, but it could not easily be reconciled with an ideology that made technology an end in itself. Some thinkers of the era believed technology must facilitate a return to the simpler life associated with republicanism. Henry Ford argued that his new invention, the automobile, would have the desirable effect of destroying both the city and what he regarded as the corruption it had brought to American life.[96] But such voices were drowned out, at least among the educated class, by those concerned that the new technology posed too great a threat to the republican tradition.

What was clearly needed was some way of reconciling the new technologies with American democratic institutions. One attempt to do this was provided by the novel *Looking Backward: 2000–1887*[97] by Edward Bellamy (1850–1898).[98] Although written and published decades earlier, *Looking Backward* was the expression, in fictional terms, of the essence of the Progressive era.[99] His major work was a triumph of the technological imagination. Bellamy had been a writer of romances, of fictional works set in different and exotic settings. It was as a romance that his most famous and successful book began, but he soon came to believe in the ideal society he had set forth therein. The plot is simply an excuse, as in most utopian fiction, for an exposition of the features of the utopian society. *Looking Backward* was an instant success when published in 1888. Groups were formed throughout the United States to promulgate Bellamy's ideas, especially in Boston, New York, and San Francisco. Bellamy had eschewed the very word "socialism," both because he did not like Marxism and because he feared the word had the wrong connotations for Americans. Indeed, the characters in his novel insisted that the various "Red scares" of the latter nineteenth century were deliberately set up by capitalist agents to discredit criticism of the system. Bellamy and his followers used the word "Nationalist" to describe their ideas. Industry was to be "nationalized" rather than "socialized." The Nationalist movement was the means by which his followers sought to spread Bellamy's message,[100] which was calculated to appeal to middle-class Americans upset about rampant industrialism and its effect on their economic and social status but not interested, indeed fundamentally opposed to, changing social relations in any Marxist direction.

It was no wonder that Bellamy's message, like that of Henry George before him, appealed especially to the traditional middle class. Another source of its appeal was the fact that it did not involve violence. Put quite simply, Bellamy believed that in the future the centralization and consolidation of American industry would eventually convert all the nation's economy into one conglomerate. At that juncture the population would do the obvious and sensible thing and have the government take it over and administer it.

This theory of revolution—peaceful and rational—was designed to appeal to the middle class's distaste for turmoil and real social change, but these positive features of a utopia simply reflected many aspects of the new rationalized existence that progressivism was already effecting. According to the theory, everyone would be enrolled in an "industrial army," and politics would virtually come to an end, manifesting the increasingly popular belief in the superiority of neutral "administration" over partisan politics. This, after all, was the beginning of an era in which nonpartisan elections and the public administration movement flourished in America. Bellamy's utopia was essentially the technocratic state of the Progressive era extrapolated and projected into the future as a utopian vision. The Progressive era denigrated traditional partisan politics and sought to rise above it. In Bellamy's military- technological society, politics completely disappears, to be replaced by public administration, a society run by an essentially co-optative gerontocracy of elite managers. In trying to conquer the machine, the society itself has been captured by it.

After Progressivism

Throughout the period between the end of the Civil War to World War I, Marxist ideas and socialist political organization flourished from time to time, especially in the early twentieth century, along with the interest in populism, progressivism, and managerialism. The Golden Age of American socialism as a political force was in the pre-World War I period. Led by Eugene V. Debs (1855–1926), a former president of the American Railway Union and a leader in the ill-fated strike against the Pullman Company in 1894, the American Socialist party tried to operate as a standard American political party. It managed to elect several congressmen and numerous state legislators and controlled the mayor's office in many American cities including Milwaukee, Bridgeport (Connecticut), and Berkeley.

The American Socialist party was effectively destroyed by the Wilson administration during World War I, since it continued to preach the traditional Marxist line by opposing the war as imperialist; it was the only Socialist party in the world that remained predominately faithful to Marx on this issue. Debs went to jail for sedition, but while still in an Atlanta prison he received 920,000 votes for president in 1922. The American socialist movement, however, made few contributions to socialist thought. Above all, American Marxist theorists and movements continued to accept the Marxist view of technology as unequivocally good in itself, representing humanity's conquest of nature, and making possible the exchange of the "realm of necessity" for the "kingdom of freedom."[101]

But if the Marxists did nothing to illuminate the dilemma that technology posed for republicanism, at least one important American thinker tried, however obliquely, to do so. Economist Thorstein Veblen (1857–1929) was born on a farm in Wisconsin. Throughout his very checkered career as an academic economist, he was a thorn in the side of the reigning political-economic orthodoxy. Veblen believed—like Marx, whom he professed to abhor—that human beings naturally were productive animals, possessing an "instinct of workmanship." For Veblen, a prodigious student of languages and anthropology, history was the story of a conflict between naturally workmanlike human beings and conquering aristocracies that despised ordinary work in favor of vapid abstractions. For Veblen, the great conflict was not between economic classes in Marxist terms, but between the rational sedentary working class and various groups of warrior nomads who had conquered them and exalted pecuniary virtues at the expense of rational ones. Veblen considered the American businessman not only intrinsically oppressive, but in the service of phantasms that got in the way of steady work. He attacked the dominant ethos at its central point—the belief that businessmen were hard-headed and practical and reformers were dreamers. On the contrary, according to Veblen, it was the businessmen who were the dreamers. His *The Theory of the Leisure Class* (1899),[102] though widely attacked, became a classic. In it he coined the phrase "conspicuous consumption," which is now in popular usage.

Veblen's theory of social change held that the real work of society was now done by the industrial class.[103] That being the case, it was only a matter of time until the workers—and above all the engineers—overthrew the mere manipulators of monetary symbols and took their rightful place as the rulers of society. His *The Engineers and the Price System* (1921),[104] was combined with some of Bellamy's ideas in the technocracy movement of the 1930s[105] and became something of a Bible for the technocrats who hoped, in vain, that the engineers would revolt against the established order and create one geared to the rationality of the machine.[106] Alas, what little revolting the engineers did was far milder in form and aim. Veblen's hope for them as a new revolutionary elite became only a fantasy.[107]

The period between World War I and the advent of the New Deal saw little change in American thought regarding technology. Progressive ideology degenerated into managerialism on a grand scale. Business ideologists sought to create a new popular faith in American capitalism and in the "the American Way," in which labor would be a happy junior partner of a paternalist, all-powerful managerial class.[108] The total acceptance of the dominance of technology during this time is symbolized by the Chicago World's Fair of 1933, glorifying a "Century of Progress" with its slogan, "Science Discovers—Technology Makes—Man Conforms."

The New Deal itself was so ideologically formless and so dominated by political opportunism that it never came to grips with the problems that technology posed for the republican ideal.[109] New Deal policies often sought to serve the welfare of the poorest, primarily urban, one-third of the nation, but its political support rested largely on the traditional big-city political machines. It divorced the Progressive ideal of good government from social reform. New Deal aims were essentially conservative and not intended to break new ground ideologically.[110] Since the New Deal was unwilling— actually uninclined—to challenge the American class structure directly, it had to be content to strive for the economic well-being of the poorest within the confines of the existing order. Its experiments in ideas, like its experiments generally, were brought to an end by the approach of World War II.

During the New Deal era, dissent against the existing order continued, inspired both by the failure of capitalism as exemplified in the Great Depression and the slowness with which the economy was responding—if at all— to New Deal ministrations. There were movements such as technocracy, but there was also some continuing interest in Marxist socialism, which might have become more popular if so many erstwhile socialists had not become involved in the revived labor movement, which the New Deal had done much to foster. Many elements of the old Populist critique came to the fore again, though now it was somewhat soured by the taint of fascism—in the form of opposition to the New Deal led by Senator Huey Long of Louisiana and Father Charles Coughlin, among others.[111]

A much more direct challenge to American technological change came from a vastly different source. A group of primarily literary intellectuals, centered around Vanderbilt University in Nashville, argued that only a return to the agrarian values of the past, endangered both directly and indirectly by the urban, industrial-centered New Deal, could restore decency, order, and prosperity to the country. It was the South, they argued, the least industrialized region of the nation and the most faithful to its original republican culture, that could lead the way toward revival. In their manifesto of 1933—entitled *I'll Take My Stand* after the words of the Southern anthem "Dixie"—the "Nashville Agrarians," as they came to be called, openly challenged both urbanism and machine civilization. They had little direct political impact, as other Southerners continued to prefer the newfound largesse of a Democratic administration in which the South wielded great power, and most of their leaders eventually turned exclusively to literary pursuits. But they and their intellectual descendants had a profound influence on the post-World War II conservative movement and upon certain elements of the contemporary environmental movement as well.[112]

Despite all these stirrings, the average American remained firmly convinced that technology brought about a wonderful existence which fully justified any costs in freedom, and that the promise of the future was even greater. Some members of the white majority were dimly aware of the human costs of slavery and the treatment of Indians. Still others felt that the growth of the factory system and large corporations had somehow constricted the lives of Americans, and they had obscure misgivings about what had happened to the republic of free human beings envisaged by Jefferson. Others, far fewer in number, had some idea of the impact of uncontrolled land development and industrial growth on the environment. But most took all of this in stride. They knew little about the critics of mainstream America, and, had they known, they would have dismissed them out of hand. They remained content with what America had become. They had faith that it was not only the greatest country on earth but owed most of that greatness to its willingness to use technology to exploit the natural resources a beneficent God had provided. What had so obviously worked in the past could and would continue to work in the future. The theme of the Chicago World's Fair of 1933 had been "A Century of Progress." The New York World's Fair of 1939 laid claim to be "The World of Tomorrow," a streamlined utopia that reflected popular tastes, and did much to shape them as well.[113] Storm clouds across the Atlantic heralding the coming of World War II did much to disquiet the fair goers, especially in 1940; but on the eve of the war Americans still had faith that technology could and would bring about a better world for themselves and their children.

Chapter Seven

MARXIST SOCIALISM, NATURE, AND TECHNOLOGY

Without question, the most important political thinker in the modern world has been Karl Marx (1818–1883). Though his ideas may have almost disappeared in many former Communist nations, they still remain a potent force in the contemporary world. It is thus important to examine carefully what Karl Marx and those who claim to be his followers have to say on the subjects of nature and technology.

To do so adequately, we must first try to put Marxist thought in perspective. Let us begin with the question of nature. Marxist ideology does not fully separate the question of people and nature from that of people and technology. A central tenet of Marxism is that people approach nature to subdue it. Indeed, as we shall see, for Marx people do not simply subdue nature through technology, in a very real sense they and their technology create nature in the first place.

Marxist philosophy is a derivative of that of the greatest philosopher of the nineteenth century, Georg Wilhelm Friedrich Hegel (1770–1831). Hegel shared Plato's idealism and belief that material reality was a reflection of the Spirit, which was for Hegel the only reality. What we think of as physical nature was for Hegel a form of estrangement of the *Geist*, the Absolute World Spirit whose coming to consciousness constituted history. Marx inverted the basic Hegelian view of the universe—and set Hegel "right side up"—by making matter the central and exclusive aspect of the world and spirit its purely derivative reflection. He retained much of the basic Hegelian language and mode of thinking about reality, especially regarding Hegel's revolutionary insistence that nothing was really stable, that being was an unreal category, and that everything was process.[1] But for Marx, humans, not the *Geist*, constituted the center of the universe, and their views of nature and of technology were human centered.

Much confusion exists about important aspects of Marxist thought. As a leading student of Marx's views on nature has written, "There is no systematic Marxist theory of nature of such a kind as to be conscious of its own speculative implications."[2] The physical universe was, for Hegel, a manifestation of the Spirit. For Marx the physical universe of nature—though in some sense existing independently of people—only has meaning in relation to them. It is really only in terms of people that Marx is concerned with nature. In Marxist doctrine, nature exists for the benefit of people, to be exploited and used by them. At first Marx accepted the Renaissance and Enlightenment views that scientific laws were only confirmed by experimental practice,[3] that we could only know what we could create. But Marx took these corollaries a step further. Nature did not *really* exist until man had worked on it.[4]

Marx had been strongly influenced in his youth by the Hegelian philosopher, Ludwig Feuerbach (1804–1872). Feuerbach had broken with Hegel over his emphasis on the Spirit, and substituted the importance of the human as a sensuous living being. He especially criticized Hegel's philosophy of religion, arguing that religion represented the projection of people's best qualities onto an imaginary being. But Marx had objections to Feuerbach and other materialist philosophers on purely philosophical grounds. Since they still regarded nature as a fixed datum and philosophy as a reflection of it, they failed to recognize the historical transition that had taken place with the movement from agrarian to industrial production. In agrarian society land existed independently of people. In industrial society, as Marx wrote in the *Grundrisse*, nature had become a "mere object for man" and had long ceased "to be recognized as a power in itself."[5]

But Marx went further. The economic interpretation of history illuminates not only the history of human society but the nature of nature itself. Nature too has a history, inseparable from people's actions on it. The story of nature and humanity is one story. As he wrote in the *German Ideology*: "We know only a single science, the science of history. History is to be contemplated from two sides, it can be divided into the history of nature and the history of mankind. However the two sides are not to be divided off; as long as men exist the history of nature and the history of men are mutually conditioned."[6]

In human history, the human, a component of nature, has in turn affected nature in its entirety. Thus nature is affecting itself. This process takes place through the use of human labor. But it is only through working within nature that we can come to know it. Since, as noted earlier, nature is of no interest to Marx except insofar as human beings can use it, the only laws of nature that mean anything are those derived through human practice. People have needed nature to survive, according to Marx. To survive within nature, they have needed to understand its laws. But as Marx holds in *Capital*, "This understanding grows out of the practical transformation of the world."[7]

But this transformation works in the other direction as well. In changing the world of nature, humans themselves are changed. To understand this fully, we must understand two key aspects of human action in the labor process. The first is stated in Marx's famous discussion in *Capital* of the difference between the activity of humans in nature and that of other animals:

> We presuppose labour in a form that stamps it as exclusively *human*. A spider conducts operations that resemble those of a weaver, and a bee puts to shame many an architect in the construction of her cells. But what distinguishes the worst of architects from the best of bees is this, that the architect raises his structure in imagination before he erects it in reality. At the end of the labour-process, we get a result that already existed in the *imagination of the labourer* at its commencement, that was therefore already ideally present. He not only effects a change in the form in the material on which he works, but he also realizes a purpose of his own that gives the law to his modus operandi, and to which he must subordinate his will.[8]

Thus humans differ fundamentally from their fellow animals in how they make use of nature because of their ability to think. Unlike other animals, humans are self-conscious in their appropriation of nature.

Technology also fundamentally changes the human. For Marx, it is above all the tool that enables a human to take over nature. "Nature itself becomes one of the *organs* of his activity, one that he annexes to his own bodily organs, adding stature to himself in spite of the Bible."[9] In other words, in creating tools, people have made part of nature part of themselves. But this very tool also shapes the human being, leading to both increased physical and mental capabilities.

There is no question that Marx and Engels, besides appreciating nature for its utility, also shared a liking for its beauty.[10] They severely attacked capitalism for degrading nature, spoiling the forests and rivers, and destroying the farmland. But they also "shared the attitude toward nature held by contemporary men of industry and commerce and by the millions of settlers migrating to new lands to struggle with the hardships of the frontier."[11] The present revival of interest in ecology as a moral cause has led to much debate about where Marx and Engels had really stood on the question, with proponents of Marxism trying to present them as friends of the environment.[12]

In fact, however, Marx was a complete believer in the subordination of physical nature to people and their purposes. As a noted contemporary interpreter of Marx puts it, any "resurrection of nature" that does not entail its complete domination by mankind "cannot be logically conceived" within the system of Marxist materialism.[13] Much of the confusion over what Marx really stood for stems from a failure fully to appreciate the extent to which Marx in his mature thought abandoned the Feuerbachian romanticism of his youth to accept a position in which nature had meaning solely as the servant of humans. As one Marxist-inclined critic has noted, "Marx had no doubt that the progressive mastery of nature was the presupposition of human emancipation."[14] Says another scholar of Marx: "In later life he no longer wrote of a 'resurrection' of the whole of nature. The new society is to benefit man alone, and there is no doubt that this is to be at the expense of external nature."[15]

Marx's attitude toward nature can in large measure also be inferred from his numerous remarks about such things as the "idiocy of rural life."[16] He was a notorious critic and indeed an enemy of the peasantry, as were his followers, at least those in Europe and North America.[17] Such an attitude is hardly compatible with idealization of unspoiled nature.

Marx's ideas on nature were not completely shared by his friend and collaborator Friedrich Engels. This divergence between Marx and Engels cannot be understood without taking into account the extent to which Engels, even more than Marx, was impressed by contemporary science and was eager to assimilate Marxism into it. Engels's main concern in the *Dialectics of Nature* was to demonstrate that the physical sciences in the modern world were returning to the concept of nature found in the early Greek philosophers and, later, in Hegel: that the world was not something that was fixed forever and "ossified" as the natural sciences of even the Greek era had claimed, but something that had emerged out of chaos and was still developing.[18]

Commentators have differed on whether Engels's ideas about nature were more coherent or subtle than those of Marx. Robert C. Tucker, for instance, held that Engels's *Dialectics of Nature* was "a *mélange* of Hegel at his worst and the materialism of such nineteenth-century writers as Haekel."[19] But Tucker goes on to say that Engels maintained faith with Marx's idea of the material world as the materialization of human beings when Engels wrote how much the world had been changed by them: "There is devilishly little left of 'nature' as it was in Germany at the time when the Germanic peoples immigrated into it. The earth's surface, climate, vegetation, fauna, and the human beings themselves have infinitely changed, and all this owing to human activity, while the changes of nature which have occurred in this period of time without human activity are incalculably small."[20] Marxist scholar Stanley Aronowitz, on the other hand, argues that Engels's views were superior to Marx's in that he insisted that human beings were part of nature and the "distinction between *homo sapiens* and the natural kingdom was not to be made into a sharp dichotomy." Engels, according to Aronowitz, restored the Greek view of nature as "a constant process of coming into being and passing away."[21]

Marx himself wholeheartedly accepted the Darwinian concept that humans are part of nature and, according to legend, wished to dedicate *Das Kapital* to Darwin, "who declined the honor, saying that he has onus enough from his own work."[22] But unlike Darwin, for whom humans were always simply one more part of nature, Marx believed that people could somehow transcend and control it. Indeed, human history has been the story of progress toward that goal.

Prometheus, as Marx wrote in the preface to his doctoral dissertation, was "the grandest saint and martyr in the philosopher's almanac."[23] Marx's attitude toward the world always retained that Promethean thrust, glorifying the human conquest of nature. Marx and Engels were optimists about nature, as, according to later apologists, befitted men of their time.[24] A modern ecological theorist has lumped Marx with Locke and Adam Smith as one of the "philosophers of the Great Frontier."[25] Both Marx and Engels accepted the Baconian dictum that to command nature, one first had to obey it. Engels was graphic in expressing the dangers involved in not really doing so when, in a famous passage, he wrote:

> Let us not, however, flatter ourselves overmuch, on account of our human conquest of nature. For each such conquest takes its revenge on us. . . . Thus at every step we are reminded that we by no means rule over nature like a conqueror over a foreign people, like someone standing outside nature—but that we, with flesh, blood, and brain, belong to nature, and exist in its midst, and that all our mastery over nature consists in the fact that we have the advantage over all other beings of being able to know and correctly apply its laws.[26]

The first Marxist theorists were concerned that humans might violate basic laws of nature. They were of the opinion that when the final triumph of socialism had taken place, nature would somehow be treated more kindly. Exactly what form this reconciliation with nature might take was never made explicit in classical Marxist texts, however. Marx himself, as we have seen, was a vocal opponent of rural values. As a contemporary sociologist has noted, "Marx regarded the onset of urbanism as one of the blessings of capitalism," while "Engels . . . was . . . anguished by creeping urbanism," and wrote that the " 'isolation of the individual . . . is everywhere the fundamental principle of modern society. But nowhere is this selfish egoism so blatantly evident as in the frantic bustle of the great city.' "[27]

Marx and Engels were in strong agreement with respect to the role of technology, its centrality to the very definition of humanity, and its overwhelming importance to the very possibility of creating socialism. For Marxism, "Complex technology must be wholeheartedly accepted because, in the long run, it will emancipate man from the repressions of historical civilization."[28]

Marxists might have had views about how technology should be controlled and administered by society,[29] but they joined liberals in refusing to criticize the basic technological constitution of modern society.[30] For Marxists, "Hard technology, aggressive and polluting, remains . . . the specific weapon of progress, which can only be materialistic, and quantitative first and foremost, without the vain scruples of a sensitive soul in the presence of Nature."[31]

This virtual worship of machine technology is clearly connected to the belief of Marx and Engels and their modern followers that humans are productive animals.[32] A major merit of socialism, they contended, is that it finally will free people's productive capabilities that are shackled by capitalism. Marx and Engels had no quarrel with bourgeois technology. Indeed, it is one reason why they hailed the bourgeoisie as a "revolutionary" force in history and paid tribute to it for having finally liberated people's productive powers.[33] The socialist utopia, they argued, will clearly be built on the physical foundations laid down by capitalism. Because of their emphasis on physical productivity above all, many contemporary Marxists actually resemble capitalists, propounding an ideology of "bigger is better" and ignoring any resulting negative social consequences.[34] Anarchist Murray Bookchin has said that Marxists "articulated the liberal project more consistently and with greater clarity than its most blatant liberal apologists."[35]

Technological determinism holds that changes in technology in themselves directly, inevitably, and independently cause changes in economic and social structure. Marx was not a technological determinist. He held rather that the social structure in which technology operates is what determines the character of a society.[36] It was this very position that prevented Marxists from recognizing the extent to which their societies did not achieve socialist ideals of equality and freedom, despite having done away with the existence of privately owned productive property.

A major, perhaps *the* major, issue raised by the rise of industrial technology was the degree of freedom the workers could enjoy on the job. Paradoxically, despite all the talk about alienation in the workplace, classical Communist theory accepted the necessity for strong authority in the production process, in diametrical opposition to the position of the anarchists, with whom they contested for the leadership of the struggle against capitalism in the nineteenth century.[37] Engels, attacking the anarchist position in an essay published in Italy in 1874, argued that combined action meant organization and asked, "Is it possible to have organisation without authority?" The answer clearly was no, because—and this is all important—the authority was not based on human will but on technology itself. In the factory, "All these workers, men, women, and children, are obliged to begin and finish work at the hours fixed by the authority of the steam, which cares nothing for individual autonomy." In a pregnant passage, Engels states "If man, by dint of his knowledge and inventive genius, has subdued the forces of nature, the latter avenge themselves upon him by subjecting him, in so far as he employs them, to a veritable despotism *independent of all social organisation.*"[38] Thus we find Engels in effect repudiating much of Marx's position on the relationship of technological factors to social organization. It should be little wonder, therefore, given this ambivalence at the heart of Marxist theory, that Marxist regimes were never able successfully to come to grips with the worker alienation that necessarily continued to exist even after the triumph of socialism.

Contemporary Marxist Theorists

The inability of Marxism to deal directly with the relationships among humanity, technology, and nature is illustrated in the writings of most contemporary Marxist theorists as well. In the United States, mainstream Marxist thought, among Socialists and Communists alike, has reflected the traditional belief that the death of capitalism will solve the problems of alienation from society and nature. This will happen because of an unprecedented increase in material production that socialism will make possible by freeing man's technological genius from the shackles of the profit motive. This is the essential moral of writers as disparate as democratic-socialist Michael Harrington[39] and the more orthodoxly communistic Paul A. Baran and Paul M. Sweezy.[40]

This inability follows from the whole tradition of Marxism as established in the early days of the Soviet Union. When Lenin took power after the Revolution of 1917, rapid industrialization became the key to the survival of the new regime. He blamed centuries of czarism for the unwillingness of Russians to work hard, often enjoining them to imitate German diligence.[41] On many occasions, he endorsed the theories of the pioneer American industrial engineer Frederick Jackson Taylor, stating that "Taylorism" was simply another technology that the Revolution could adopt from the capitalists.[42]

But Lenin also made what would turn out to be a fatal mistake for his version of socialist democracy in believing that it would be possible to separate technical control of productive processes from full managerial control over the economy. Following the well-established doctrine of Marx and Engels (actually borrowed from Saint-Simon) that after the Revolution the "government of men" would be replaced by the "administration of things," Lenin held that any worker who could read and count could manage a factory. Thus professional managers would no longer be required. But Lenin made an exception for technical experts such as agronomists and engineers. They were still needed, yet would pose no danger to the socialist regime. Such people had worked for the czars for pay and would also work for the new socialist state.[43] In time, as we know, the sheer logic of modern industry sowed the seeds for technocratic control by a new bureaucracy that would rule over the masses in their name.

Not only Lenin but also Leon Trotsky believed that people must conquer nature through technology. Indeed he was, if anything, truer to the original Promethean Marxist inspiration than was Lenin. Under socialism, he wrote, man will become a "superman," altering the course of rivers, leveling mountains, changing nature in accordance with his own superior tastes, and ultimately altering his own nature as well.[44] Though Joseph Stalin despised Trotsky and had him expelled from the party and the country, and eventually had him assassinated in his Mexican refuge, there is no reason to think that he did not share the same beliefs on nature and technology, if perhaps with less messianic fervor.[45]

Stalin's recent successors in the former Soviet Union were forced to take a different position. Environmental disaster was too obvious to be ignored any longer in favor of maximizing production.[46] Initial recognition of the problem was hedged for a long time as theorists maintained the priority of human needs and desires, as defined by mainstream Western history. Thus one wrote that "to see the future in a rejection of 'Western technologism' in favour of 'Eastern submergence in nature' means to absolutize the split in mankind and to perpetuate backwardness."[47] All of this, like so much else of the past, is now being swept aside. The traditionalist opposition to the Communist regime, embodied in groups such as Pamyat, made restoration of nature part of their crusade for returning to the past.[48] But even aside from such extremists, a major environmental movement has arisen, spurred on by such events as the atomic disaster at Chernobyl.[49]

Given the basic premises of the Marxist system, science and scientism always enjoyed great prestige in Soviet society.[50] But Soviet ideology, with its emphasis on production, had disastrous environmental consequences. As one observer put it, "Soviet economic and political institutions seem designed to produce environmental deterioration and resource depletion just as inexorably as their American counterparts."[51] Throughout the former Soviet Union, the air is dangerous to breathe, rivers are polluted, and soils are dangerous to till. The great inland Aral Sea has been reduced to a fraction of its former size by careless use for irrigation of the rivers that fed it. Like its former satellites in Eastern Europe, the former USSR is a major environmental disaster area of unprecedented scope. Like them, its citizens face the dreadful problems of how to save nature without starving or going without energy in the interim.

The plight of the former Soviet Union—now the Commonwealth of Independent States—is not only tragic but ironic. Not only had orthodox Marxists always believed in the power of the technology of the industrial revolution to solve all problems, but only a decade or so ago they were looking to postindustrial technology as the new savior of the Communist dream. This new hope was embodied in the briefly flourishing theory of the Scientific Technological Revolution (STR). The concept of the STR originated with the Western Marxists J. D. Bernal and Victor Perlo.[52] Though a major intellectual product of the ill-starred Dubcek era in Czechoslovakia,[53] it survived that regime's suppression by Soviet tanks to become a major theme for both Soviet and East European ideologists. The theory argued that automation, the computer, and nuclear energy were bringing about a new era that would make the elimination of work in the old sense finally possible.[54] Largely through the influence of Georgy Arbytov, it became for the 1970s a major postulate of official Soviet thought. Thus at the 1971 party congress, Prime Minister Leonid Brezhnev spoke of the need "organically to fuse the achievements of the Scientific and Technological Revolution with the advantages of the socialist economic system."[55] Much less was heard of the STR in later years, indicating the Soviet's growing realization of the extent to which they lagged behind the rest of the world in most technologies. Its brief period in the sun, however, indicates how deeply the Soviet Communist state was wedded to the basic Marxist notion of using technological progress to bring about a kingdom of freedom on Earth.[56]

In considering Marxist thinkers since the time of Marx and Engels, we have primarily looked at those in the mainstream of orthodoxy. But Marx was such a seminal thinker that various schools of thought have arisen, all claiming to represent what "the master really meant," or what he would have espoused had he been confronted with contemporary problems. With regard to nature and technology, the most important of these schools is the so-called Frankfurt school, which gave rise, in the 1960s, to the New Left both in Europe and in America.

Frankfurt school philosophers such as Max Horkheimer, while agreeing with Marx that the domination of nature took its ugliest form in capitalist society, denied that it was simply a side effect of the class struggle. As far as they were concerned, it was the root malady of the era, in a real sense transcending the class struggle. Thus the ground was laid for a fundamental reorientation of Marxist thought.[57] The main theme of the Frankfurt school was the need to synthesize Marx and Freud, a daunting task indeed. Freud was essentially a pessimist about human nature—his theory of the id has been compared to Augustine's theory of original sin—and for both of them human beings are born evil. For both only some form of rigorous personal and social discipline can prevent society from dissolving in an orgy of rape and murder. Marx, on the other hand, as we have seen, accepted both Rousseau's Romantic view that man was basically good and corrupted only by society, and the Enlightenment corollary that a better society could be created in which humanity's natural goodness could flourish.

Yet, accepting both Marx and Freud as the latest prophets, and desperately wanting to worship all their gods at the same altar, the German theorists were determined to conflate these two opposing philosophies. They were encouraged by the rediscovery of Marx's early works, revealing a Romantic concern with alienation as a psychological problem and an emphasis on human fulfillment. Various members of the Frankfurt school had differing ideas, but they finally arrived at a synthesis, which held that modern society, based on a scientific rationality that denied fundamental human realities, was both psychologically and economically alienating.

A major expression of this position is found in the work of philosopher Herbert Marcuse, an emigré from Germany who became a leading figure in the American New Left. Like Freud, Marcuse argued that a certain amount of repression was necessary, but that industrial society created a "surplus" repression that interfered with human creativity and spontaneity more than was justified. In this formulation he not only glorified basic human drives within a still quasi-Freudian framework, but enabled his proponents directly to villainize modern industrial society itself, rather than merely the social and economic arrangements of capitalism. [58] Ending capitalism is not enough, he argued, all of modern society must be radically changed.

Marx, Engels, and Lenin held that capitalist control was just as firm, if not more so, in democratic societies. Marcuse continued in this tradition in his view that the power of technology is just as great in such societies, if not more so, because technology's values are internalized. A new irresistible system of power has been created: Technological progress, extended to a whole system of dominion and coordination, creates forms of life (and of power) that appear to reconcile the forces opposing the system and to defeat or refute all protest in the name of the historical prospects of freedom from toil and domination. [59]

Marcuse, concerned with Freud as well as with Marx, found an inner connection between technological mastery of nature and the mastery of the human psyche. The control of external nature requires the control of internal nature also; this, indeed is the "revenge of nature," which forces us to dominate ourselves in return for dominating nature. The problem then becomes one of how to liberate people from this control of their own selves caused by modern science's control of nature. The solution Marcuse seems to turn toward is a new *liberating* technology. His views of science are therefore necessarily ambivalent.

Marcuse starts from the point of view of the early Marx, the "nature romanticism" that the mature Marx repudiated. He looks toward a situation "in which a new science would discover friendly and helpful qualities in nature overlooked by the domineering attitude toward nature characteristic of ordinary science."[60] This would be the basis of a new technology and would make possible a wide variety of benefits to society: "Some aspects of the new technology can be delineated: the complete rebuilding of cities and towns, the reconstruction of the countryside after the ravages of repressive industrialization, the institution of truly public services, the care for the sick and the aged."[61] But exactly what this new technology will consist of, just how it will differ from existing technologies save in intent, Marcuse never tells us. Faithful to his version of Marx, Marcuse sees all science and technology as historically relative, and holds therefore that an entirely new science and technology would be possible in a liberated society, a science which "might discover new laws of nature, and a new technology [that] might extract nature's resources in a less violent way."[62]

In any case, the new technology must be able to do all that the current technology does in conquering scarcity, since man can never enter into the "kingdom of freedom," the promised land of Marx's earthly eschatology, unless scarcity no longer exists. Yet as one student of Marcuse's thought argues, this is a catch–22 situation, since "there can be no liberation without abundance, but there can be no abundance without alienation. The means for abundance hurt the ends for which abundance is itself the means."[63] Thus in the last analysis, Marcuse's thought remains primarily an abstract, theoretical aspiration, however noble or ignoble.

Standing in pointed opposition to many of Marcuse's positions is the German philosopher Jürgen Habermas, a fellow member of the Frankfurt school and arguably the leading Western Marxist theorist of our day. Unlike Marcuse, Habermas does not believe that science is basically conditioned by society, but that it represents an objective picture of a permanent reality outside man.[64] In a direct response to Marcuse's claim that modern science and technology are a specific project of capitalism, Habermas argues that "the achievements of modern technology are a project of the human species as a whole."[65] Habermas is involved, however obliquely, in nothing less than "the rejection of the Frankfurt School theory of the domination of nature as the basis of human domination."[66] He holds that science must yield not simply aesthetic satisfaction but knowledge with its potentiality for manipulation and control.[67] For Habermas, any "resurrection of nature" that does not entail its effective domination "cannot be logically conceived within [Marx's] materialism."[68]

Indeed, Habermas goes back beyond Marx, to the Enlightenment itself, of which Engels claimed Marx to be the final culmination. In the words of Joel Whitebook, he seeks the successful completion of the Enlightenment project and "takes rational autonomy as a perfectly adequate idea of selfhood, and feels no need to formulate an alternative emancipatory concept such as a new sensibility."[69] But Habermas is famous for his theory of "communicative ethics," which are independent of the means of production.[70] As Whitebook also points out, "Communicative ethics is thoroughly anthropocentric. By virtue of his communicative abilities, man is the only value- bearing being."[71] Thus nature cannot have rights of its own, and must be dominated by science.

Though both members of the Frankfurt school, Marcuse and Habermas diverge widely in their views on nature and science. They also diverge in their views on social action, though both remain consistent with the basic Frankfurt approach by which Marx must be understood in the light of Freud. There was also a practical political consequence, for Marcuse at least, in this approach. If psychology was just as basic as economics, one could argue that oppression in modern society was not primarily economic in the old sense— the capitalist bourgeoisie dominating the blue-collar proletariat—but primarily psychological, a rational-scientific elite dominating all of society. This, of course, relates to the concept of technical reason that Marcuse and Habermas have debated. In this case the source of revolt against the existing order would not necessarily be the old working class, but could well be other socially oppressed groups. Indeed, this was one of the central tenets on which the New Left, especially in America, differentiated itself from the Old Left. The enemy was no longer capitalism as such but rational bureaucracy. In this new struggle, the old working class no longer constituted the cutting edge.

The phenomenon of the increasing affluence of the working class and its increasing acceptance of the existing capitalist order has long been noted. It led many critics of classical Marxism, both anti-Communists and "Revisionist" socialists such as Eduard Bernstein, to argue that Marx had been wrong and that the Revolution was not going to take place because the proletariat, its natural protagonist, had ceased to exist. The middle class, which Marx had predicted would collapse into the proletariat, was not collapsing but changing, transforming itself into a new technical intelligentsia into which many former working-class members were being integrated. What was crucial in this transformation was, of course, modern technology, which had vitiated the predictions Marx had made during the early industrial age.

But what if the new technical intelligentsia became members of the proletariat like the old blue-collar working class? Then the Marxist revolutionary project would still be a going concern. That the technical intelligentsia are part of the proletariat was one of the central contentions of the New Left, especially in Europe. Discussion of the concept of a "new working class" began in the existentialist Marxist journal *Arguments* in 1959.[72] The leading proponent of this theory was the French theorist Andre Gorz, himself an existentialist Marxist and disciple of Jean-Paul Sartre.

Gorz held that "the technical intelligentsia—which had become the main bearer of the forces of production—had now become a new working class."[73] He also contended, along with another existentialist, the Marxist Serge Mallet,[74] that the new working class was not only part of the working class, but its avant-garde. Its demands were more basic, and it could not be bought off with higher material benefits.[75] A similar thesis was proposed by French New Left sociologist Alain Touraine.[76]

What did the new working class want? Nothing more or less than the total reorganization of work and society. The major means of social change would be through what Gorz called "autogestion"—worker control over employment and the work process itself. This would strike directly at the main evil of modern capitalism. To Gorz, "the chief evil of capitalism was not exploitation but the alienation of the workers' creativity."[77]

Like the American New Left, Gorz emphasized the important role of students in the revolutionary process. Their education gave them intellectual independence. But at the same time, they were also involved in the reproduction of the labor force, since their schooling prepared them for a "proletarian role in industry."[78] In May 1968 students took the leading part in a revolt that shook Paris for a brief period. At one time even the Communist trade union, the CGT, supported the general strike despite the fact that the Communist party had not been sold on the idea of a new working class.[79] Most organized French workers, including hypothetical members of the new working class itself, took part. The strike became something of a great festival of nonalienation. Ultimately it failed because most of France supported the De Gaulle government and stability, and so it was not the cause of major changes in French society; today the revolt is little but a memory. Undaunted, Gorz has continued to attack the old Marxist concept of the working class and revolution, but not without an attempt to combine Marxism with such fashionable causes as ecology. Gorz's *Ecology as Politics*[80] is a breathless jumble of dire warnings about the environment—statements about how human beings now want not simply economic emancipation but true freedom across the board (including freedom from traditional drudgery); it is also checkered with traditional Marxist jargon.

The ideas espoused by Gorz and others tempts one to agree with the judgment of the anarcho-socialist Murray Bookchin, a keen and not unsympathetic student of Marx, that "a theory that is so readily 'vulgarized,' 'betrayed,' or, more sinisterly, institutionalized into bureaucratic power forms by nearly all its adherents may well be one that lends itself to such 'vulgarizations,' 'betrayals,' and bureaucratic forms *as a normal condition of its existence.*"[81]

He would appear to be right. Despite the tremendous extent to which Marxist theorists have recognized the importance of the relationship of nature and humanity, they still have not advanced beyond the mainstream of the Enlightenment from which Marxism arose, and have little to say to anyone who does not accept the exploitative premises of modern liberal capitalism.

Chapter Eight

NATURE AND TECHNOLOGY IN ISLAM

Islam is on the move again. After centuries of economic, political, and cultural subordination to the industrialized world of Europe and America, the Islamic nations are regaining much of their self-confidence. Over a millennium after their legions were stopped by the French, a large and growing Muslim minority of immigrants to France are challenging French ethnic and secular identity. Britain, long the overlord of much of the Near East, is now home to large numbers of Muslims from its former imperial possessions, and perhaps as many people attend the mosques every week as the well-established Church of England. Even in the United States, which has never ruled over Muslim areas except for a small part of the Philippines, in a few decades the Muslim population may well exceed the Jewish population.

It is not only in the traditionally non-Muslim world that the power of Islam is rising, but in the Middle East itself and in the former Soviet Union. The Muslim-dominated oil fields of the Middle East, on which many Western nations depend for energy, are a major source of international instability, and new, former Soviet republics composed of Muslims are emerging on the international scene. Confounding the mainstream of development theorists who have long held that modernization would wipe out traditional religions such as Islam, virtually every Muslim nation in the world has a large and growing Islamic fundamentalist movement. Because almost a billion of the world's population hold allegiance to the tenets of Islam, its outlook on nature and technology is of the utmost importance to the future of planet Earth.

Though we have spoken of Islam and the West as two different civilizations, not all would agree. Some argue that Islam is in effect a heresy from the Christian tradition. Historian Arnold Toynbee considered Islam, like Christianity, a " 'deviationist' Judaic religion."[1] Lynn White Jr. speaks of it as "a Judeo-Christian heresy,"[2] a view that is vigorously rejected by a leading contemporary Muslim scholar.[3]

Be that as it may, Islam traces its origins to A.D. 613, when a native of the trading city of Mecca, Muhammad (ca. A.D. 570–632), began preaching a new religion to his polytheistic fellow tribesmen. He spoke of Allah, who, he claimed, had previously revealed himself through the Jews and Christians, but they had forgotten or neglected his teachings.[4] Muslims revere Abraham, the prophets, and Christ (whom they regard as another prophet), but for them the final and definitive revelation was made by Allah to Muhammad. The new religion spread rapidly, impelled both by voluntary conversion and by the sword, first through Arabia, then North Africa, and even as far away as France. There, the Muslim armies were defeated at Tours by Charles Martel, and they retreated behind the Pyrenees. Islam also threatened and almost toppled Byzantium despite the Arabs' poor equipment and small numbers— a tribute to the fighting spirit over military professionalism.

Not only did the Muslims threaten Western Christendom militarily, they threatened it intellectually as well. The thinkers of the West were still under the spell of the ancient Greeks, above all Aristotle. Now the Muslims who had retained the originals became the source of many of the ancient texts that now appeared, and were eagerly devoured, in the West. Some works of Aristotle were first translated from Greek into Arabic by Muslims, and then, later, from Arabic into Latin by Saint Albert the Great. These translations influenced how Aristotle's ideas were interpreted. When Albert's pupil and successor Saint Thomas Aquinas sought to synthesize Aristotle and Christian scripture, he also had to wrestle with the ideas of Arab philosophers such as Avicenna (980–1037) and Averroës (1126–1198), who were among the great commentators on Aristotle.

Ironically, while the Arabs were preeminent in preserving and spreading the scientific work of the ancient Greeks, they did very little to extend it. At first they did seem to be on the way to overtaking the Greeks in science. During the four centuries from 750 to 1150, according to one Muslim scholar, "Islam held the lead in scientific activity," with original works appearing especially in mathematics, optics, astronomy and medicine,[5] and geography and history were well developed as well. Despite these promising beginnings, in a few centuries Islam lagged far behind Europe in scientific knowledge and technological application. How did this happen?[6]

"Even in the early Middle Ages," writes one historian, "the parts of Europe adhering to the Latin Church began to show a technological dynamism superior to that of the generally more sophisticated cultures of Byzantium and Islam." The problem of accounting for the "relative passivity toward technology" is accentuated by the fact that Islam and the Western world were closely related societies. But, in Islamic thought, "Science was not separated from philosophy, nor philosophy from theology, a position which in the long run could be counterproductive from a scientific point of view."[7] A Muslim scientific expert says that "Islam refuses to break the unity of thought in the face of economy and politics, science and technology, religion and society: the epistemology of Islam is the matrix that webs all the elements in a single orientation, based on the human soul."[8]

The Islamic position not only requires that science and religion be kept together, but automatically precludes the Baconian use of science to conquer nature. Science and religion cannot conflict as they did in the West. Science in Islamic philosophy is, like everything else, judged on moral grounds, and real science is by definition almost good, since it must exist for the sake of social good.[9]

The primacy of moral considerations dominates the Islamic view of science, unlike that of the mainstream West. This position has won the admiration of those in the West who are unhappy with the alleged value-neutrality of modern Western science.[10] Yet another non-Muslim observer argues that the Islamic attitude is simply a heritage from the classical and medieval worlds.[11]

It is impossible to understand the Islamic attitude toward science without understanding the Islamic attitude toward nature. Nature was created by God not simply for man's use but as "man's testing ground. Man is enjoined to read its 'signs.' Nature has therefore been created both orderly and knowable. Were it not so, were it unruly, capricious and erratic, it would be a 'ship of fools' where morality is not possible." But nature is not merely an object of knowledge. It is a sacred trust from God. "Man has, of his own accord, accepted nature as a trust (*amana*) and a theatre for his moral struggle. . . . Man is but the deputy of God possessing no authority save that of a steward."[12] This view, of course, differs basically from that often attributed to Christianity.

According to the Islamic view, a transcendent god "does not necessitate debased creation: *de-divinisation* need not imply *de-sacralisation*. Indeed, Islam holds that there is no such thing as a profane world. All the immensity of matter constitutes a scope for the self-realisation of the spirit. All is holy ground. As the Prophet so beautifully puts it: 'The whole of the earth is a mosque.' "[13] Yet despite their alleged differences about the value of nature, Islam and Christianity—at least a major part of medieval Christianity—are united in stressing that contemplation rather than manipulation should be the primary way in which nature is approached. It was this basically contemplative attitude toward the unity of science and religion that enabled the Muslim world to accept much from the sciences of others.[14] It also makes it so difficult for contemporary Islam to deal with a modern Western scientific tradition that rejects the unity of knowledge and the primacy of contemplation and moral judgments in scientific activity.

Just why did Muslims, after a distinguished beginning, fall behind in the sciences? The Islamic scholar Fazlur Rahman attributes its decline to what he calls a "fateful distinction" made between "religious sciences" (*ulum shariya*) on the one hand and "rational or secular sciences" (*ulum agiya*) on the other. Philosophers argued that because life is so short, preference should be given to the sciences that could help men save their souls rather than to secular sciences. The rapid spread of the mystical cult of Sufism, which opposed spiritual life—which is stressed—to rationality, played a role as well. Rahman also notes a sociological reason: those who had religious degrees could obtain jobs as qadis or muftis (judges in religious courts), while philosophers or scientists could work only for the royal court.[15] In any case, as he puts it, "The Muslim attitude toward knowledge in the later medieval centuries is so negative that if one puts it besides the Qur'an one cannot help being appalled."[16] All this occurred despite the fact that historically there has been no Islamic counterpart of the "religion versus science" debates that has troubled the Christian West, especially in the nineteenth century.[17]

The Western economic historian David Landes is more direct in blaming the Islamic religion and its leaders for the Muslims' scientific and intellectual decline. "In the Muslim world," he writes, "it was religious rather than national or ethnic pride that posed an obstacle to the importation of knowledge from the outside. From the start, Islamic culture was at best anxiously tolerant of scientific or philosophical speculation."[18] The great Western scholar of Islam, G. E. von Grunebaum, largely agrees with Landes about the extent to which the sciences were never quite at home in a Muslim world that was primarily religiously oriented. But the failure of science was tolerable, since while "the loss did indeed impoverish Muslim civilization . . . it did nothing to affect the livability of the correct life and thus did not impoverish or frustrate the objectives of the community's existence as traditionally experienced."[19]

The eminent historian of Chinese science, Joseph Needham, has a different perspective. He notes that Muhammad often had words of praise for merchants and few for agriculturalists, and that the early Arab cities were highly mercantile in character. But when the caliphate was established in Baghdad, the society became more bureaucratic on the Persian model, and in time much like that of the Chinese. Needham concludes: "So perhaps what began in Islamic civilization as a mercantile culture, ended by being thoroughly bureaucratic, and to this might possibly be ascribed the decline of Arabic society and particularly of the sciences and technology."[20] Thus, "Moslem thought froze into a fixed mold just at the time when intellectual curiosity was awakening in western Europe—the twelfth and thirteenth centuries A.D."[21]

What was true of science was also true of technology. Here also, after a brilliant start, a long period of stagnation ensued. As in philosophy and other fields, the Muslims were at first ahead of Europe, and also an agency for the transmission to Europe of new knowledge and techniques. Much of Muslim technology was borrowed directly from China. From the Chinese the Arabs, like the Western Christians, received gunpowder and the compass, and learned the art of papermaking.[22] Still, the Muslim world was several centuries slower in accepting printing from the Chinese "because learned prejudice long forbade reproduction"; it was not until the eighteenth century that printing surfaced in Islam.[23]

But if the Muslims were wary of printing, they remained fascinated by inventions that seemed to imitate activities of life: ". . . inventions in the service of the miracle. . . . What created a sensation in the late eighteenth century was not the new spinning machinery, but the manlike automatons who walked, played instruments, spoke with human voices, wrote or drew."[24] Technology was not for practical use as much as it was for celebrating the wonder of the world.

Muslim attitudes toward technology, like Muslim attitudes toward science, were shaped by the importance Islamic thought placed on the reverence for nature and service to man. As a Muslim scholar describes it, Islamic philosophy already sets up the framework for assessing technology: "Such concepts as *adl* (all-pervasive justice), *istslah* (public interest), *khalifah* (trusteeship) and *iotisad* (moderation) and Shariah injunctions, for example in environmental areas such as *ihy* (land reclamation), *harim* (conservation areas) and *hima* (public reserves), can accurately map the circumference of technological activity."[25]

In addition to philosophical convictions about respecting nature, by the late Middle Ages the Muslim powers displayed an increasing resistance to technological change that extended even to military technology, usually the first area in which such change is accepted. "There was roughly a forty-year lag in Islam's adoption of the European cannon" during the Middle Ages.[26] Ironically, the decisive defeat of the Mamelukes, the originally alien military class that ruled Egypt, by the Ottomans in 1514 was due to the Mamelukes' even more conservative militarism. They had accepted the cannon a full sixty years before the Turks, but for various reasons refused to use ordinary firearms in battle.[27]

There is disagreement among non-Muslim and Muslim scholars as to the contemporary attitude about science and technology in Islam. Europeans and Americans tend to assume that the Koran (Qur'an) and modern science and technology are basically incompatible. One Western scholar typically asserts that "a regime seeking to be true to the Koran had difficulty in coping with twentieth-century technology since those who mastered the technology of the West were unlikely to remain fanatically faithful to Muhammad's revelation." Many Westerners go on to hold that the very elaborateness and rigidity of the Muslim way of thought and life has inhibited attempts to adjust it to the West, and that the only recourse for most Muslims who do not wish to renounce their religion outright—as most do not—is to react by "a rigorous compartmentalization of mind."[28]

Many Muslim scholars would strongly reject this whole perspective, however. Starting with the last century, "modernist" Islamic scholars have argued that true Islam is not incompatible with modern science and philosophy. Some have argued that since the Koran advocates learning for its own sake, a Muslim can safely enter into the spirit of modern science, indeed he must do so.[29] Others say that there is no point in trying to reconcile the Koran with the theories of science since even modern Western scientists admit that theories are only tentative, and "It is therefore possible for a Muslim scientist to explore the universe without first of all accepting an anti-Islamic concept as an absolute criterion."[30] Still others argue that it is necessary to return to the spirit of the Qur'an: "What is of primary importance is the cultivation of the spirit of scientific inquiry as demanded by the Qur'an," and that one must be concerned with "Imbuing higher fields of learning with Islamic values."[31]

There is ambivalence about whether and to what extent Islam can absorb the West's technology without destroying the "traditional integrity" of its society.[32] One author argues that "there is no way a Muslim society can overlook or, indeed, escape the consequences of values inherent in Western technology. . . . *All* technologies available to a developing country are designed to support the various keepers of power."[33]

Paradoxically, Muslim-oriented thinkers about science and technology identify with the West's concern with nature and the environment. The widespread publicity that has been given to the environmental crisis in the Western and world media in the 1970s and 1980s has convinced many Muslim scientists that they were right about science's responsibility to society and man's stewardship of nature, after all. This despite the fact that the actual record of Muslim countries was, if anything, worse than that of the rest of the world, in part because they have been "importing Western technology uncritically and blindly," since "the leaders of many wealthy Muslim nations have become completely separated from their religion and their roots in the land."[34] Ironically, one observer notes, given the terrible problem of "desertification" in many Muslim countries, "it is the tiny state of Israel which is coming up with virtually all the new scientific developments in turning the desert green."[35] Note is taken again, in a contemporary context, of the extent to which Islam has always made people not the masters but simply the stewards of nature.[36]

Muslim scholar Ziauddin Sardar attacks Lynn White Jr.'s thesis about Islam being, like Marxism, a Judeo-Christian heresy and "equally responsible for the debasement of nature." Sardar argues that Islam has a code of environmental values that would prevent such debasement, and the fact that the Islamic nations are no better than the West in terms of the environment reflects the circumstance that they have ignored the Shariya law—the legalistic part of the Qur'an—for the last three or four hundred years, having assimilated the value system of the Occidental colonizers. Islam, as he describes it, has many concepts to show how the environment, "in all its kaleidoscopic richness, must be preserved."[37]

As Muslim nations continue to turn to their religious heritage for guidance, one can expect these concepts to become more refined and play a major role in the planet's future.

Chapter Nine

NAZISM ASSAULTS THE WEST FROM WITHIN

The mainstream of Western thought accepted the Enlightenment and the industrial revolution that resulted from it as unalloyed goods. There was, however, one major dissenting voice in this view in the twentieth century. The German Nazis repudiated the values by which the modern West had grown.

There has long been a dispute over the extent to which the Nazi regime was an aberration or a phenomenon that arose out of the major themes of German cultural history. There seems to be little question that Hitler tapped sources deep in the German psyche. The precedents of Nazi ideas—though not in the extreme form to which they were taken by Hitler—can be found throughout the preceding centuries.[1] Germany was, above all, the locus of the Romantic movement, which downgraded reason in favor of emotion. Its German manifestation in writers such as Herder and Goethe emphasized the idea that Germans had a unique "soul" that arose from their direct relationship to the German land and history. This specialness was reflected in German-Norse mythology and celebrated in the music of Richard Wagner—a notorious anti-Semite[2]—whose Ring Cycle especially was inspired by the myths of German struggle, triumph, and eventual collapse. Hitler was a great admirer of Wagner and reportedly saw one of his operas over thirty times.[3] Yet Nazism also had origins in the science of the nineteenth century and in its attendant emphasis on evolutionary struggle and mass irrationality.[4]

The fascist movement as a whole—of which Nazism must be regarded as a part even though the Nazis emphasized the race rather than the state as the primary source of human greatness—has always been ambivalent about technology, regarding it both as dangerous to the soul yet still capable of being infused by it. We must keep in mind that the fascist impulse was widespread in Europe among intellectuals, especially at the turn of the century, in reaction against the pieties of bourgeois liberalism and the industrial revolution.[5] But "Marinetti and the futurists in Italy, Wyndham Lewis and Ezra Pound in England, Sorel, Drieu la Rochelle, and Maurras in France were all drawn to right-wing politics partly out of their views on technology."[6]

Closely tied to this new view of technology was a strong attachment to nature: "The 'blood and soil' motif was a fairly general one. . . . However, German National Socialism was unique in a very crucial aspect: in it, the rebellion against the Judeo-Christian concept of the role of man vis-à-vis nature achieved expression in a genuine religion of nature."[7] The belief that Germans had a special affinity to the German land was basic to the *volkisch* ideology prominent in Germany during the nineteenth century and upon which the Nazis to a large extent drew,[8] following the precedent of German conservatism generally.[9] This ideology held that the German *Volk* were different in kind and superior to other peoples because their ties to the land and the past had not been corrupted by modernity.

The Weimar Republic was born in German defeat at the end of World War I and was therefore thoroughly detested by most of German society, including the intellectuals. It was during the Weimar period that the debate over technology—*der Streit um die Technik*—was finally consummated.[10] Some accepted the growth of new technologies with enthusiasm. But even more viewed what they called *Amerikanismus*—mass production and consumption, Taylorism, and the whole rationalization of industry—as "a plague threatening the German soul." The prominent periodical *Die Tat*'s synthesis of nationalism, "middle class" anti-capitalism, and anti-Americanism was representative of the popular view.[11]

The debate over technology and society involved most of the major intellectual figures in Weimar Germany. A major philosophical precursor of National Socialism was Carl Schmitt (1888–1985). A student of Max Weber, Schmitt believed that the combination of an authoritarian state and advanced technology could restore political dynamism to a bureaucratized society. He agreed that "political romanticism demanded a break from what he viewed as the passivity and escapism of nineteenth-century German romanticism,"[12] and he held that technology enhanced its users' domination over human beings and nature. Opposing parliamentary discussion, he felt that "the *Geist* of technology, once separated from liberal and Marxist notions of progress and rationality, possessed an elective affinity for authoritarian politics."[13] As a noted constitutional lawyer he came to argue that parliamentary democracy could not meet the needs of the new industrial society, and he pushed the idea of strong leadership, or the *Führerprinzip*.[14]

Of special importance in paving the way for Nazism among German intellectuals were three figures: the sociologist Werner Sombart, the historian Oswald Spengler, and the novelist and poet Ernst Jünger. Of these only Sombart openly accepted Nazism, with the publication of his book *Deutscher Sozialismus* (1934). Spengler refused to accept Nazi racism, and was therefore ostracized. Jünger eventually became an opponent of Nazism. But all of them contributed to the climate of opinion among Germans that made the intellectual as well as political victories of Hitler possible.[15]

Sombart was a major figure in a subtype of socialist tradition that considers the finance capitalist, not the industrialist, the villain. He identified Jews as having market rationality and commercial greed and Germans as having productive labor. His *Deutscher Sozialismus* has been described as "an explosive mixture of sympathy for National Socialism, enthusiasm for 'German technology,' and disgust with the supposedly bygone liberal-materialist-Jewish era."[16] In his essay "Technik und Kultur" (1911), Sombart held that technology was having an adverse effect not only on general culture but on the skilled craftsman, and blamed this and other German economic problems on the domination of technology by the "economy." Later, in *The Quintessence of Capitalism*, he referred to technology as an externalization of the "Germanic-Roman spirit" and contrasted it with the speculating Jewish spirit. After World War I he began to speak of the unsung hero, the self-sacrificing engineer surrounded on all sides by the "meaninglessness of our material culture." But he eschewed any "philosophy of technology," holding that it was "culturally neutral, morally indifferent" and hence could be placed at the service of either good or evil;[17] thus the need for a strong state to control it.

In his prime Oswald Spengler was a world-renowned figure, known for his gloomy reflections on the future of mankind. A largely self-taught high-school teacher, he burst upon the world in 1918 with the publication of *The Decline of the West*.[18] Spengler essentially agreed with Sombart and Schmitt that technology was becoming dominated by money: "The ancient wrestle between the productive and the acquisitive economies intensifies now into a silent gigantomachy of intellects, fought out in the lists of the world-cities. This battle is the despairing struggle of technical thought to maintain its liberty against money-thought."[19]

Spengler was somewhat ambivalent about technology and its impact. *The Decline of the West* was criticized by many as overly antitechnological. In 1931 he wrote *Man and Technics* "to establish his protechnological credentials." He argued that technology was not a purely rational process; it involved intuition and will.[20] Though not himself a Nazi, Spengler had set forth the basic theme of Nazi attitudes toward technology, endorsing it for its own practical sake by removing it from the world of the Enlightenment, which both the Nazis and Spengler despised, and making it part of the *volkisch* soul. In this context, it could be allowed to go its own way.[21]

Perhaps equally important to Spengler among Weimar-era intellectuals paving the way for Nazism was Ernst Jünger. Though he came to reject Nazism outright and expounded a mystical pacifism, his novels about World War I, in which he had served as a much-decorated soldier, were immensely popular in Germany right after the war, and they glorified war itself. Not only was war itself noble, war was an expression of how technology could be joined to comradeship. Jünger "celebrated the will over 'lifeless' rationality." Like Spengler, Jünger believed that technology could be assimilated to *volkisch* ideas.[22]

Jünger hated the market system as much as any Communist, but his ideal was not the Soviet worker councils but the *Gemeinschaft* of the war. This paved the way for his belief that technology could be integrated with the human will, a cornerstone of Nazism. For Jünger, according to Herf, "Germany had a special mission among the modern industrialized nations to demonstrate that technology and culture, modern means and traditional values, are not necessarily in conflict. Technology in the service of a 'German war' would foster the 'victory of the soul over the machine.' "[23]

Jünger was a complete technological determinist and believed that far from being neutral, "technology was inherently in conflict with democracy. Authoritarian technology required an authoritarian state." Nor did he object to this condition. "Technology is our uniform," he wrote, and those who wear it have a "second and colder consciousness" that has the "capacity to view itself as an object," and thus, in one commentator's words, "place the body beyond pain and pleasure."[24]

The post-World War I years witnessed a major change in German thought. In the words of Walter Benjamin, a prominent left-wing intellectual and member of the Frankfurt school, "In the language of battle, the right abandons its enmity to technology." From now on those who rejected the rationalization of German society would paradoxically embrace new technology rather than look back to a bygone rural era.[25]

Prominent in this whole debate were the engineers, among whom there were many partisans of the right wing. As one historian phrases it, "A reactionary modernist tradition . . . was developing *inside* the German engineering profession," which was drawn to Nazism because "they shared a considerable number of points in both theory and practice."[26]

In fact, Nazism drew upon a major desire among Germans to combine the best of the past with the requirements of modern life, to "turn history into myth," and "to recreate through modern planning and technology the modern version of an organic medieval community."[27] This desire, I must emphasize, was not limited to those sympathetic to Nazism, but permeated most of German society; some elements even leaned to the Left, such as the Bauhaus school of architecture.[28]

National Socialism, in its extreme manifestations, would seek to revive the pre-Christian pagan attitude of the German people toward nature, in fact to institute a virtual religion of nature. But here also it drew on long-standing cultural impulses, such as the *volkisch* movement, that were manifest even during the Weimar republic. Before World War I, a widespread youth movement called the Wandervogel sought to create an elite devoted to the German nation and German traditions.[29] During the armed political unrest that wracked the Weimar republic from its outset, the German government, deprived by the Treaty of Versailles of a substantial regular army, often had to draw on the *Freicorps*—armed organizations of right-wing militants, usually war veterans—for support against the far Left. The Wandervogel supplied a whole regiment of *Freicorps*, wearing their emblem, the swastika, upon their helmets.[30] The Wandervogel also used the greeting "Heil" and believed in the abolition of usury, in reinstitution of the guild system, in "Aryan" values and high breeding, and in a special German socialism that opposed rule by big business. Given these similarities with what soon became the Nazi movement, it is no wonder that the Wandervogel movement was sometimes confused with Nazism.[31]

More than a few elements of Wandervogel philosophy survived in Nazism. Many in the Wandervogel movement, especially after World War I, made a special virtue of the primitiveness of the youth hostels they used—so much more "natural" than hotels or homes. But it was not only among those on the political Right that renewed stress was placed on nature. Various forms of mysticism and ritual found their way even into the rationalist movements fostered by the Social Democrats, and the sunbathing and nudist clubs that flourished numbered among their members many prominent figures of the Left.[32]

What was true of the return to nature in general was also true of the worship of nature in particular. The famed World War I general, Erich Ludendorff, was involved in a pagan cult that sought to revive ancient, pre-Christian German values.[33] Thus we have a major paradox. Nazi Germany was so wedded to modern technology that Hitler's minister of armaments and war production, Albert Speer, could blame runaway technology for the barbarism of the Nazi regime.[34] At the same time, the regime encouraged the revival of a primitive religion of nature. Hitler himself, according to Speer, was "downright antimodern."[35] "Industrialization," Hitler held, "has made the individual completely unfree . . . it is a workmill in which any originality or individuality is totally crushed."[36] After victory, those in Hitler's circle believed, Germany should return to a craft economy that aims at quality and rejects the assembly line. Speer regarded this outlook as pure romanticism, but it was what Hitler wanted.[37]

Yet we cannot deny that despite their basically antimodern instincts, the Nazis gained much from their support of certain forms of technology. Civil engineering, which had produced giant aircraft hangars and the autobahns that crisscrossed the nation, excited those who found construction itself admirable.[38] The German physics community continued to thrive as well.[39] Not only did the aeronautical engineers try to use the jet engine (a British invention) to revive the Luftwaffe in its dying days, but their cousins the rocketeers, through their invention of the V-1 and V-2 rockets, laid the groundwork for the future generation of Inter-Continental Ballistic Missiles.[40]

Hitler was not only a fervent ideologue who murdered millions to fulfill his beliefs, he was also a skilled political leader. No matter how great the instruments of terror at his disposal, he still had to pay some attention to where the German people and nation were in terms of their own society and ideas. Hence the Nazi leadership always paid close attention to German public opinion, through secret polls. As a result, the German churches had to be kept in line, and their continued support of the Nazi regime, based on their traditional Augustinian respect for established authority, was encouraged. Within the Nazi movement there necessarily evolved a dual set of doctrines— one for the inner elite, and one, more conventional, for the German population at large.[41] Hitler approved of this division, despite his personal skepticism about some of the more bizarre attempts of some Nazis to revive pagan rites to express pre-Christian German religious traditions.[42]

It has even been argued that the Nazi religion of nature represented trends that were widespread in modern Western society.[43] One scholar contends that the Nazis ignored the supernatural belief in a deity but "were strong believers in a religion of nature *heavily* mystical in content."[44] A major architect of the National Socialist religion of nature was the party ideologue Alfred Rosenberg, who wrote *Der Mythus des 20. Jahrhunderts.* Hitler admitted that he could read very little of this tedious book with its detailed descriptions of Nordic as opposed to Judeo-Christian concepts, and stated at one point in 1942 that it should not be regarded as the official doctrine of the party.[45] But Rosenberg's view was one that "either because of or in spite of him, was, *widely held* by most committed National Socialists."[46] Rosenberg was much more concerned than Hitler—who was still a political pragmatist in many respects—with creating a mythical structure that could serve the purposes of religion, but both were united in believing that nature was the touchstone of morality and had made some men superior to others.[47] National Socialists were concerned with creating a *Weltanschauung* that could combine religion and science and, by deifying nature itself, "allow for a bridging of the gap between spirit and matter."[48]

Hitler had always deified nature. In Hegelian fashion he said that "man is God in the making."[49] At the same time, he was highly skeptical of attempts on humanity's part to control nature. He felt that people could only conform to nature's laws, and it was the purpose of science to discern them. Indeed, he was highly critical of those like Rosenberg who tried to revive old Germanic customs, and he was contemptuous of astrology, which he regarded as a "swindle."[50] Yet despite this attitude on Hitler's part, the Nazi movement developed a major commitment to the revival of ancient, sometimes almost occult ideas.[51]

This was not entirely illogical, given the thrust of Hitler's views. Herman Rauschning, an early supporter and confidant of Hitler who later rejected Nazism, wrote of the movement's belief—beneath the surface of "scientific naturalism"—in "natural mysteries" and the coming of a new age to be ushered in by some kind of "great world transformation," and of Hitler's "biological mysticism."[52] In the school for SS leaders at Bad Tolz, the basic text said, "We are the bearers and shapers of eternal life."[53] The leading analyst of National Socialist religion sums it up this way: "Amidst the crudely pantheistic and proto-existentialist verbiage, one idea emerged with some clarity . . . that National Socialism was based upon a 'life-affirming' principle which was the principle of nature itself."[54] All the world was one. As Nazi philosopher Ernst Kriek put it, "There is no inorganic nature, there is no dead, mechanical earth. . . . The great mother has been won back to life."[55]

But part of affirming the final significance of nature as an arbiter of human life was the belief that nature made some people superior to others, a kind of Social Darwinism with moral underpinnings. But by that very same morality, National Socialism was found wanting, since it was losing World War II. In a final irony, the religion of nature was declaring its proponents bankrupt, unable to meet its demands.[56]

Chapter Ten

THE ORIENT, NATURE, AND TECHNOLOGY

Languages reflect the worldviews of those who use them. It is thus extremely difficult to look to Eastern philosophies for answers to "Western" questions about nature and technology. Yet the questions must be asked, even if only to give Westerners a new perspective on the meaning of the questions and their own answers to them.

The early Greeks and Romans were aware that Asia was somehow different. But for them, Asia was really West Asia, above all Persia. Contact with East Asia was almost nonexistent, and it was completely lost after the fall of Rome,[1] not to be resumed again until the Middle Ages. After the travels of Marco Polo, it suddenly occurred to the Western consciousness that there is another world out there. This other world rivaled—indeed surpassed—Christendom in opulence and sophistication; in addition, it had an entirely different way of looking at life. Ever since then, some in the West have argued that not only is the East more civilized, but also deals more wisely with the relation between matter and spirit. In recent decades this sensitivity has also meant to many that the East has a better appreciation of nature and technology than the West. One Western philosopher, speaking for many, has said that "it is . . . apparent that a great deal of the ecological crisis stems from Western-based technology and the metaphysical and axiological positions that have sustained and nurtured that technology."[2]

This tendency to generalize about "Eastern" philosophy and culture is exemplified in the argument of the distinguished Yale scholar F.S.C. Northrop, who has contended that there is simply one Eastern tradition, albeit with several branches. He summarizes what he considers to be the essential common denominator of Eastern civilization: "The meaning of Oriental civilization—that characteristic which sets it off from the West—may be stated very briefly. The Oriental portion of the world has concentrated its attention upon the nature of all things in their emotional and aesthetic, purely empirical and positivistic immediacy."[3]

We are thus faced with two basic questions. Is there simply one Eastern culture or many? How does the East regard the problems of the relationship of man to nature and to technology that are of concern to us? We will leave aside formal consideration of the question of unity and look at each major Oriental culture in turn. As we do so we will find many similarities in outlook—at least vis-à-vis the West—but important differences as well. We shall look first at India, the purest of the allegedly "spiritual" cultures. Then we shall examine China, about which the relevant literature is more vast. Finally we shall deal with Japan, which now, despite its traditional past, is a major factor in deciding how the modern world will deal with nature and technology.

India

Of all the major cultures of the East, none is as far removed from that of the West than the culture of India. While China and Japan, in their major philosophies as well as everyday life, are concerned with things of the practical world, India has, above all, been concerned with the "spirit."[4] Recent investigation has unearthed vast new evidence about the earliest Indian civilizations, but much remains obscure, including the causes of their disappearance. But we do know that the death blow came with the invasion of India by tall, fair-skinned nomads from Central Asia in the second millennium B.C., destroying all in their path but leaving no physical artifacts of their own.[5] These people called themselves Aryans, "the noble ones," and soon controlled the subcontinent, driving the indigenous inhabitants, the darker-skinned Dravidians, south and east into subjugation. Their language, Sanskrit, is the basis for all subsequent Indo-European languages, and the one in which the great epics of Hindu mythology, the Vedas, are written.[6] A fundamental characteristic of Indian religious civilization, in contrast with the sacred books of the Hebrews involving a historical covenant with Yahweh, is that there is no sense of time. The Hindu indifference to time seems part of its original impulse, reflecting an indifference to a major dimension of human life.

In order to comprehend the attitude of Hindu thought with regard to nature and technology, we must try to understand their religious and philosophical ideas as such. The essence of the problem is that Eastern thought not only differs from that of the West with regard to answers, it differs with regard to questions. In order to grasp what the Hindu thinks about nature we will have to investigate what the Hindu thinks about the world as a whole. The Hindu, like most orientals, is not really concerned with nature as such as we usually think of nature. The concern of most Eastern philosophies and "religions" is, paradoxically, this-worldly, with humanity and its day-to-day conduct. "From the very beginning," one Western student of oriental ideas asserts, "the speculations of India's sages were aimed at solving life's basic problems. Their philosophy grew out of their attempts to improve life."[7]

Even religion itself is viewed somewhat differently. Hinduism, like India itself, has been capable of absorbing alien influences and incorporating them, not simply by dissolving them in an already existing religious substance, but syncretistically accepting them side side by side with existing teachings. Even so, it does have its own essential characteristics. One article of faith that all Hindus accept is "that man will work out his destiny through the interaction of karma (the law of cause and effect that determines his station in life), dharma (the duties incumbent upon him in that station), and reincarnation (rebirth in another life)."[8]

For the Hindu, man is imprisoned in his body and the world of illusion. In this system the only ultimate reality is spiritual. It is, in Western philosophical language, an essentially idealistic philosophy, in which the appearances apprehended by the senses are not the ultimate reality. In this context the Hindu religion is quite logical. Hindus believe that there is one ultimate Godhead—not a personal Supreme Being in the Western sense, but a divine essence that underlies all others. The various gods and goddesses of the complex Hindu pantheon are simply "masks" behind which this supreme God hides. The world is a result of God emptying himself (*atma-yajna*) in an act of self-sacrifice. It will end when people, following the same pattern, give up their own lives to become one with God again. As a sympathetic Westerner puts it, "The basic myth of Hinduism is that the world is God playing hide- and-seek with himself."[9]

In such a worldview, ethics takes on a new meaning. One performs "good deeds" not for their own sake but so that in one's next reincarnation one will be more able finally to shake off life entirely and become liberated by returning to God. Good and evil have no ultimate meanings. They are merely different sides of the same coin, different ways in which God manifests himself. The role of the Hindu is simply to accept them both as equally necessary, and valid, components of the universe.[10]

In this context the Hindu view of nature is also easier to grasp, since nature, like mankind, is an expression of the Godhead, not something created but something poured out of the divine essence. As a result it too is "good," and should be treated with reverence.[11] Even though people are ultimately destined for liberation from the body, in the meantime they have to live in it as best they can and accept their natural environment. Since nature is basically at one with oneself, to struggle against nature is to struggle against oneself."[12]

Buddhism

The basically accepting attitude toward nature found in Hinduism is also found in Buddhism, a major Eastern religion that grew out of it. Buddhism arose in the sixth century B.C. through the Guatama, the "Awakened One," who died about 545 B.C.[13] Legend has it that the Guatama Buddha, a sheltered youth of wealth and high position, was so horrified when he first saw suffering and decay that he sought an explanation. He first turned to the Hindu Upanishads, but found them wanting.[14] Then he entered a regimen of impoverished wandering, meditation, and asceticism. He found this approach also unsatisfactory. Then, in a burst of inspiration, he decided that all seeking is only an illusion. Many were converted to this system of belief even in his lifetime, and became monks or nuns, following him and seeking to share his awakening and enlightenment.

The Buddha did not present a didactic system. Rather, he invited his followers to embrace a path of living: "Buddhism is essentially a way of life; it is not a philosophy about life." He died at age eighty, commanding his disciples with his last breath to "work out your own salvation with diligence."[15]

What does "salvation" mean in Buddhism? How does one work it out? How does this endeavor condition one's attitude toward nature and technology? In an era is which many leading social critics such as E. F. Schumacher have called for a return to "Buddhist economics,"[16] the last question especially is a vital one.

Buddhism is highly humanistic. The practical problems are the problems of living human beings. And what is the most important human problem? We have seen from the Buddha's own life that one must account for poverty, disease, and death.[17] Why is there suffering? For Buddhism, suffering is caused by desire. As one philosopher notes, *"It is the craving of a self that gives rise to suffering."*[18] Buddhism solves the problem of the suffering self by denying the very existence of a self. As Alan Watts puts it: "It is fundamental to every school of Buddhism that there is no ego, no enduring entity which is the constant subject of our changing experiences."[19]

What kind of morality can exist in such a system of thought? For Buddhism, since there is no self that is the subject of moral duties, there is no real person to be punished or rewarded. One's objective in life is to become enlightened, to become free of the illusion of self, to attain *nirvana*, that merger with the total universe in which fulfillment consists.[20]

Yet despite this aversion to morality as it is known in the Western world, the metaphysics of Buddhism provide the basis for an identification of people and nature that precludes despoiling nature for transitory human advantage.[21] To begin with, there is no God-directed task to conquer the natural world, which would, like all human strivings, be folly in itself and an obstacle to liberation. For Buddhists, nature is *dharmata*, a dependent arising, something that comes out of phenomena.[22] Thus a Buddhist culture is one that lives in harmony with nature.[23]

By the same token, a Buddhist culture will not make a god out of technology. "In Buddhist cultures man has not been subordinated to things and machines," one observer contends,[24] and since people are regarded as controllers of their own fate, they can choose whether or not to be ensnared by their own ignorance and decide what forces, if any, can be their master.

Thus both of the major religious movements in Indian culture, Hinduism and Buddhism, preach a way of life that does not radically differentiate people from nature, and does not insist that people exploit nature for their own benefit through technological means. But this is religious and philosophical theory. What has the actual technological practice of Indian society been?

The common belief is that science and technology did not flourish in India, or in the Orient generally, not only for philosophical reasons but for social ones that derived from these philosophical priorities. Scientists were not regarded as important contributors to the real task of society, which was enlightenment, and, like the ancient Greeks, the Indians looked down upon those who worked with their hands.[25]

The case is not quite that simple, however. Indian science and technology, though no match for that of the West at the time of India's conquest by the British, was not without its contributions. The Gupta Empire (467–320 B.C.), though Buddhist, excelled in science and technology as well as in the arts, and it has been described as "the most advanced country of its time."[26] Indian astronomers knew that the Earth was round and that it rotated on its axis. Indian mathematicians invented the concept of zero and the so-called Arabic numerals, imported to the West many centuries later by Arab mathematicians.[27] When the Gupta Empire was destroyed by invaders from central Asia, India plunged into turmoil, but this cannot gainsay its achievements in science and technology within a Buddhist culture.[28]

Indeed, most of the antitechnological impulse of traditional Indian society arose not from religion but from social structure. Buddhism flourished briefly in India and then was reabsorbed by Hindu society. The primary social feature of Hinduism is the caste system, within which the lower castes are those who do the "dirty" work, with their hands. Just as there was no real social pressure in ancient Greece and Rome to develop labor-saving technology in a society where the brute physical work was done mainly by slaves, so there was little incentive in India where the lower castes traditionally performed these tasks.

Yet at the time it came into contact with the modern West, India was not hopelessly outclassed in technology. It was the world's greatest exporter of textiles in the eighteenth century.[29] Even when India was not ahead, it was still a player in the game. But if India, contrary to what one might have predicted based on philosophy alone, was not left at the post in a technology designed to produce goods, it did have problems. India was never able to relate well to nature directly through its technology, above all to the soil, which is of overwhelming importance in a primarily agricultural country. India was destroying its natural heritage despite—indeed, largely because of— a low level of agricultural technology.[30] The regular, devastating floods of India, like those of China, bear witness to the fact that simple lip service to harmony with nature at an abstract philosophical level is no answer to the problems of creating and maintaining such harmony on a day-to-day basis.

What role has politics played in the Indian relationship to nature and technology? The vast irrigation networks that have underlain the use—and misuse—of the soil for centuries were created and maintained by the central government.[31] But during the course of the conquest of India, British imperialism destroyed the traditional economy, and the traditional society as well, to the greatest extent possible. Karl Marx welcomed this destruction as a necessary step forward in human progress,[32] but many Indians did not, for ideological as well as immediate practical reasons. The great leader of the Indian independence movement, Mahatma Gandhi, deplored the overuse of modern machinery.[33] Gandhi looked toward an Indian utopia from which most of technology would be removed.

But after Gandhi's assassination, India rejected his views and sought, under Nehru and his successors, to modernize through large- scale industrialization. Most of India is still a dirt-poor agricultural society based on the traditional village. But there is within it the equivalent of a modern developed nation that is the equal in population and living standards of, say, Italy, and industrial and scientific progress proceeds apace.[34] Ironically, India excels in such areas as nuclear physics; the ancient Brahman distaste for getting one's hands dirty therefore does not extend to using the blackboard to deal with equations as abstruse and removed from ordinary life as the most convoluted and recondite teachings of traditional Hindu philosophy.

China

Throughout most of its history, China has effectively been isolated from the Western world. The West has always been aware of China, however. The Roman Empire had marginal contact with it.[35] When China was rediscovered by the West in the late Middle Ages, Westerners became fascinated with it. All educated people were aware that a vast, prosperous, and civilized empire existed that operated on principles radically different from those of the Christian West. What might they learn from it? Although the early reports about China were so full of praise and appealed particularly to those who were less than enamored by Western Christendom, in time this admiration lessened as it became evident that China was much weaker economically and militarily than the West. Soon it too became a victim of imperialism. But why? What was inherent in Chinese civilization that made it less able than the West to use science and technology to harness national power? That question has fascinated scholars for over a century, and it relates directly to our own concerns about how the Chinese regarded nature and technology.

The great early seminal period of Chinese history was the Chou dynasty (1115–221 B.C.),[36] the time of Confucius (551–479 B.C.) and the semilegendary sixth-century B.C. philosopher Lao-tzu, founder of Taoism and supposedly an elder contemporary of Confucius.[37] It is impossible to understand Chinese thought without understanding the "religions" of Confucianism and Taoism. The great scholar on China, Joseph Needham, referred to what he called the *philosphia perennis* of China as "an organic naturalism."[38] Confucianism and Taoism, according to another observer, "can be characterized as man-centered, and thus world-centered."[39] Neither suggests "a postulated scientific or a doctrinally formulated, theological object."[40]

Both Confucianism and Taoism had a common source in the famous *I Ching*, or the *Book of Changes*, which dates from somewhere between 3000 and 1200 B.C. and lies at the very foundation of Chinese culture.[41] Though apparently quite different in their emphases, Confucianism and Taoism are complementary. Like virtually everything in traditional Chinese culture, they reflect the fundamental belief of the Chinese, symbolized even visually by the *yin* and the *yang*, that the universe is not made up of opposites that are in conflict but opposites that, like male and female, are necessary to each other.[42] Confucianism is primarily a philosophy for those in the midst of life, while Taoism, according to one scholar, is generally "a pursuit of older men."[43]

Confucianism is above all humanistic, and is barely concerned about the physical universe of nature.[44] It emphasizes the development of humanness, which also includes an emphasis on tradition.[45] Confucianism wants to maintain the social order as it has always existed. If harmony has been achieved, why disturb it? Among other things, one must have reverence for the wisdom of elders and for elders—above all one's ancestors—themselves. Along with this attitude comes a profound distrust of innovation, indeed a disinterest in looking at nature for clues as to how people might live better. But this acceptance of tradition and the unwillingness to look to nature for ideas has not prevented the Confucians from dominating nature. Rivers especially were to be "disciplined," in accord with their task to serve the common good of society.[46] But politics has played a major role in Confucianism as well. By 206 B.C. it had become the official philosophy of the Chinese Empire, and for some two thousand years the road to power came through Confucian study,[47] which inhibited progress in science and technology. One scholar writes that "the intellectual climate of Confucian orthodoxy [was] not at all favorable for any form of trial or experiment, for innovations of any kind, or for the free play of the mind."[48]

Unlike Confucians, the Taoists, though still essentially humanistic, have paid great attention to nature.[49] Tao means "the Way," taking its name from the first word of the Taoist book, *Tao Te Ching* (The Classic of the Way and Its Power). The way to live is according to nature—external nature and one's own. The utmost stress is placed on spontaneity, and Taoists reject all attempts to live according to external codes.[50] According to Taoists, there was no God who produced the world from outside by making it (*wei*); the world just grew, it was a product of non-making (*wu-wei*).[51] Taoists follow this principle of nonmaking in their own lives, and are content to sit back and grow like the universe of nature itself. This is best done when we are convinced, as the Tao teaches, that we are a part of this simply growing nature. We achieve this state through some form of absorption in nature itself. The means by which we attain this union with nature is the same means we must use to run our own lives. We must reject the idea of a separate ego. Here Taoism's compatibility with Buddhism—not yet introduced into China—is clear.

The impact of Taoism upon the Chinese attitude toward nature must be seen in context. In the Bronze Age, the Chinese, like most other peoples, created various forms of nature-religions, involving shamans and blood sacrifice, worshiping not only the sky-god T'ien but a vast number of spirits animistically identified with entities in nature.[52] Even after the primitive forms of nature-religion declined, the emphasis on conformity with nature remained. It was vital that all aspects of life be attuned to natural patterns.[53] Indeed, unusual natural disasters were a sign that the regime itself was not virtuous, and that the emperor had lost the "mandate of heaven."

Nature has remained a major theme and presence in Chinese poetry and literature throughout the nation's history.[54] Chinese and Japanese landscape painters "celebrated wilderness over a thousand years before Western artists."[55] This attitude was nurtured by Confucianism, which treated human and physical nature alike: "The aim is not conquest, but a life in conformity with it."[56] The identity of human and physical nature was also the position of the Taoists, who, however, gave priority to the study of physical nature in achieving this joint understanding. They "felt in their bones . . . that until humankind knew more about Nature it would never be possible even to organize society as it should be organized."[57] Despite their interest in understanding nature, however, the Taoists were unable to do so in any scientific fashion because of their philosophical convictions. The Taoists also differed from the Confucians in their attitude toward subduing nature, especially rivers, which the Chinese considered important. "*Wu wei*, the idea of moving with the flow of a stream and doing nothing contrary to nature, meant opposing structures that too rigidly confined rivers or diverted them from their course."[58]

Central to Chinese thought from the beginning had been what Y-Fu Tuan calls "geopiety." "In China," he writes, "from remote antiquity to the modern period, the gods of soil and grain were worshipped side by side with the ancestors." Nor, as we have seen, did geopiety permit any distaste for wilderness.[59] This attitude of respect for nature allegedly pervaded all aspects of Chinese activity. Even Chinese monuments, unlike those of the West, did not challenge nature even though the Chinese had the engineering skill to do so; they believed in being in harmony with nature.[60]

Despite their theoretical respect for nature, the Chinese, like the Native Americans, mistreated nature as much as the Westerners did. Above all the misuse was apparent in agriculture, which has always been the mainstay of Chinese existence. China also destroyed its forests; indeed, the Yellow River gained its name from the products of soil erosion it carried to the sea.[61] China's land use practices were not really in harmony with basic Chinese philosophy.

If in theory at least the Chinese did not share the Western belief that nature was an adversary to be crushed, how did this attitude affect their technology? The evidence is strong that for many centuries Chinese technology was not only equal to but superior to that of the West. But at a crucial point it failed to continue to develop, primarily, according to most scholars, because it had no scientific basis. As we saw, Chinese veneration for nature did not completely keep them from seeking to manipulate it, as in the case of rivers. But perhaps they were unable to master nature fully because they were unable to understand it in the same scientific terms as the West.

Recent research has established that the Chinese were for long far ahead of the West both in priority of inventions and in their application to human life.[62] Indeed, Needham argues that it is a fundamental misunderstanding to maintain, like Alan Watts, that in traditional Asian cultures "it was felt easier for man to adapt himself to Nature than to adapt Nature to himself." Directly quoting and refuting Watts, Needham asserts "this thesis is falsified by twenty centuries of Chinese scientific and technological history."[63]

Chinese technology and science were the wonder of the world when extensive contact with the West was begun in the twelfth and thirteenth centuries. Many important inventions such as the magnetic compass, gunpowder, and paper were borrowed from China. During the Han dynasty (A.D. 0–220), paper was invented and a primitive compass was developed, largely for use in deciding the most propitious places on which to erect temples. Printing from woodblocks began during the T'ang dynasty (A.D. 618–907), though movable type was not used until after the thirteenth century and did not completely triumph until modern times because of the ideographic nature of the Chinese language.[64] Gunpowder was invented and used in pyrotechnical displays also in the T'ang dynasty. In the reign of Emperor Li Lung-chi in the eighth century, an iron suspension bridge was erected over the Yellow River and a water-powered astronomical clock was developed.[65] Indeed, Needham credits the Chinese, through their invention of the clock, with making possible the mechanical world-picture that underlay the development of Western science.[66]

Nor did Chinese civilization, despite its strong Confucian class bias, eschew labor-saving devices. This attitude contrasts with that in Europe, where many machines were ignored because of indifference to the well-being of workers, or fears of unemployment due to technology.[67] The traditional Chinese culture also created its own system of medicine, long despised by the West as hopelessly primitive, but recently regarded more favorably because of such empirically based features as acupuncture.[68] The Chinese developed a form of immunization through vaccination centuries before the West.[69]

Given its early head start in science and technology, how did China fall into a position of subservience to the West? Why did it fail to move on to new scientific triumphs, as the West was able to do? Various answers have been given to this question. Needham stresses the philosophical aspects. He argues that the essential breakthrough in the West came only with Galileo and the mathematization of science. He does not accept the idea that Chinese class and political structure inhibited science, at least initially. Above all, he paradoxically gives credit for the growth of Western science to the inherited idea of natural law and a divine lawgiver, an idea he does not himself accept. As we have seen, this idea is foreign to all the basic Chinese philosophies, which reject the idea of a personal God who created the universe out of nothing and laid down laws from above.[70]

The Chinese, according to Needham, did accept the idea of some kind of natural law in the form of social rules of conduct, as in Confucianism. But after overthowing the brutal rule of the Legalists in the fourth century B.C., they did not like the idea of rigidly codifying such laws. Not accepting the idea of a Supreme Being, having depersonalized God, they could not search for his commands in nature.[71] The paradox comes in the fact that, according to Needham, their ideas are more consonant with contemporary physical science than those of the Renaissance philosophers of science. But having intuited them prematurely, as it were, they skipped a necessary stage of error on the way to that modern science.[72]

But Needham, though he emphasizes the philosophical aspect, does not completely deny social causes in the failure of Chinese science. He attributes the lack of growth of modern science to the same factors that prevented China from developing a mercantile class: the domination of society by the entrenched scholar gentry, those whose knowledge of the Confucian classics gave them the right to hold political power. When traditional Chinese feudalism decayed, it led not to capitalism but to a kind of bureaucratic feudalism, in which the scholar-gentry "effectively prevented the rise to power of or seizure of the State by the merchant class, as happened elsewhere." There were stirrings of democratic ideas in traditional China, but they were not linked to a rise to power of the merchant class and to technological change.[73]

Other scholars lay the blame for China's failure to develop a technology directly at the feet of Confucian orthodoxy, reinforced by the politically based stranglehold of the elite scholar-gentry on Chinese society. After he pays homage to Needham for dispelling old notions of a scientifically and technologically backward China, economist Rudy Volti writes: "Technological innovations could be regarded with appreciation, but actual involvement with them was regarded as beneath the dignity of a scholar." Yet Volti does accept Needham's thesis that hostility to mercantile activities played a major role in "stifling the development of science and technology."[74]

A noted historian of Chinese Confucianism, Joseph Levenson, agrees with the idea that Confucian orthodoxy stifled science, and he contrasts the empirical orientation of Chinese science with the systematic interrogation of nature espoused by Bacon and Descartes. But he rejects the idea that has increasingly been pushed in China since the Communist revolution that any real Chinese scientific tradition exists: "A number of historians of modern China have tried to find an impressive Chinese pedigree for modern science; and their efforts seem, paradoxically, a subjective response to the fact that none exists."[75] A more economically minded historian, Mark Elvin, combines economic and philosophical reasons for China's failure to meet the challenge of developing modern science and technology, and for its economic stagnation.[76]

It is only in the context of lack of interest in new technology and continued economic stagnation that we can understand how the Chinese could have provided such a poor market for European technology in the earliest centuries of contact. During the period when mercantilism dominated Europe, Chinese porcelain, silk, and cotton had to be paid for in gold and silver. This was not because the Chinese shared the mercantilist belief in the singular importance of precious metals, but because European textiles and pottery were very inferior to their own, and the Chinese had no interest in using European technological products such as clocks, optical devices, and even firearms in daily life, regarding them simply as toys or curiosities.[77]

Yet precisely because China refused to accept modern science and industry, it was finally forced to do so. Like other stubbornly traditional societies, its technological weakness engendered military weakness as well, which made it helpless to resist Western thrusts to force it into economic subservience. The British, unable to find other products the Chinese would buy willingly, in order to balance its trade with China forced the Chinese to open up the country to the sale of opium, forbidden by law. After the Opium War (1839–1842), China was forced to accept Western trade activity and make concessions of sovereignty, not only to Britain but eventually to other foreign powers as well.

The shock, of course, was tremendous. The Chinese had always regarded themselves as the only real power in the world and all others as inferior barbarians. Now they were being put under the barbarian yoke. This led to an orgy of soul searching that has not ended to this very day. It is of interest how China could maintain its sense of superiority and restore its national sovereignty without giving up the culture on which its sense of superiority was based in the first place. Many have simply insisted that it could not be done. Others have offered a solution that has been advocated in other non-Western nations, that one would have to distinguish between the material and the spiritual. The latter insist on the superiority of traditional Chinese principles, above all Confucian ones, while accepting Western technology for practical purposes. Thus one can still maintain the position that the traditional ways are superior, because the spiritual is superior to the practical.

Even before the Opium War, the handwriting had been on the wall. The debate was joined early in the nineteenth century by various Confucian scholars who were de facto Westernizers, the so-called school of "self-strengtheners." Notable among them were Lin Tse- hsu (1785–1830) and, above all, Chang Chih-tung (1837–1909). Chang was a great advocate of railroads and heavy industry and argued that Chinese learning could be maintained for substance (*t'i*), and Western learning introduced for function (*yung*). The *ti-yung* school emphasized military technology above any other. But, as one scholar notes, "Soon the list of indispensable superior techniques lengthened, to cover industry, commerce, mining, railroads, telegraphs . . . and essential traditional attitudes were almost casually dissipated by seekers after the useful techniques which were to shield the Chinese essence."[78]

In fact, Chinese leaders slowly tried to modernize their country, step by step, in the nineteenth century, beginning with the military.[79] But attempts to internalize the new outlook into the heart of Chinese society failed. In 1864 Li Hung-zhang proposed that the new categories of knowledge be included in the all-important traditional civil service examination. His proposal was rejected. Those who rejected it were not entirely wrong in doing so. Unlike the "self-strengtheners," they realized that traditional China could not endure in a new world of railroads and factories, but clung to the old ways even if to do so meant ultimate defeat.[80]

The Taiping Rebellion (1850–1864), largely inspired by Protestant ideas, had sought to modernize, but was defeated.[81] The republic of 1912, which finally brought an end to imperial China after many millennia, continued to witness the struggle between the modernizing desire to exalt Western science and technology and the reactions of those still loyal to the past.[82] The great watershed came with the rise to power of the Communists, and above all the leadership of Mao Zedong.[83] Mao represented the Promethean urge within Marxism personified. Despite the fact that the main current of Chinese thought, as we have seen, affirms nature as good, Mao struggled against it in various ways, almost desperately.[84]

This struggle against nature has been coupled with an attempt to rewrite the history of science in China. Science, in Marxist thought, is of course an unmitigated good, therefore a proud Communist state—and the Chinese Communists are above all nationalistic—must have a scientific past. Drawing on recent research, past discoveries are exalted, and even the ancient Taoists, now that they are safely out of the picture, can be glorified as pioneers in the naturalistic tradition.[85]

Some have held that Mao was somewhat wary of technological change. Certainly he was, but not of technology itself. He was rather concerned with an imbalance between large-scale technology and revolutionary China's needs, hence the stress on small-scale technology, on "walking on two legs," associated with the "Great Leap Forward." For Mao this policy was a matter of Marxist ideology as well as national self-sufficiency. His major concern was that the advance of Chinese industrialization not depend unduly on foreign importations nor, importantly, on domestic expertise. With regard to the latter, he was more orthodox Marxist than the Stalinists in this respect, and he rightly feared the example of the Soviet Union, where a small cadre of technical experts had taken over the revolution and real power in the state.[86]

Mao's successors have repudiated most of his policies with regard to science and technology, emphasizing increasing interaction with the West, including the advanced capitalist West, and the training of native experts, as well as competition and a certain measure of economic freedom. In the course of doing so they have also repudiated the Maoist emphasis on frugality and followed the West down the path of pollution through industrial growth— even spectacularly so. At the same time, anxious to retain a monopoly of power for the party and its ideological leadership, they are seeking to limit the spread of Western ideas in the areas of culture and politics. This seems to be the old debate of the nineteenth and early twentieth centuries all over again: to what extent can China accept Western technology and yet maintain its own culture?

Japan

Despite its current industrial prowess, Japan has long been regarded, and regards itself, as a civilization in which nature has a special place of honor. Like all peoples, the Japanese originally worshiped nature and natural objects and forces.[87] The major indigenous Japanese religion, Shinto, arose in this context.[88] Shinto, it has been said, "was a form of nature worship that deified mountains, forests, storms, and torrents in preference to fruitful, pastoral scenes since the wild was thought to manifest the divine being more potently than the rural."[89] The main feature of Shinto regarding nature is the concept of *kami*. *Kami* are difficult to describe in Western terms, but they may be said to be sacred spirits that can take on various forms but are usually found in natural objects.[90]

The intimate relationship between creation, nature, the *kami*, and the Japanese people is at the heart of the Japanese attitude toward nature: "As seen in Shinto mythology, the *kami* express their sacredness and power through their embodiment in nature. This idea contrasts sharply with the Judaic and Christian traditions, which tend to emphasize the distance between God and man and the inferiority of nature."[91] This attitude toward nature is not unique to Shinto, but is found in all the major Japanese religious traditions. Like the Chinese, the Japanese, unless they are Christians, do not regard religions as mutually exclusive. Shinto, the indigenous religion, soon became combined with Buddhism, the imported religion,[92] and today most Japanese practice both at the same time, priority depending on the occasion and context.[93]

What is Japanese Buddhism like? Shinto, as noted, teaches that nature is inherently divine and all natural objects partake of sacredness. Buddhism originally did not necessarily mean some form of nature worship, but it was not totally incompatible with it and was soon headed in that direction in nature-loving Japan.[94]

The major facet of Japanese Buddhism is Zen, which has been and still is widely influential throughout Japanese culture.[95] It has also enjoyed a great vogue in the West, especially in the United States since the 1960s. A vast literature about Zen, directed primarily at Americans, exists.[96] So great was the vogue at its height that many whose religious convictions were other than Buddhist have accepted Zen techniques as useful for spiritual growth.[97] Zen has shown a special attraction for Americans interested in environmental causes,[98] and the Pulitzer Prize-winning poet and noted supporter of radical environmental causes, Gary Snyder, is a disciple of Zen.[99] Thus it is vital to be as clear as possible about what Zen Buddhism is and what exactly it teaches about nature.

Zen actually arose in China, a result of the impact of the central principles of Mahayana Buddhism upon Taoists and Confucians.[100] The Mahayana school taught that *all* men could follow the path of enlightenment. (It is noteworthy that all the schools of Oriental religious and philosophical thought we have been examining are directed primarily if not exclusively to men; women were always so outside the society that little heed was paid to them.) Many of the terms used to describe Zen are Chinese in origin. Even *zen* itself is a version of *chan*, a term describing the new system developed in China, and the sudden flash of insight, *satori*, which is central to Zen, is analogous to the Chinese *ten wu*.

The golden ages of Zen in China were the last two hundred years of the T'ang Dynasty, from about 700 to 906. The Rinzai school was introduced into Japan in 1191 by the Japanese monk Eiiai (1141–1215), who set up monasteries at Kyoto and Kamakura under imperial patronage. The imperial connection is important and does much to explain the fact that Zen provided the Japanese warrior class, the *samurai*, with their code of *bushido*. Zen teaches that each person should be aware of what he or she is doing and why, following nature, paradoxically, in a conscious fashion. Just as fish naturally behave in a certain way, all groups, in accord with Zen's basic Taoist origins, have their "way." *Bushido* is the "way" of the warrior.[101] Zen appealed to the military class because of its practicality, earthiness, and simplicity. There is an anomaly here, since Buddhism is essentially pacifist and *bushido* often scandalizes fellow Buddhists. But it has been argued that this too is an illusion. *Bushido* "seems to involve the complete divorce of awakening from morality. But one must face the fact that, in its essence, Buddhism is a liberation from conventions of every kind, including the moral conventions."[102]

What does Zen "teach" about nature? The word "teach" is put in quotation marks because the most distinct feature of Zen is that Zen, even more definitively than the Buddhism of which it is a part, does not have teachings in the sense of any system of propositions, but insists on the all-importance of self-awareness and self-enlightenment.[103] In Zen this self-enlightenment is acquired through anecdotes presented in question-and-answer form, called the *mondo* system. Usually the questions, such as the famous "What is the sound of one hand clapping?" have no answers as such; their purpose is to make the student aware of the absurdity of the very terms of the question. The Zen tradition is passed on not by indoctrination but initiation, and from the point of view of epistemology resembles skills such as swimming, which can only be learned by initiation and practice rather than through didactic discourse.

The essential tests through which a Zen student becomes accepted as truly adept and capable of becoming a master himself are the *koan*, an exercise that requires demonstrating "understanding" of the *mondo* of the old established masters. But all this is, in a sense, contentless. What one looks for is the sudden flash of insight, *satori*, which is by definition incommunicable, a matter of shared experience. It is this feature of Zen that makes it accepted by many who do not share the basic metaphysical tenets of Buddhism.

As far as nature is concerned, many Zen scholars might insist that Zen is not "pantheistic."[104] But what that really means in a "system" that has no clear teachings about God is difficult to discern.[105] Zen does not postulate any ontological identity of humans and nature, but rather a psychological/mystical union at the level of human consciousness. That is why Zen is so much better expressed in art, including the arts of daily living, than in words.

Zen has influenced the Japanese love of nature not only directly through its "teachings," but indirectly as well. Most Zen monasteries are located in the mountains, and the monks are necessarily in intimate contact with nature.[106] Zen has had an effect on Japanese life both through its attitude toward nature and for its creation of the basis for the Japanese love of simplicity, as evidenced in the hermitage or the tea house, and the tea ceremony itself. In any event, whatever its philosophical nuances, Zen did not create but exemplified and reinforced a love of nature already present in Japanese culture from the earliest days.[107]

Some argue that this Japanese feeling toward nature continues to this day, that indeed it is something of a "safety valve" during times of Japanese economic and political stress.[108] Professor R. Byron Earhart, a specialist on Japanese religion, however, notes that "the traditional world view is being questioned."[109] He calls attention to the important discrepancy between words and deeds when he says that "Nature may be seriously polluted, but it is still revered and extolled, particularly in graphic art and literature."[110] Mainstream Western attitudes toward nature are clearly making major inroads into Japanese culture. Traditional attitudes have been "rapidly fading in Japan since the hasty introduction of modern science and technology."[111] Attempts are being made to counterattack against Westernization and the consequent destruction of the environment, but it is an uphill battle.[112] Masao Watanabe, professor of the history of science at the University of Tokyo, holds that not only will Japan have to reject the "science fetishism" of the West, but return to Buddhism.[113] In 1974 he was concerned that the hour for decision was getting late, as "still immersed in nature itself, the Japanese people do not quite realize what is happening to nature and themselves, and are thus exposed more directly to, and are more helpless in, the current environmental crisis."[114]

Not all would agree that it is quite that simple. One Western scholar contends that the Japanese have always been ambivalent toward nature. In Japanese mythology, he argues, the divinities of nature are born from the lower orifices of the female deity and all those related to culture from the head of the male deity, so "sex stands on the side of nature, whereas culture is represented by the processes of purification." He goes on to say that the basic ambivalence consists in the fact that nature "though it looks beautiful . . . is also the realm of change, decay, and putrefaction, to which is opposed the purification of culture." This strand of distrust of nature remained underground throughout Japanese history, masked by the cultural conquest of nature which, " 'if left alone simply decays.' . . . It may be said that the traditional interpreters of Japanese culture have failed to see this point, blinded as they were, perhaps, by a Western romanticism which is out of place." He advances the idea that when in 1868 the government officially separated Shinto and Buddhism, the effect was to break the traditional bond between people and place and lay the groundwork for a total reorientation of Japanese attitudes and practice toward nature. Thus, despite its religious tradition, at the end of the century Japan "opened to the West in a catastrophic manner: rejecting much of its past, it . . . assimilated in no time the idea that nature is something to be controlled, rather than man's activities in it."[115] As a result, in modernization, "Japan has become a land destroyed and polluted."[116]

Technology

In the case of technology, as in the case of religion, Japan was not much of an innovator but a master at borrowing and using for its own ends. As in the case of religion, Japan's major source of outside borrowing in science and technology was China. These areas were not totally unrelated, especially in terms of the eventual impact of Confucian science on Japan. But it was a long process. There was a wave of Chinese influence from about 600 to 894, followed, in a pattern found several times in Japanese history, by a period of semiseclusion from 894 to 1401, when Japan shut itself off from outside contacts. Another period of Chinese influence took place from 1401 to 1854.

This whole era of assimilation of Chinese learning was interrupted by a brief flirtation with the West. Portuguese adventurers reached Japan in 1542, and between 1549 and 1551 the Jesuit priest, St. Francis Xavier, established a mission in Japan that flourished for half a century. Like Father Matteo Ricci, the sixteenth-century Jesuit missionary to China, the Jesuits in Japan tried to impress their hosts with the latest in Western learning and science. Ricci's mission to China eventually floundered, largely because of Vatican distrust of his attempts to merge Catholicism with indigenous Chinese culture. Things never got that far in Japan. The Jesuits, and all other Christian missionaries, were suddenly thrown out of Japan in 1587 by a government that was afraid of losing Japanese cultural integrity.

It was difficult for Japan to adopt Western science, above all its mode of thinking, because of the nation's Chinese intellectual heritage. The Japanese were more willing to accept specific technologies, especially in the area of military technology, where they clearly recognized the superiority of foreign techniques. In nonmilitary technology, the case of printing is noteworthy. Woodblock printing began in China in the seventh century, and entire Buddhist scriptures were being printed by the ninth century. The technique was developed in Japan in the eighth century and was widespread. Movable type made of wood, porcelain, and copper appeared in China around 1030, whereas printing with movable type did not appear in the West until the middle of the fifteenth century. It was first used in Japan in the sixteenth century, initially primarily for Chinese classics and Buddhist scriptures. But by the late sixteenth century it was also being used to produce secular materials.[117]

There was one technology, however, that Japan accepted from outside, and then apparently rejected for several centuries: the gun. Its adoption has become a textbook case in the relationship between technology and society.[118] In Japan the gun took the form of the arquebus, introduced by the Portuguese in 1543. It was soon not only accepted but manufactured by the Japanese themselves, who already had an advanced metallurgical industry that exported to all of Asia.[119] Guns soon dominated the battlefield, and Japan became frequently involved in civil wars. With the gun, not only tactics changed, but it may also have contributed to the strengthening of central authority, since it was easier to monopolize the gun than the sword.[120] But when the Tokugawa clan came to control the shogunate and hence Japan, the Japanese not only began a new period of isolation from the outside world but apparently also lost interest in the gun.

How did this happen? The samurai, who were at the top of the social ladder in prestige, regarded the gun as beneath them and far inferior to the sword, a gentleman's weapon. Guns were still prized and used in hunting, but "it was their use on the battlefield that was called into question, for aesthetic and moral reasons. To oversimplify the matter, the gun was considered too inelegant a way to kill an enemy."[121] Not only did the era mark a great revival of chivalry, but there was fear that the widespread possession of guns by the *heimin*, or common people, would lead to social instability. Therefore, in 1588 Hideyoshi enacted a law to prevent peasants from possessing firearms. After war with China and Korea came to an end, guns were virtually eliminated from the nation, and an apparently idyllic age ensued.[122] Yet a commentator notes that the 260-year-old dynasty was marked by over a thousand peasant uprisings, and that "the gun was not rejected out of any sentiment of nonviolence, but to preserve a cruel and despotic social structure that favored one particular group." Eventually the feudal system came to an end in Japan—though later than in Europe—with the samurai defeated by an army of common people who had somehow obtained guns.[123]

Throughout the period of isolation following the ejection of Westerners from Japan, the scientific tradition imported from China continued to remain dominant. But many aspects of this tradition were being criticized long before Perry forced Japan to open itself up to the West. Change was being widely advocated behind the scenes.[124] Many Japanese scholars, though still steeped in the old learning, were becoming aware of its inadequacies and turning to new ideas from the West, though there was argument about how to respond to encroaching Western science.[125] In the early nineteenth century traditionalists had claimed, as in the parallel case of China, that "Western science, which they both feared and despised, was really Confucianist or at least fully compatible."[126]

When the Togugawas isolated Japan from the world in the seventeenth century, they had the military power to make it stick. It was obvious after Perry's fleet appeared in Tokyo Bay in 1854 that this was no longer the case. Japan was relatively so weak that to persist in isolation would only invite the fate encountered by other colonies.[127] Foreign pressure soon led to the collapse of the Togugawa shogunate. The emperor resumed power in the Meiji Restoration of 1868.[128] Now Japan finally had to face modern science in its fullest Western development.[129]

The proud Japanese, though impressed with barbarian power, were determined that they would not fall under Western sway. They had witnessed other empires fail to try to meet the West on its own terms, above all in military technology. The Japanese had abandoned the use of guns as major weapons of war, but now, despite arguments about whether foreign ideas should be allowed to infiltrate the kingdom, they quickly developed a consensus that the only way to defend Japan against possible foreign occupation was through the production and use of Western military technology.[130]

Thus Japan embarked on one of the major feats of controlled social change of our time, adopting Western science and technology while struggling to retain its political and cultural integrity.[131] As in the case of China, Japan attempted to combine "Western technology, Eastern ways," but unlike China, it has been largely, and uniquely, successful. Japan was able to absorb Western technology without major societal disruption.[132]

Since the time of the Meiji Restoration, Japanese science has been almost exclusively supported and directed by the government.[133] How successful it has really been is a matter of dispute. Some argue that in the years before 1889 a major scientific revolution occurred in Japan similar to that of seventeenth-century Europe.[134] In any case, from the outset—and more so even than in early modern Europe—science and technology in Japan were wedded to the power of the state.

The reason for this marriage was that traditional science in Japan, because of its Chinese origin, was closely tied to Confucianism. But as a Japanese scholar notes, Japanese and Chinese Confucianism are radically different. For example, faith and bravery are virtues in Chinese, Japanese, and Korean Confucianism, but benevolence is a virtue only among Chinese and Korean Confucianists; loyalty is common to both Japan and Korea, but it is not considered a prime Confucian virtue in China; Chinese Confucianism is humanistic, and Japanese Confucianism is nationalistic; the religion of the Japanese imperial court was Shinto, but the actual ideology of the administration was Confucian; finally, China became a civilian Confucian country, but Japan became a military Confucian country.[135]

These differences were crucial to the way the two countries greeted Western technology. China's Confucian bureaucrats were primarily interested in literature, Japan's in war. As a result, "Whereas China's bureaucrats showed a solid opposition to the science of the West, Japan's governments from the Tokugawa Bakufu through to the imperial government which followed the Meiji Revolution showed nothing but an enthusiastic desire to acquire this same science." And acquire it they did, and rapidly. Because Confucianism was intellectual and rationalistic rather than mystical, Japan did not endure the same opposition to science as had early modern Europe.[136] In addition, the Confucian bureaucracy was well adapted to become the manager of modern industry—not on an individualistic capitalist model, but in accordance with the bureaucratic structures based on loyalty that are traditional in Japanese society.

The government pushed science throughout modern Japanese history, and even exempted from military service in World War II those students studying the physical sciences. The scientists who emerged from the wartime expansion of science faculties were not ready in time to contribute to the war itself, but they became important in Japan's postwar economic boom.[137]

Whether Japan will really make it scientifically is still in doubt. Both the Japanese themselves and outsiders fault them for lack of creativity in the basic sciences. This may be simply a matter of relative neglect.[138] Or it may, as one Japanese observer suggests, be a more fundamental matter of rejecting the scientific spirit itself.[139] By the early 1990s this latter view seemed increasingly less tenable.

Whatever the state of science and technology in Japan, there is no question that the Japanese have become a major world economic power in the current age of science-based technology. But at the same time they have paid a price in various ways. Not only are they crowded and harried; not only is their fabled consensus in danger;[140] but the nature they have always claimed to cherish has become the victim of pollution.[141] They even have a special word for pollution broadly defined, *kogai*.[142] The Japanese, having chosen Western technology, have chosen the Western dilemma as well.

Chapter Eleven

TECHNOLOGICAL CORNUCOPIANISM

Despite the pessimism engendered by World War I, the following decades were witness to the intensifying belief that science could solve society's problems. Technology, it was still widely believed, was a cornucopia that could in time provide complete material abundance enabling humanity to forget about scarcity and problems of allocating scarce resources. A major figure in popularizing this viewpoint was the British novelist and publicist, H. G. Wells.[1] His widely read *The Outline of History*, first published in 1923[2] and subsequently republished in large editions, was a major force in sustaining the popular belief in the possibilities of modern industrial society. His *Modern Utopia*, published in 1905,[3] had already been widely admired as literature as well as speculation. It became a modern classic, having a wide influence throughout the ensuing decades.

The twentieth century has been, above all, the age of the machine and machine-generated plenty. So strong was the impact of modern technology that it spawned a revolution in art as well as in other aspects of life. A new movement in the arts and literature called "futurism" arose in Italy after World War I. It was highlighted by a "Futurist Manifesto" (written by the poet Filippo Marinetti) that was noted for its dictum that "a roaring motorcar that seems to run on shrapnel is more beautiful than the *Victory of Samothrace*."[4]

Nowhere was the influence of the machine, and its uncompromising functionalism, as strong as in the United States.[5] The period after World War II saw a continuing faith among Americans that the power of science and technology could bring salvation not only to America but to the whole world,[6] and this view remains strong in many quarters today. It persists despite the criticisms that are increasingly leveled against science and technology

154

because of their possible detrimental effects on democratic institutions and on nature and the environment.[7]

Proponents of salvation through technology assert that technology is in itself a humanizing force rather than a dehumanizing one, as its critics often charge. Many testimonials are made to the aesthetic and intellectual pleasures one derives from knowing and building, and even from beholding the work of others.[8] One supporter of technology, the engineer Samuel C. Florman, holds that "analysis, rationality, materialism, and practical creativity do not preclude emotional fulfillment; they are pathways to such fulfillment. They do not 'reduce' experience . . . ; they expand it."[9]

The 1960s were a period of great cultural and political ferment, especially in the United States. Many of the new ideas rejected parts of the civilization that had been created by industrial technology.[10] This led Florman to fear that a wave of antitechnological sentiment was sweeping the world: "Anti-technology, which for a while seemed to be a rather harmless—possibly even wholesome—undercurrent of intellectual rebellion, is suddenly a rushing tide." Florman attacks the 1930s Technocracy movement, calling it a "fiasco," and holds that the widespread fear that there is an unorganized, informal, technocratic rule in modern America is nonsense. He attacks not only the belief that technicians rule our society, but also many of the value preferences that underlie this belief. These values include the idea that self-reliance is the most important aspect of life. It is good in itself, he says, but "much more important, in my view, is the *mutual* reliance that is the basis of civilization."[11]

Florman is not only an engineer. He also holds a graduate degree in literature, and his ultimate reason for glorifying technology is literary. He sees the world as essentially tragic. Technology, which calls us to the fate of Prometheus, destroyed by *hubris*, is justified *because* it appeals to human pride: "Yet pride, which in drama invariably leads to a fall, is not considered sinful by the great tragedians. It is an essential element of humanity's greatness. It is what inspires heroes to confront the universe, to challenge the status quo."[12] What Florman does not sufficiently emphasize is that Prometheus's pride represented a revolt against the gods, an attempt to reject their control over the natural world and place it in human hands. This was an act of defiance similar to that of Adam and Eve and their partaking of the fruit of the Tree of Knowledge of good and evil. Prometheus's punishment parallels their expulsion from the Garden of Eden. In both cases, the offense was that humans sought to become like God in controlling the universe. Modern technology, according to its critics, commits the same sin.

Since in order to control the universe humans must also control themselves, it is only logical that the progress of scientific control over physical nature should run in tandem with an increasing conviction that people can control their own nature and that of society. This conviction is reflected not only in the changing literature of utopia, but also in its first cousin, science fiction.[13] The question then increasingly becomes: "Now that we have mastered physical nature and human society, what comes *next?*"

Beginning in the late nineteenth century and under the influence of Darwin, people were finding a common answer. Humanity, they reasoned, will reach a higher—indeed, a more godlike—plane of existence in which the conquest of the physical universe will make possible its repudiation in a manner not unlike that of many Oriental religions. The whole incarnational basis of the Christian tradition—that God literally became man in the person of Christ and human life itself can be holy—is rejected. The tradition is doubly rejected since humanity is not saved by redemption through Christ but by itself, through its own technology.

There were already hints of this development in such utopian classics as Edward Bellamy's *Looking Backward*, wherein the minister's sermon speaks of the next, higher stage of human evolution made possible by society's conquest of poverty, conflict, and itself. But it did not come to full fruition until the science fiction writing of the twentieth century, most notably in the works of Olaf Stapledon and Arthur C. Clarke. In Stapledon's classics, humanity evolves through almost inconceivable millions of years into new and usually higher forms, some of which barely resemble humans physically.[14] It even acquires the ability to influence the past. But even humanity fails. The solar system—in which humanity is trapped, in Stapledon's view—disappears, and people with it. Stapledon's outlook is essentially pessimistic, and his work ends on a strongly elegiac note.

Arthur C. Clarke, famous to Americans especially for the movie *2001*, is more optimistic. He is not only the author of numerous science fiction books but of an important nonfiction work, *Profiles of the Future*.[15] His position on humanity's future is best expressed in his classic novel, *Childhood's End*.[16] Here the planet has come in contact with a species that is superior almost beyond imagination. Chosen humans are gradually permitted to share some of the powers of the aliens and step beyond the bonds of bodily existence. This outcome epitomizes Clarke's basic view of the universe. Spirit and matter finally become one, the goal of the cosmos since the beginning of time.

This fundamental hatred of the bodies given to us by nature is not unique to the writers of science fiction. The ancient Gnostic view is found at work also in the thinking of other visionaries, such as E. F. Estfandiary. He writes:

We must not accept the human body. Never again be content with it.
We must modify the body, redesign and redo it completely.
We must de-animalize ourselves.[17]

Estfandiary may be extreme in his views, but he is not alone in his rejection of the natural order. Some reject the idea of accepting and following nature blindly. Ethics, they argue, is a human creation and as such antinatural. T. H. Huxley, the famous biologist and friend and defender of Darwin, held that "the ethical progress of society depends not on imitating the cosmic process . . . but in combating it."[18] Although not all those who repudiate the classic belief in human limitations are so grandiose in their aspirations as to overshoot realistic human possibilities completely, many come close.

Biotechnology

One area in which technology is being used to alter radically the traditional concepts of human biological nature is the growing field of biotechnology. Not all who work in or support its efforts would support Estfandiary completely. But they share with him the belief that man has inherited the powers of God the Creator, and can and should use scientific knowledge to reshape the world of living things.

Since the dawn of history, humans have altered the living world. Plant species have been bred since the beginnings of agriculture; corn as we know it could not exist without human intervention. Cattle and horses and dogs and cats have long been shaped by animal breeders. But all this has involved trial and error based on empirical knowledge. Not until the discovery of the laws of heredity by the Austrian monk, Gregor Mendel (1822–1884), in the mid-nineteenth century could breeding be put on a scientific basis and the modern science of genetics arise. Science since then has learned how to locate and physically move genes from one species to another and, through genetic engineering, to alter the characteristics of plants and animals. As a result, "Man now possesses power which is so extreme as to be . . . godlike."[19]

So far the major applications of genetic engineering have been in the field of agriculture. Plants are being created that are more resistant to disease, insects, or extreme temperatures. Cows are altered so they can produce more milk. Economically, the stakes are vast, and some envision a day when food can be produced even without the use of land.[20] Visionaries even look to a future in which genetically engineered bacteria can become building materials, and artificial photosynthesis will supplement the natural mode.[21]

But the real frontier of course involves human beings. Ever since the publication of Aldous Huxley's science-fiction classic, *Brave New World*, in 1932, the subject of human cloning has been a source of scientific and moral controversy. Every living cell in every living thing contains its complete genetic blueprint. Very early in the life of the plant or animal, cells become specialized and the blueprint becomes dormant. But it is possible to take living cells and to use them in the laboratory to create copies of the whole creature. This has been done with plants and small animals, such as frogs, but not so far with mammals. James Watson, Nobel Prize-winning codiscoverer of DNA, predicted in 1971 that a human being would be successfully cloned between 1990 and 2020.[22] Not all scientists would agree even today that this is technically possible, but it is the ethical aspects that have drawn the most comment.

Should humanity resort to cloning? This would involve the final rupture between sex and reproduction. Sex without reproduction has long been possible. Primitive birth control techniques are as old as human society. The industrial revolution made the artificial condom possible, and birth control pills have long been on the market. Now cloning makes it possible to have reproduction without sex. For those concerned with the eugenic future of the human race this is a highly desirable development. Nobel laureate Joshua Lederberg contends that "if a superior individual . . . is identified, why not copy it directly, rather than suffer all the risks of recombinational disruption, including those of sex? Leave sexual reproduction for experimental purposes; when a suitable type is ascertained, take care to maintain it by clonal propagation."[23] For those who feel the future of the race is an abstraction that clouds the issue, and that natural processes should be respected, human cloning would not be desirable at all. Protestant theologian Paul Ramsey holds that "the practice of medicine in the service of life is one thing; man's unlimited self-modification of the genetic conditions of life would be quite another matter."[24]

Human cloning at present is still a hypothetical matter. Other biotechnological interventions in the natural process of birth are not. Human beings' desire to have children at all costs has led to the widespread use of artificial insemination (even using sperm donors other than the husband) and surrogate motherhood, in which the fertilized ovum is carried to term by a woman other than the producer of the ovum. Who the real "mother" is in these cases is a matter of great dispute and the subject of bitter and highly publicized court cases.[25] Once again, as in the case of cloning, ethicists disagree as to how far human beings should use technology to change the natural order. Cases such as these, added to the problems associated with modern medicine's ability to keep alive in a vegetative state those who would have died in earlier eras, has led to a new profession, that of the bioethicist, who is often consulted in cases of doubt.[26]

Though the questions of genetic alteration of human beings through genetic engineering and the revolution in reproduction most dramatically pose the issues about human nature and technology, advances in biotechnology pose other problems. Probably the most socially important consequence that faces us is genetic screening. Medicine has been busy isolating the genes that cause or predispose humans to hereditary diseases such as Huntington's disease, certain forms of cancer, and alcoholism. There is widespread support both among scientists and politicians for the Human Genome Project, a federally financed attempt to map as rapidly as possibly the 50,000 to 100,000 different human genes.[27] But what happens when science is able to tell who has which "defective" genes? When the information gets out, as it almost certainly will, will the individuals in question find it difficult if not impossible to get jobs or insurance coverage? On a broader level, as S. E. Luria of the MIT biology department has asked, "Will the Nazi program to eradicate Jewish or otherwise 'inferior' genes by mass murder be transformed here into a kinder, gentler program to 'perfect' human individuals by 'correcting' their genomes in conformity, perhaps, to an ideal 'whiter, Judeo-Christian, economically successful' genotype?"[28]

What could happen is already evident from the use some have made of the related technology of ultrasound, which enables physicians to tell much about a baby by examining the fetus while it is still being carried by the mother. Designed to check the future health of the child, the machines are used in India almost exclusively to determine its future gender. If it is to be a girl, it is frequently aborted because women are not valued in Indian culture. Sex ratios of newborns in India are already being drastically altered because of this technology.[29]

Biotechnology is clearly an area in which those who want to use technology to change human nature to a form they deem superior will become manifest.

Conventional Utopianism

But this is an extreme. There are many quite conventional contemporary utopian thinkers who simply project the progress of the past indefinitely into the future for centuries, or even for millennia.[30] Others focus on particular aspects of society, but within an overall stance of believing that things are inexorably becoming better.

One influential thinker has been the psychologist B. F. Skinner. Skinner taught at Harvard for many years and has been a major force among American behavioralists. He held, in essence, that man is no different from other animals and can be conditioned to behave "virtuously." For Skinner there was quite literally nothing except that which could be directly observed: mental states are fictions because the mind is fiction. Yet, somewhat paradoxically, there is something that could be conditioned, in people as in animals. He explored these issues in his classic utopian novel, *Walden Two*,[31] and in his manifestoes, *Beyond Freedom and Dignity*[32] and *Science and Human Behavior*.[33]

Skinner's basic point was that, because freedom is an illusion, we cannot escape being conditioned by various forces in our environment. It would be better if we ourselves did the conditioning. This means of course that some human beings are to be granted direct control over our society—like the conditioners, the leaders of the *Walden Two* experiment—and this eventuality has led many critics to view Skinner as a totalitarian who would not only destroy human dignity (which he did not believe in, as he clearly said), but subject us to a new tyranny.[34] Skinner has been quite influential. His books have sold widely and have been the subject of much comment. In the 1960s a commune based on his ideas was established in Virginia, and it survives to this day.[35]

On what basis would judgments be made as to good or evil in the new society? Skinner's answer was deceptively simple. What is advantageous to human society is a matter of what is found through experience to be conducive to its survival. Nature itself teaches us right and wrong. Skinner apparently believed that in this he was being faithful to the implications of Darwin's theory of evolution. In a sense he was. What Skinner did not seem to realize— indeed, the very thought would have horrified him—is that he was not only aligning himself with the conservative beliefs of Edmund Burke but also with the tradition of natural law. This tradition, which goes back from Stoics such as Cicero to the ancient Greeks, holds that the "ought" can be derived from the "is."[36] In a supreme paradox, the optimistic belief that unbridled science can save modern society rests on a profoundly prescientific foundation.

Another major prophet of scientific-technological optimism in the contemporary world has been the late R. Buckminster Fuller. Fuller was a largely self-trained polymath, a man who operated in many different fields in which he had no formal credentials. Fuller was widely respected as an inventor; the famous "geodesic dome" is probably his best-known trademark.[37] His message is as simple as it is grandiose. It is that man is capable of anything, and the universe is a treasure house of resources that only awaits the opening of the human mind to reality so it can be used for whatever one may need. In Fuller's thinking, technology is not a means by which human beings master the universe. He finesses the problem of whether and how humanity should control technology by assuming that the universe, humanity, and technology are essentially one: "Universe is technology—the most comprehensively complex technology. Human organisms are Universe's most complex local technologies." Ultimately, as in the fantasy of Arthur C. Clarke, humanity and the physical universe will merge into one: "Humanity can only evolve toward cosmic totality, which in turn can only be evolvingly regenerated through newborn humanity."

Aside from his general optimism, perhaps the most important thing about

Fuller was his reaffirmation of humanity's desire to escape from its Earth and time-bound existence and conquer the universe: "Humanity at this present moment is breaking the critical-proximity barrier that has programmed him to operate almost entirely as a part of the ecological organisms growing within the planet Earth's biosphere."[38] Despite his obfuscatory prose style, Fuller's message is crystal clear. Through technology, people can escape from being human and conquer the total universe and merge with it, in effect becoming God. Prometheus acted correctly. Indeed, the serpent's promise to Eve was an honest one, and its honesty will be revealed in the near future. It is little wonder that Fuller became so popular. His thinking was at the root of much of the desire of the United States—and the former Soviet Union as well—to conquer outer space, the great human project of the last half of the twentieth century. As one space scientist wrote, "The single purpose of evolution is the perpetuation and eternalization of life; its entire strategy is focused on that goal."[39] He went on to say that "science and spiritual-religious philosophy both hold the belief that man and his creation, the fourth kingdom of the machine, will find in the universe infinity and eternal life."[40]

Not nearly as grandiose in his claims and intentions, but still a technological cornucopian at heart, was the publicist and controversialist Herman Kahn. Originally trained as a physicist, Kahn burst upon the public consciousness in the 1950s with the publication of *On Thermonuclear War*.[41] In this book he argued that we must learn to "think about the unthinkable"[42] and that such a war would be indeed terrible but not quite as infinitely horrible as most people seemed to assume. After all, *some* would survive. Much controversy ensued, of course. Undaunted, the now famous Kahn in the following years turned to the study of the future in general.[43] In time he founded, and until his death directed, the Hudson Institute, a "think tank" that soon turned from early interest in military-related topics almost exclusively to optimistic speculation about the future. Working with colleagues, Kahn produced *Things to Come*[44] in 1972, and in 1976, for the American bicentennial, *The Next 200 Years*.[45]

It is in this work that the most grandiose—and at the same time the most insidious—aspects of Kahn's thought are revealed. He and his associates "predict" that by the year 2176 the gross world product will be 300 trillion dollars (in 1976 dollar values), give or take a factor of five.[46] Not only is this wildly optimistic in comparison with most other economic forecasts, but the "give or take" clause renders it virtually meaningless, to say the least. Yet Kahn claimed generally that this kind of objection did not really matter, since what he was doing was not "predicting" but presenting "scenarios," descriptions of what might or might not happen, depending on prior developments.[47] Thus in effect he eats his cake and gets to keep it as well.

Kahn's predictions, however, have a fatal flaw. Though in many individual cases his optimism may be justified, collectively it cannot be. It may, for instance, be true that we can produce more food by irrigating more land. It also may be true that we can produce more coal by using water on a vast scale in the production process. It is manifestly impossible that we can do both at the same time with the same water. Kahn was not simply counting on making an inside straight; he was counting on repealing one of the fundamental laws of physics by having the same material in more than one place at the same time.

One way in which Kahn's disciples, and Kahn himself on occasion, could reject the ideas of those who felt that cornucopianism was basically erroneous was to argue that in fact the Earth was not the closed system many ecologists seem to regard it as being. Not only does the Earth receive at every moment vast energy inputs from the sun, but mankind can, as in the visions of Fuller and others, leave it to find new resources elsewhere. Space travel will make all the universe a storehouse for man.

Prominent among the prophets of space, besides Arthur C. Clarke, was the late Princeton physicist Gerard K. O'Neill, who did much to publicize and lobby for space colonization,[48] and founded the Space Studies Institute (SSI). Space colonies would have many functions, according to their advocates. First, they would solve the Earth's resource problems, providing both abundant energy from the sun and the possibility of mining the moon and the asteroids for minerals.[49] Space industrialization could make possible the production of goods without pollution, at low cost, in zero-gravity environments.[50] But O'Neill is much more ambitious and far reaching in his aims. The "establishment of permanent, self-sustaining colonies of humans off-Earth" will have several vital results, he contends. It "will make human life forever unkillable, removing it from the endangered species list, where it now stands on a fragile Earth overarmed with nuclear weapons"; "the opening of virtually unlimited new land area in space will reduce territorial pressures and therefore diminish warfare on Earth," and the small scale of space colonies will make possible experiments in direct democracy reminiscent of the New England town meetings of the American heritage.[51]

Much attention has been given to the question of what kind of societies would evolve on space colonies.[52] O'Neill seems to assume that they will be self-governing. Some claim that they would provide an opportunity to try out new life-styles that would be beneficial to humanity.[53] Some are concerned that space colonization would be an "invitation to disaster," since it might lead to space wars and piracy.[54] It is also argued that in space, people might have the same polluting effect they have had on Earth and in time create many of the same environmental problems.[55] The same kinds of dispute arise over what might happen when, as many take for granted, eventual contact will be made with extraterrestrial cultures,[56] for it is widely held that even if none exist in our solar system, man in time will be able to venture far beyond it.[57]

The basic thrust of all of these discussions is clear. Man's voyage to the moon is only the beginning of his conquest of the whole universe.[58] Those who are dubious are silenced by the argument that, had Columbus not sailed the oceans, the New World would never have been found. But others find the analogy disquieting. As one says, "The ocean, terrifying as it can be—is not space—we came forth from this ocean. . . . Alien as the ocean may seem to some of us it is a part of our heritage, the outer darkness of space is not."[59] The darkness of outer space may not be our heritage, but it will be that of some of our descendants. It is clear that for some proponents of space colonization, the probable result, which they are quite willing to accept, is a change in the nature of the colonists. Says O'Neill: "In a longer time the effects of genetic drift will show, as human groups separated by great distances evolve into noticeably different forms of humanity."[60]

What is really involved is an attempt to create a new world through technology, a humanist challenge to the physical universe, now conquered by humans.[61] This conquest can be looked upon as a manifestation of a physical mysticism that—as in the views of men such as Clarke and Estfandiary—defies traditional Western religion by seeing human evolution as leading to the final spiritualization of matter, in which mankind, by becoming the universe, in effect becomes God. But this conquest-transformation need not appear in an openly atheist guise. One of the most influential figures in popularizing the transformation in the twentieth century has been the French Jesuit priest-paleontologist, Teilhard de Chardin. Teilhard's idea, though clothed in Christian imagery, is essentially that the universe-in-the-making needs the cooperation of man's spirit for its completion.[62] He quotes, with approval, his friend Julian Huxley's dictum that "man is nothing else than evolution becoming conscious of itself."[63] So optimistic was he that for him even the invention of the atomic bomb was a good thing. Its explosions herald the coming of the new spirit that will dominate mankind, the "Spirit of the Earth."[64]

With Teilhard we reach the zenith of the optimistic belief in what science and technology can do to improve the human condition. The claims of nature as it came from God's hands are now completely ignored. The total human conquest of nature through technology is not only possible, it is inevitable. It is not only inevitable, it is a divine command. Through material power we find salvation.

Chapter Twelve

CONTEMPORARY CRITICS
OF TECHNOLOGY

It is unrealistic to believe that mankind could thrive—or even exist—without the use of tools, especially modern mankind. As a noted philosopher has said, "Clearly, then, it would be silly for anyone to announce that he is 'against' technology, whatever that might mean. We should have to be against ourselves in our present historical existence."[1] Some observers argue that there is a danger in letting the use of technology mesmerize people so that they fail to pay attention to what really makes them human and sacrifice their identity at the altar of technology.

Prominent among these thinkers is the modern German philosopher, Martin Heidegger (1889–1976). Not only is Heidegger a giant among modern philosophers, but his work on technology stands almost alone in the field.[2] Heidegger was a philosopher's philosopher who, like Kant before him, devoted his whole life to philosophy and, like Kant, rooted in one small geographic area of Germany. Heidegger, described as "the most stubbornly regional character in modern culture," made the "homelessness" of modern man a major philosophical theme.[3] This very passion for local roots may have been one of the things that attracted him to Nazism, with its emphasis on a nationalistic appeal to regionalism as opposed to cosmopolitanism.[4]

Heidegger was a Jesuit novice for a short time, studying theology at the University of Freiburg, but he soon turned to philosophy.[5] From the first he was attracted to Aristotle's concern with being,[6] but soon went beyond Aristotle in his almost fanatic concern with the "thingness" of things. His first major work, which indeed was the core of all he ever had to say, was *Sein und Zeit*, published in 1927. Heidegger felt philosophy had taken a wrong turn after

Plato by rejecting the centrality of being, and he was determined literally to start all over again, to rethink the entire tradition from the outset.

The whole long labor of Western philosophy had been to bring science into existence, "to establish the world of objects over which the subject, mankind, has now to assert its mastery."[7] Mankind now existed in a Nietzchean world, in which the will to power was everything. Heidegger "sees in Nietzsche's philosophy the completion and consummation of metaphysics, and that must mean also the consummation of the essence of technology. Nietzche's overman might be said to be technological man *par excellence.*"[8] In an essay, "European Nihilism," written in 1940 but not published until 1961, "Heidegger argues that modern technology could only arise in a world which has become nihilistic or forgetful of Being. The nihilism of Nietzche's 'will to power,' Heidegger claims, is the culmination of Western subjectivism and leads to the pure 'will to will' of the technological age."[9] For Heidegger the way out was to start the whole labor all over again from the beginning, by changing the relation of the objective and subjective worlds.

The estrangement from Being, which Heidegger sees as the root of the total failure of the Western tradition, is fundamental. The basic error, he holds, begins with Plato's simile of the cave in *The Republic*, wherein Plato made the truth subject to ideas, dependent on verifiability in a positivistic-scientistic fashion.[10] For Heidegger, existence itself "means to stand outside oneself, to be beyond oneself." Heidegger takes man to "be a field or region of being." This field he calls *Dasein*, which in German literally means "being-there." *Dasein* "is his name for man." By looking at things as individual objects, subject to man's ideas, the Greeks destroyed our ability really to relate to them. Heidegger's ideas have many resemblances to Taoism, in that he holds that the world of being is a total system that we must learn not to use but to participate in. We must seek not to manipulate Being but to contemplate it. His "final answer to Nietzche" is that "Western man has got to fetch Being back from the oblivion into which it has fallen. Man must learn to let Being be, instead of twisting and dislocating it to make it yield up answers to our need for power."[11]

For Heidegger, it is in wondering and reflection that we find the truth, not in sensory data. We must become more passive regarding truth, and let it reveal itself to us. We must become open, "vulnerable" to truth as in the "dialectic of prayer." Man "is not the enforcer, the opener of truth (as Aristotle, Bacon, or Descartes would have him), but the 'opening for it,' the 'clearing' or *Lichtung* in which it makes its hiddeness manifest."[12] Our attempt to impose our ideas on the world, rather than being open to the world itself, is the thrust underlying our whole technological effort to make the world over in the image of our ideas. Because this is philosophically wrong, it is bound to be self-defeating. "The essence of technology is danger," in Heidegger's words. When technology becomes total, it "lifts mankind to a level where it confronts problems with which technical thinking is not prepared to cope."[13] Technology, Heidegger believes, may already have taken us beyond the point of no return.[14]

Heidegger's concern with technology is not based on its ravages in society and nature or grounded on social or political concerns; it is based "on rigorously philosophic ground."[15] *Techne* has been debased. Modern technology, which seeks to master the earth rather than cooperate with it, is something radically different than traditional technologies. For Heidegger, "Technology is . . . no mere means. Technology is a mode of revealing." Freedom does not consist in choice, but in the process of revelation of what the world is really like, the process of unconcealment.[16] The essence of modern technology lies not in technology itself but in the purposes to which it is put, what Heidegger calls "enframing." This is where the danger lies.[17] For Heidegger, "Humanity has in some measure been impelled by the nature of things . . . to embrace enframing and the aggressive form of technology it calls forth."[18]

Not technology itself, but how it is looked at is the problem. The Platonic mistake compels us to use technological knowledge in an improper way: "What is dangerous is not technology. There is no demonry of technology, but rather there is the mystery of its essence. The essence of technology, as a destining of revealing, is the danger."[19] What Heidegger seems to be trying to say is that it is not technological artifacts that are dangerous, but the belief of humans that technology can explain ultimate reality. Our problem is that we have made the world a mere object of our desires: "Nature appears everywhere—because willed from out of the essence of Being—as the object of technology."[20] The ultimate technology is the ultimate danger. According to a contemporary sympathizer, for Heidegger "the cybernetic model of man is the *exemplar* of today's Western thinking that is dominated by logistics, calculation, and technology." The solution, of course, is radical rethinking of our concept of Being, beginning with our view of our own selves. Heidegger "envisions a new dawn of human thought with the destruction of technological rationality."[21]

What are the implications of Heidegger's ideas and what has been his influence? We must first confront his commitment to Nazism. There is no question that it was not a mere "flirtation" during a trying time in his life. Some claim Heidegger "broke" with the Nazis largely because he began to realize that they were not as opposed to the wrong kinds of technology as he had at first supposed. But he never—even after the war—recanted or sought to clarify any of the philosophical propositions on which he agreed with them. Nor did he fully repudiate the ideas they shared.[22] Despite philosophers' apologetics, the issue of Heidegger's Nazism has resurfaced in the past several years, especially in France, where his following among philosophers was great.[23] No question: Heidegger was a Nazi. His ideas, like theirs, reflected the twentieth- century German ambivalence toward technology. Despite this, his ideas on technology are so powerful and have been so widely echoed that they must stand—or fall—on their own merits.

According to one critic, Heidegger's ideas have had a major role in shaping neo-Marxist criticism of technology, especially that of Marcuse: "The neo-Marxist idea of a new 'humanized technology,' of a return to harmonic concordance between human needs and the laws of production, is pure Heidegger."[24] This is ironic. Heidegger really appeals to a more primitive view of the world.[25] He is "an agrarian through and through."[26] Yet even critics of modernity who are not committed agrarians can find him congenial. Taking their cue from his belief that "enframing" does not stem from technology alone, but from its associated social context they, in a sympathetic observer's words, "have been opposed . . . not so much to what technology *is*, but rather to its nontechnological essence, a technocratic mentality, compatible with what he calls 'enframing.' "[27] Heidegger continues to be a major focus of discussion wherever the "philosophy of technology" is debated. Despite his political past, there seems to be no reason to expect his influence to lessen.[28]

Jacques Ellul

While the ideas of Heidegger remain a major component of the debate about technology and society, far more directly influential is the contemporary French thinker Jacques Ellul. Compared to Heidegger, his style makes him much more accessible, or at least he seems to be. Jacques Ellul is a radically different kind of philosopher and person than Heidegger. Trained in law and theology, he participated in the French resistance against Nazism before a long career teaching law and sociology.[29] He has been described as "at the deepest level a theologian—that is, a reflective Christian."[30] His first books were about theology. He noted that it was a query from the great theologian Karl Barth asking what he meant by "technique" in *The Presence of the Kingdom*[31] that led him to write the book, *La Technique; ou L'enjeu du siècle* (1954). Even before it was translated into English as *The Technological Society*,[32] the book made him famous and led to major conferences on technology and society in the United States,[33] followed by a vast outpouring of works by Ellul.[34]

Ellul has views on the direction of modern society that seem to be so negative as to be highly controversial, and a huge secondary literature has responded.[35] Opinions on his ideas vary; they largely, but not completely, reflect the critic's own ideas about technology and society. No doubt the fact that Ellul is not only a theologian but a Calvinist (which not all critics know) and thus predisposed to determinism affects his ideas.[36] Many who share his negative ideas about modern technological society are unhappy with what they regard as his belief that nothing can be done about it. Lewis Mumford, who shares Ellul's belief in what Mumford calls "the evils of megatechnics," accuses him of "fatalism."[37] Another critic speaks of Ellul's alleged " 'malevolent' technological determinism."[38] Bernard Gendron, surveying various contemporary attitudes toward technology, classifies Ellul as a "Dystopian" critic in his extensive discussion of Ellul's ideas,[39] and Witold Rybczynski carries on a running battle with him in defense of modernity in the pages of *Taming the Tiger*.[40] Yet it may not be that simple.[41]

Ellul's fears about technology do not stem from a belief that technology represents a disturbance of the natural order of things. For Ellul, "the natural does not have an eminent and normative value."[42] Yet as a theologian, he quite clearly rejects the interpretations of Genesis that give people complete dominion over the earth, empowering them to act as Demiurge to complete the work of God. Creation as it came forth from the hand of God was "*perfect and finished.*" Technique (by which he *usually* means technology) came into existence only after the Fall, when the descendants of Cain raised flocks, made music, and forged iron and bronze tools.[43] It is an aspect, albeit a necessary one, of mankind's fallen state.

But these new technologies fundamentally change man's relationship with nature. "After the break between God and humanity work as such appears . . . as something necessary and harsh. . . . Work in the biblical narrative is never presented as a joyous accomplishment, a human blossoming, but as a reality which is difficult, exhausting, and painful."[44] But not all work is necessarily a good thing. Not only does man not have dominion over the earth as a result of an original divine injunction to Adam, but there are strict limits on how he can legitimately deal with creation.[45]

Ellul's main concern is not with what people through the use of technology have done to nature, but that the technological environment has *replaced* the natural environment as the world in which people find themselves situated. This is not only Ellul's central contention and insight, but the source of most of the obscurity in his thinking and the criticism it has elicited. If *everything* is technique, how can the existence of technique become an explanatory principle? In his introduction to the American edition of *The Technological Society*, Ellul defines technique as the "*the totality of the methods rationally arrived at and having absolute efficiency* (for a given stage of development) in *every* field of human activity."[46] But what, one might ask, does such an all-encompassing concept leave out? Only those actions that are random or irrational would seem not to meet his criteria for being part of "la technique."

One sympathetic observer, William Ophuls, makes much of the fact that Ellul distinguishes between "tools," more or less neutral instruments that can be used by people for purposes they choose, and technique, "which imposes a certain behavior on men."[47] Yet what is it that is really "imposed"? There are two questions involved here. One is whether or not all modern activity is really as rationally means-ends oriented as Ellul seems to believe. That is an empirical question of fact, which can be debated. But even supposing that Ellul is correct, and that people always behave in that way, what is he attacking? Nothing less, it would seem, than the rational nature of modern society itself. If that is the case, what would he put in its place—a deliberately irrational society?

Before we can come to grips with Ellul's ideas, we must understand his intellectual methodology. As one critic of *The Technological Society* notes: "The subject and thrust are sociological, but the method is phenomenological. Ellul uses data, both historical and sociological in origin, less as clues to a pattern than as the artillery of his mounted offensive upon the reader." He subscribes, the same critic avers, to Engels's law "that changes in quantity can lead to change in quality." The great increase in technologies has caused them to change their character.[48] Actually, Ellul's way of dealing with ideas was strongly influenced by Marx. He had first read Marx at age nineteen, while working to support himself in school; he became a Marxist and devoted himself to the study of Marx's ideas, though he never joined the Communist party. At twenty-two he began to read the Bible and was converted to Christianity with, as he puts it, a "certain brutality." He soon recognized that Christianity and Marxism were irreconcilable. But he also "saw clearly that one could not deduce directly from the Biblical texts political or social consequences valid for our epoch. It seemed to me that the method of Karl Marx . . . was superior to all that I had encountered elsewhere."[49] Things become more confusing because, although Ellul insists that there is a "difference between the concept of *technique* and the concept of *technology*" and that it is a "a grave error" to confuse them, he explicitly has allowed translators to render *technique* as technology.[50]

What is Ellul really trying to tell us about technology? First, he says that although human beings have always used technologies, they were local and limited and part of a whole rather than a defining environment. Now the technological system is everywhere, and technology is dominant in our society. Technology is, moreover, determining.[51] Technology not only is determining, it is autonomous: it is *independent of man himself.*[52] It is "self-determining in a closed circle. Like nature, it is a closed organization which permits it to be self-determinative independently of all human intervention."[53] "Neither man nor group can decide to follow any road that is not technological."[54]

Technology constitutes a total environment, replacing nature. Technology "is in fact the environment of man." "Man once lived in a natural environment, using technical instruments to get along better in it, protect himself against it, and make use of it. Now, man lives in a technological environment, and the old natural world supplies only his space and his raw materials."[55]

Technology is not simply a matter of one or more particular technologies, adding their impact to one another. It is above all a total system. This is central to Ellul's thought: "The technological system is a qualitatively different phenomenon from an addition of multiple technologies and objects."[56] It is a system like nature, from which man cannot escape.[57] This system, like all systems, has its own norms. The dominant one is efficiency, defined on technology's own terms. Technology creates its own moral standards: "Independent of morals and judgments, legitimate in itself, technology is becoming the creative force of new values, of a new ethics." So pervasive is the technological norm of efficiency that everything becomes an efficiency-oriented technology, even sex and religion.[58]

What are the effects of technology on human beings? Ellul is ambivalent on this issue. Technology does not necessarily destroy the highest human capabilities, as many contend, but it in some sense negates them by making them part of itself.[59] Technology is at fault above all for destroying human freedom. It does of course remove us from our unfree dependence on nature. But this freedom is simply exchanged for a new bondage, to technology itself. This new world created by technology, Ellul tells us, is one "into which freedom, unorthodoxy, and the sphere of the gratuitous and spontaneous cannot penetrate. . . . The more technical actions increase in society, the more human autonomy and initiative diminish."[60]

Ellul's ambivalence toward contemporary technology is manifested in his attitude toward the computer. The computer may or may not contribute to the apparently desirable end of decentralization, but it is significant in helping reduce people to a lesser position in the face of technology: "Technology is a process of exteriorizing human capacities. And the final step has been taken. Man is faced with another being capable of doing everything that man used to do, but with greater speed, accuracy, etc." Moreover, the computer "performs a task *inaccessible* to man."[61] It has come along to save the technological system in the nick of time.

If the triumph of technology destroys human freedom in an existential sense, what of human freedom in the political sphere? Ellul's initial major work, *The Technological Society*, was soon followed by two books, *Propaganda* and *The Political Illusion*, which argued that political freedom was necessarily an illusion in a technological society. Freedom in technological society is not threatened by technocracy or anything like it. "There are very few technicians who wish to have political power," he says, but "we are watching the birth of a technological state, which is anything but a technocracy." Ironically, the notion advocated by some—that the way to solve the problems posed by technology is for the political state to control it—would itself lead to technocracy, since "a state qualified to dominate technology can only be made up of technicians."[62]

The problem of the impact of technology on political freedom is far more subtle and complex. Within existing democratic states, especially when the population has come to depend on the material happiness it believes technology will bring them, "the power of the politician is being . . . outclassed by the power of the technician. . . . Only dictatorships can impose their will on technical evolution."[63] In fact, Ellul argues, the technical solutions are virtually self-justifying:

> When the expert has effectively performed his task of pointing out the necessary ways and means, there is generally only one logical and admissible solution. The politician will then feel himself obliged to choose between the technician's solution, which is the only reasonable one, and other solutions, which he can indeed try out at his own peril but which are not reasonable.[64]

Note that it is not really the technician as such who has power in Ellul's view but technique itself. Note also his assumption that there is always simply *one* right way of doing things, a naiveté reminiscent of the dreams of the Progressive reformers of late nineteenth-century America. Government itself begins to fade away as technology replaces it. In an echo of the Marxist utopian belief—derived of course from Saint-Simon—that after the revolution the government of men will be replaced by the administration of things, he writes that "the political power is no longer precisely a classical state, and it will be less and less so. It is an amalgam of organizations, reduced because, in the interplay of techniques, decision making has less and less place."[65]

The Technological Society was written over a generation ago. Has Ellul changed his beliefs about the dominance of government by technology in the intervening years? Not really. We are still told that "technologies now make it possible to shape desire, and public opinion forms on that basis." Technicians still "are at the origin of political decisions."[66] By his stress on the overarching nature of technology in the modern world, Ellul, despite methodological disclaimers, has constructed an intellectual system in which freedom is necessarily lost because the only real freedom would be the nonexistent ability to reject the total modern social world.

From the perspective of Ellul, the problems raised by rampant technology are not quite those seen by other observers. As already noted, he is not concerned with what technology has done to nature, since he does not consider nature to have any special value in itself. Nor is he concerned with anything like pollution, since all such problems created by technology can and will be solved, if only by technology itself. But for him this is no escape, since although because these new technological fixes for the problems created by technology might make it possible for people to have a "pleasant and livable life," still "this is nothing more than substituting an artificial system and a technological fatefulness for the natural system and the fatefulness of the gods. There is no retort, no original invention by man."[67]

Technology has become all-powerful in society. How can we deal with this omnipresent, omniscient system? Shall we embrace it as the will of God? Ellul explicitly refers to the ideas of his fellow Frenchman and theologian, Teilhard de Chardin, who as we have seen, embraced technology as the means to salvation; but Ellul does not critique Chardin's ideas, regarding them as one of many theories that were "superficial" and impractical.[68] What really are our choices, if any, in regard to technology, in Ellul's views? Not many. For Ellul, as a critic writes, the question " 'Can man control his own technologies?' has no real meaning, as the autonomous character of technique denies man even such an option."[69] Our only hope is to accept the reality of technology as it exists while denying it a mental hold over us. For Ellul,

> It seems to me that the only means to mastery over Technique is by way of "de-sacralization" and "de-ideologization." This means that *all* men must be shown that Technique is nothing more than a complex of material objects which have as their sole result a modicum of comfort, hygiene, and ease.[70]

He is the first to admit how difficult it is to grasp what he thinks is wrong with our technological society. He writes:

> I realize I may be asked: "But if man can develop all his potentials through technology, what more do you want?" A tough question to answer. How can we point out that highly technicized sex is not love? That playing with complex or fascinating apparatus is not equivalent to a child's playing with bits of wood? That the nature reconstructed by technology is not nature? That functionalized nonconformity is not existential?[71]

Ellul, despite his rejection of technological rationality as a norm, is no supporter of mere rebellion against it, no partisan of the "greening" of the modern world. The system is far too strong for that, and the worldwide countercultural movements of the 1960s were futile: "It is not the sensualist, irrationalist explosion that is going to shake the system per se, even the slightest bit." Nor can withdrawal into communes do the job; they too can exist only in the context of a technological society: "Little can be expected of them . . . for they lead to setting up a marginal society, with no influence on technology, and with a way of life that is highly dependent on outer technologies."[72]

Yet despite its triumphs, including the extent to which its norms have been internalized among people, the technological system is not really all-powerful after all. It has its own inner problems. There are limits: "Psychologically, ideologically, man cannot put up with *everything*. . . . The malleability, the plasticity of the social organism are not indefinite. Hence, choices must inevitably be made among technologies, among innovations. We cannot, today or tomorrow, do everything that technology overabundantly proposes."[73] But these choices among technologies—which are necessary and will become more so each day—are still not real choices. They can take place only within the technical society.[74]

Is Ellul right about technology and its power? A sympathetic observer notes that despite the fact that some hostile critics say his analysis is "so general that it leads to withdrawal from political life," Ellul has not, paradoxically, done so.[75] But the paradox is a real inconsistency. Despite his activism, Ellul essentially shares Heidegger's whole rejection of the modern age. In judging his ideas, one cannot help thinking that the problem of society's relationship with technology is not quite as clear-cut as he thinks, and we are still potential masters of our destiny.

Technology Is Not the Answer: More-Moderate Critics

Many believe technology can solve all of our problems. Some, like Heidegger and, despite disclaimers, Ellul, regard it as the major villain in modern life. But there are many others who, though not quite as negative, still share a basic skepticism about modern technology and what it has done to nature and society, and they contend that, if left uncontrolled, it threatens the continuance of the human race and its culture.

An important contribution to the political criticism of the cornucopian vision comes from the veteran American anarchist, Murray Bookchin, an active participant in current debates within the American Green movement. Bookchin starts out from a Marxist perspective, which he has gradually abandoned in most of its conclusions, yet which he cannot seem to escape in its methodology. But he warns that while science and technology can provide us all with a good life, they can only do so if we respect natural limits. Thus he allies himself in practice with the ecological movement, though he has little but contempt for "environmentalists," whom he regards as too timid to really attack the base of the system of modern industrial capitalism.

For Bookchin, the way in which physical nature maintains itself despite lack of common coercive direction is the perfect analogy for the society that anarchists have always envisioned, and the two visions are one. Thanks to modern science and its undoing of scarcity, we are now able to implement this vision: "Post-scarcity society . . . is the fulfillment of the social and cultural potentialities latent in a technology of abundance." Bookchin holds species fulfillment to be just as important as species survival: "Man's desire for un-repressed, spontaneous expression, for variety in experience and surroundings, and for an environment scaled to human dimensions must also be realized to achieve natural equilibrium." But—and here he parts company with major critics of technology—both fulfillment and survival have been made possible by science and technology. The time is now ripe, he argues, for "the absolute negation of *all* hierarchical forms *as such*." This liberating revolution is made possible not only by technology in the physical realm but by ecology in the intellectual realm; it has shown that "balance in nature is achieved by organic variation and complexity, not by homogeneity and simplification." It also tells us that we are not in sole charge of nature: "Ecology clearly shows that the *totality* of the natural world—nature viewed in all its aspects, cycles and interrelationships—cancels out all human pretensions to mastery over the planet." That anarchism and ecology are so congruent as to be virtually identical is Bookchin's constant refrain. "I submit," he says, "that an anarchist community would approximate a clearly definable ecosystem; it would be diversified, balanced, and harmonious."[76]

Bookchin is in fact something of an agnostic on technology as such. He makes "no claim that technology is necessarily liberatory or consistently ben-eficial to man's development." But he directly attacks the belief of Jacques Ellul and others that man is destined to be enslaved by technology and technological modes of thought.[77] While aware of the extent to which early Marxism and socialism were generally obsessed with the struggle against nature, he believes they are now outmoded. Bookchin has become a cheer-leader for total automation, including replacing not only workers' physical skills but the mental ones as well, completely freeing them from toil.[78] But while machines will do the work, people will at the same time be close to nature. Indeed, there is "a need to make man's dependence upon the natural world a visible and living part of his culture." Agriculture will in some sense be paramount, and "Nature and the organic modes of thought it always fosters will become an integral part of human culture. . . . Culture and the human psyche will be thoroughly suffused with a new animism."[79]

The ultimate synthesis of Bookchin's ideas is found in his book, *The Ecology of Freedom: The Emergence and Dissolution of Hierarchy*. Bookchin believes that while we must seek to guide the natural world in which we live, our hand must be light. He approvingly quotes ecologist Charles Elton's observation that managing the future of the world will be less like playing a game of chess than steering a boat, and says that "what ecology, both natural and social, can hope to teach us is the way to find the current and understand the direction of the stream." He argues that the modern use of the word "technology" is radically different from what the Greeks understood by *techne*: "The goal of *techne* is not restricted to merely 'living well' or living within limit. *Techne* includes living an ethical life." Technology, he realizes, "does not exist in a vacuum, nor does it have an autonomous life of its own." Bookchin is especially concerned to refute the view of some critics that technology can be judged simply in terms of its intrinsic characteristics such as size and scale, which "deflects us away from the most signficant problems of technics—notably, its ties with the ideals and social structures of freedom."[80]

For Bookchin, "a liberatory technology presupposes liberatory institutions." Technology itself is therefore not deterministic. The most significant feature of the modern factory system was not its tools, but "the technics of administration."[81] In sum, he writes, condemning the ideas of E. F. Schumacher, Amory Lovins, Ivan Illich, and Karl Marx, among others, in one paragraph,

A libertarian society must be created that can absorb technics into a constellation of emancipatory human and ecological relationships. A "small," "soft," "intermediate," "convivial," or "appropriate" technical design will no more transform an authoritarian society into an ecological one than will a reduction in the "realm of necessity" of the "working week" enhance or enlarge the "realm of freedom."[82]

"Post-scarcity," he emphasizes, "does not mean mindless affluence; rather, it means a sufficiency of technical development that leaves individuals free to select their needs autonomously and to obtain the means to satisfy them." He condemns those who would have the world's poor wait for their needs to be satisfied because of alleged ecological and resource problems without saying anything about "the artificial scarcity engineered by corporate capitalism." At the same time, he is quite at odds with most other ecologically oriented critics of the existing order in being highly skeptical about the possible exhaustion of scarce resources before substitutes can be found.[83]

Bookchin ultimately bases his vision of the new society on a naturalistic ethics, in which nature is the "source for an objectively grounded ethics."[84] Indeed, in the final analysis, his projected ecological society is itself based on nature: "What we call 'human nature' is a biologically rooted process of consociation, a process in which cooperation, mutual support, and love are natural as well as cultural attributes." He is not so naive as to believe, however, that cooperation is the only human attribute; there are other impulses as well. But they need not lead to the kind of society we have now: "It is by no means a given that individuality, autonomy, and willfulness must be expressed in domination; they can just as well be expressed in artistic creativity."[85]

Although some physical scientists have broken ranks with cheerleaders for unlimited progress through technology and have expressed concern about problems of our contemporary society, the leadership in criticism has come, significantly enough, from the biological sciences. Ecology, after all, is a branch of biology, not chemistry or physics. A leading critic of certain aspects of technological civilization is the microbiologist and geneticist Garrett Hardin. Hardin's major—and virtually sole—criticism of our current society has to do with the fact that, in his view, there are simply too many people in the world. Technology, he argues, can no longer solve the problem.[86] It is excess people that create the pollution that menaces us, and they threaten us with a lowering of our standards of living as well. Virtually all social and political problems are amenable to solution, but can be solved, and solved only, through the limitation of population. Hardin's almost monomaniacal concern with population leads him inexorably into endorsement of a wide variety of political positions, which include not only possible (indeed probable) eventual coercive controls on human birth, but restriction on foreign aid and strict immigration controls.

Though long active as an academic biologist, Hardin's career as a social critic did not blossom until the publication in 1968 of "The Tragedy of the Commons,"[87] which immediately caused great controversy.[88] After beginning by restating the Malthusian approach to population, that "a finite world can only support a finite population; therefore, population growth must eventually equal zero," he argues that liberal capitalism cannot provide the answer: "We can make limited progress in working toward optimum population size until we explicitly exorcise the spirit of Adam Smith in the field of practical demography." In a commons, where each herdsman grazes his cattle on the common without paying an individual price per head, it is the rational interest of each to graze the maximum number, even if the commons itself is overgrazed. The same principle applies to the creation of pollution: "The pollution problem is a consequence of population." The only solution to the population problem is coercion. Here Hardin sounds much like the early social contract theorists Hobbes and Locke, with his commons taking the place of their state of nature.[89]

Hardin is not a technological optimist. He regards technology with skepticism. With reference to the "Technological Imperative," he asks: "Must we use everything we invent? Who says so?" Much of the drive behind technological change stems from belief in "progress," but that also is not among Hardin's gods. He writes: "The man who says 'You can't stop Progress' is a genuinely religious man. But is this the best religion on board a spaceship?"

He exults at length about the halting of the SST and asks, regarding our future needs, "Can anyone think of a technological invention that is *needed* to improve the quality of life? Even one? I think not?"[90]

The attack on Hardin's dim view of the world, his belief that sharing the wealth of the rich nations with the poor is not only unnecessary but counterproductive, has come from many quarters, above all from the Third World and the Left.[91] Most of Hardin's critics start by granting the validity of his basic theory of the "commons" but react against the ethical conclusions he draws from his analogy. Yet even physical and social scientists are far from agreeing with him. As several of them put it, "research carried out in the 21 years since Hardin's article often lead to conclusions that challenge the conventional wisdom." Studies of the actual behavior of fishermen in Maine and the New York area show that Hardin "failed to take into account the self-regulating capabilities of users." They argue that "common-property resource management is not intrinsically associated with any particular property-rights regime." They conclude that "studies after that of Hardin have shown the dangers of trying to explain resource use in complex socio-economic systems" in terms of "simple deterministic models."[92] When all the dust settles, Hardin proves to be not simply inadequate on moral grounds but scientifically as well.

Virtually all contemporary critics of technology and its consequences owe a debt—sometimes acknowledged, sometimes not, sometimes unconscious— to the work of the late Lewis Mumford. This polymath social critic wrote extensively about many things. But his greatest contribution was as a pioneer in making us conscious of the relationship between technology, which he usually styled "technics," and civilization. Indeed, his changing attitude over his lifetime, from early enthusiasm for technology to later skepticism, provides something of a barometer of shifting feelings among part of the American intellectual elite. Over a long career his ideas were modified and extended. But they never wavered in their essence, which was that technology and civilization act upon each other; but while technology is all-important, in the last analysis it is people who decide. Unlike Ellul, Mumford did not flirt with the defeatist dogma of technological determinism.[93]

Though Mumford wrote about the machine in numerous essays, his primary works on the subject are four books: *Technics and Civilization*,[94] *Art and Technics*,[95] *The Myth of the Machine: Technics and Human Development*,[96] and *The Pentagon of Power*.[97] Critics have detected in his thought a movement from the ambivalent[98] or optimistic[99] tone of *Technics and Civilization* to a much more hostile attitude in the two volumes of *The Myth of the Machine*. There is some truth in this. Certainly *Technics and Civilization*, which appeared in 1934, was written in an age when optimism about technology was general, and Mumford shared at least some of it. It ends with a plea for a humanistic socialism to make the machine more amenable to human ends than he holds is possible under capitalism.[100] *Art and Technics* advocates that the machine be integrated into the arts, and still resonates with the optimism of *Technics and Civilization*. It was not until *The Myth of the Machine* that Mumford really began fully to manifest the pessimism about technology with which he is usually associated.

Mumford was no partisan of a simplistic "return to nature," in a sense relativizing it by arguing that it was a historical phenomenon, not possible as long as nature was uppermost, but that eventually "the cultivation of nature for its own sake, and the pursuit of rural modes of living and the appreciation of the rural environment became in the eighteenth century one of the chief means of escaping the counting house and the machine." Some of this urge was relieved for several centuries by the expansion of Western civilization into new, wild lands; but after this "safety valve" ended, the "lure of more primitive conditions of life, as an alternative to the machine, remains." But Mumford himself does not embrace it, quite the contrary: "The advocates of these measures for returning to the primitive forget only one fact: what they are proposing is not an adventure but a bedraggled retreat, not a release but a confession of complete failure." Not only is it a failure which he rejects, but a dangerous one, since "if such defeatism becomes widespread it would mean something more than the collapse of the machine: it would mean the end of the present cycle of Western Civilization."[101]

In *The Myth of the Machine*, Mumford's writing takes on a new tone. Mumford now asserts that things had gone wrong almost from the very outset of human history. Contrary to his earlier periodization of the history of technology, the machine—in its broadest sense—dated from the first authoritarian civilizations of the Near East. These machines—automatically acting—were tightly controlled aggregations of human beings.[102] It is this human megamachine, largely independent of technology, which has survived to our own day and which makes technology itself such a threat to human existence and happiness.

Mumford was not convinced that any basic change toward a higher cosmic destiny can save us. He explicitly rejected the thesis of Teilhard that, through the creation of a worldwide information society, the noosphere, we are necessarily evolving toward such a state. Indeed, Mumford believed not only that it is not fated but not ideal as well. In any case, he holds, "that planetary supermechanism will disintegrate long before 'the phenomenon of Man' reaches the Omega point." Yet Mumford was not without hope for the future. The solution, he reiterated, is the replacement of the mechanical, the inorganic, by the living. This new model "will in time replace megatechnics with bio-technics" and lead to an economy that no longer absorbs humans into the machine but enables them to further their "incalculable potentialities for self-actualization and self-transcendence."[103]

Creating such a new regime will require a revolution, but it will be primarily a spiritual one. Mumford's revolution is actually quite selective. The system cannot be attacked frontally by mass organizations and mass efforts at persuasion, Mumford argues, since "these mass methods support the very system they attack." The only methods that have been seen to work, indeed the only ones that can work, are in effect guerrilla actions "initiated by animated individual minds, small groups, and local communities nibbling at the edges of the power structure by breaking routines and defying regulations. Such an attack seeks, not to capture the citadel of power, but to withdraw from it and quietly paralyze it."[104]

Such a worldwide guerrilla revolution was already taking place, Mumford argued in 1970, but he was far from sure of its success: "One dare not become over-optimistic even though the first stir of a human awakening seems actually to be taking place," he wrote then. Yet he remained confident of one thing: "If mankind is to escape its programmed self-extinction the God who saves us will not descend from the machine: he will rise up again in the human soul." He remained equally sure that the choice was not predetermined, but remained in human hands: "The gates of the technocratic prison will open automatically, despite their ancient rusty hinges, as soon as we choose to walk out."[105]

Chapter Thirteen

WOMAN AS NATURE AND GOD IN NATURE

A major faction among modern feminists has argued that the attitude that humans not only have the right but the duty to control or even despoil nature through technology was and is a masculine sentiment. Indeed, they contend, it is part of a whole attitude toward a world of which their own subjugation is an element. According to this feminist reasoning, men have always looked upon women as part of nature while believing they themselves to be above nature, and the domination of nature and the domination of women are part of the same phenomenon. Furthermore, according to this argument, it follows that the end of one cannot take place without the end of the other. The redemption of nature and the liberation of women is necessarily one unified goal.

Ecofeminism, as this school of thought calls itself, argues that the equation of nature and the female as the subject of domination goes back to the roots of human prehistory, to the origins of primitive religion itself. The earliest development of religion shows a movement from conceptualizing God as being of one sex to conceptualizing God as being of another—a shift from a feminine deity to a masculine one.[1] Just why this major change took place we do not know. It seems to have been correlated with changes in the relative social and economic power of the sexes, and feminist prehistorians have spilled much ink in debating its origin.[2] But take place it did. Erich Neumann tells us that "because the patriarchal world strives to deny its dark and 'lowly' lineage, its origin in this primordial world, it does everything in its power to conceal its own descent from the Dark Mother and . . . considers it necessary to forge a 'higher genealogy,' tracing its descent from heaven.[3] As a result, the female sex, identified with the Earth Goddess, must be looked down upon and kept under control lest men be dragged down to Earth with it. Thus from the outset, nature and woman were thrust aside by the male emphasis on the extranatural.

But it was not religion alone that downgraded the position of women in the thought of the West.[4] Philosophy also played an important role. Aristotle taught that women were essentially inferior to men. The philosophy of Aristotle, though it was in partial eclipse for a long time, was revived in the Middle Ages through the agency of Thomas Aquinas, who continued and reemphasized the Aristotelian belief in the inherent inferiority of women. The beginnings of modernity did nothing to alter the role of woman in Western thought. She was, like nature, still the "other," capable of both good or evil, but requiring to be dominated. Though the position of woman was low at the onset of the modern era, it became even worse as modernity triumphed in the new intellectual and social changes associated with the Baconian revolution and the rise of modern industrialism.[5] Hence, according to ecofeminists, the recovery of the equality of status of women is linked to the overthrow of the ideology of modernity.[6]

The villain, or at least the symbol, in the process of further demeaning women's status is Francis Bacon. This is vividly illustrated in the witch trials that swept Europe. The ancient lore of women was forced to retreat into witchcraft, where the new science pursued it and exacted its toll on it. Bacon's own methods were, indeed, akin to those of the inquisitors of witches, as shown by his descriptions of them.[7] There is no question that the Baconian Royal Society was persuaded that women were inferior in their understanding and dominating of nature; indeed, they were part of the problem. As the society's secretary, Henry Oldenberg, wrote, *"What is feminine had necessarily to be excluded from the Society's true philosophy."*[8]

So far did the denigration of woman go in this era that even her role in creating new life was downgraded. As a result of these developments, "Western bourgeois man, made in the image of God the Father and having chosen to cut himself adrift from 'mother earth,' had therefore left himself with no alternative but to appropriate the physical world—God's 'great automaton'— mechanically and sexually."[9]

But this modern vision of the world could not last. It is now under attack by the conjoint forces of ecological consciousness and feminism. Carolyn Merchant has discussed the parallels between feminism and ecology.[10] Many feminists have postulated a universal devaluation of women based on the emphasis on masculine domination of nature.[11] "Eco-feminist theory seeks to show the connections between all forms of domination, including the domination of nature," writes Ynestra King.[12] A similar position is stressed by activist feminist Elizabeth Dodson Gray in her *Why the Green Nigger?*

Re-Mything Genesis.[13] Feminist science fiction writer Marge Piercy has dealt with this theme in several novels.[14] The ecofeminist connection has been written about extensively. Some feminists argue that it is real,[15] but others are more dubious.[16]

Typical of ecofeminism is the view of Susan Griffin: "The fact that man does not consider himself part of nature, but indeed considers himself superior to matter, seemed to me to gain significance when placed against man's attitude that woman is both inferior to him and closer to nature."[17] Feminist theologian Mary Daly is almost livid on the subject, blaming the current pollution of Earth and the danger of nuclear war on male superiority and attitudes, holding that "phallic myth and language generate, legitimate, and mask the material pollution that threatens to terminate all life on this planet."[18] Many feminists such as Daly have come in recent years to reject traditional Western religion, with its male-centered ideas and images of God.[19] They have moved toward a revival of witchcraft—their revenge on Bacon— which they call by the traditional name of Wicca. They consider it to be the only really female expression of religion available. Their version of Wicca is essentially benign, as they claim Wicca traditionally was.[20] Not all feminists have rejected traditional religion as completely as Mary Daly, however. While embracing an essentially ecofeminist position, theology professor Rosemary Radford Ruether still considers herself a Catholic, though often in opposition to the organized church.[21] But generally, continued male domination of the Christian and Jewish religious communities makes alternatives look more appealing to many if not most feminists.

Ecofeminism is not the only feature of the contemporary feminist move-ment. Actually, it was not quite as prominent in the early days of feminism as it later became. From the outset, many feminists rejected biology as the basis of identity. Indeed, many believed only if childbirth could be escaped could women really become free and equal. This theme is strongly present in the popular and influential work of Shulamith Firestone, *The Dialectic of Sex.* Though she often speaks favorably of ecology as such, Firestone advocates using technology to its limits to free womankind from what she clearly regards as the chains of nature. For Firestone, it is *through* technology that women can and will be freed. *"Pregnancy is barbaric,"* she writes, and therefore concludes that "artificial reproduction is not inherently dehuman-izing. At very least, development of an option should make possible an honest reexamination of the ancient value of motherhood."[22]

Many contemporary radical feminists, however, are less loath to reject biology altogether. Says one: "Women's special closeness with nature is be-lieved to give women special ways of knowing and conceiving the world. Radical feminists reject what they see as an excessive masculine reliance on reason, and instead emphasize feeling, emotion, and nonverbal communication." They differ with the Western tradition on epistemology, emphasizing the spiritual, the intuitive, and communion with nature. However, she goes on to admit, women "have not yet been able to construct systematic alternatives to the present masculine science and technology."[23] This is a major dividing line within the feminist movement. Those who argue against "masculine" science and technology and for the proposition that women are somehow closer to nature are saying, in essence, that women are fundamentally *different* from men. Though she is not concerned with nature as such, this is also the thesis in the acclaimed work of psychologist Carol Gilligan. She argues that women's values not only develop differently but *are* different from those of men.[24] The mainstream of the early feminist movement in the nineteenth and early twentieth centuries—and many if not most feminists still today—would hold that men and women are essentially the same. It is this fact that renders immoral any different treatment of them. But this newer school holds that women are *not* the same. Therefore true feminism cannot be victorious until this difference is accepted and the female perspective is given equal weight in human affairs.

In such a reading, Plato cannot be regarded as an early feminist, as many have portrayed him to be, since the female guardians in *The Republic* were embraced as equals only because they had accepted masculine values.[25] Similarly, the rise to power of women such as Golda Meir, Indira Ghandi, and Margaret Thatcher—indeed, the increasing presence of women near or at the top echelons of some American corporations as well—is not really a victory for woman as such, since these women have had to become "masculinized" to get there. The controversy over the "gender gap" in American politics is a reflection of this question: do women in fact react differently than men on fundamental public issues, or respond differently to public personalities?[26]

Why do many argue that men objectify nature and seek to dominate it, while women tend to identify with it and regard it as normative? Most see these tendencies arising in the course of human, or at least Western, history, with roots deep in the traditions of Western religion and philosophy. Some contemporaries, however, look to psychology, particularly Freudian psychology, for answers. Norman O. Brown has argued in *Life Against Death*[27] that the desire to dominate nature is based on a fear of death.[28] The contemporary Marxist theorist Isaac Balbus argues that male desire to rule nature arises because child-rearing is dominated by the mother, who represents nurture and hence nature, and therefore boys grow up wanting to free themselves from her and triumph over her, while girls, on the other hand, identify with her and with nature.[29] If Balbus is right, then—as he himself is aware—

societies will have to make vast changes in their child-raising practices for changes in attitudes toward nature to become universal.

Balbus's willingness to entertain such a basic psychological reason for the problem of the human relationship toward nature is unusual among Marxists. Basically Marxism assumes that the domination of nature by humans is as good as it is necessary. The same rift between Marxists and ecologists on the meaning of the man/nature relationship also applies to feminism as well. The basic conflict between Marxist feminists and other feminists is over whether the domination of woman by man is rooted in capitalism and will disappear with it, as Marxist feminists claim, or is more primeval and will endure even under socialism. Some Marxist theorists today would downplay the original position, but even they do not claim that Marxism provides a solution to the problem of the relationship between the man/woman hierarchy and the human/nature hierarchy. Thus Stanley Aronowitz writes that "there is no linkage within Marxist-feminism between the domination of nature and the oppression of women," and recognizes the radical feminist impatience with Marxism on that score.[30] But he provides no formula to reconcile the Marxist and radical feminist perspectives.

Feminists of all persuasions are troubled by the issues raised by the position of women in science and technology as these become more and more the levers of power in Western and, indeed, in world society. Many feminists have taken the position that there are basic gender differences in approaches to the world, and that science as we have come to know it in the West is an essentially masculine enterprise.[31] This is despite the fact that feminist research is increasingly uncovering considerable activity by female scientists in the past.[32] Much controversy hinges on whether or not women do as well as men in mathematics, and if not, whether the differences are genetic or social in origin.[33] The issue is crucial, since mathematics is the key to all of modern science. But even if it is true that science and technology as usually thought of are in some sense "masculine" in nature, this may represent not cause for alarm but rather opportunity. As one feminist, Liz Fee of the Johns Hopkins University, has put it, "The problem is not one of making women more scientific, but of making science and technology less masculine," which is also "the issue of making it more completely human."[34]

So far, one feminist notes, "women have not yet been able to construct systematic alternatives to the prevailing masculine science and ideology."[35] But this may not be entirely necessary. Women, according to a noted male scholar on technology, have always been dominant in the technologies that really keep the world going, what is often referred to—sometimes disparagingly—as "women's work": cooking, housekeeping, child care.[36] Another student of technology, an ardent proponent of its virtues not simply as a means to human physical survival and advancement but as an intellectual enterprise in itself, takes a somewhat different position. He argues that whatever the truth of the arguments about women's inherent skills—a question on which he is deliberately agnostic—the fact that women are not well represented in his beloved engineering profession despite their gains in other areas is due to social factors. But the social factors are not questions of discrimination, but rather a realization on the part of bright young women that engineering is really—alas—a powerless profession. This is a shame, he holds, since "technology lies at the heart of our contemporary culture . . . [and until] women share in the understanding and creation of technology—which is to say, until large numbers of women become engineers—they will suffer from a cultural alienation that ordinary power cannot cure."[37]

But alienation, to some extent, lies in the eye of the beholder. Many might say that our technological civilization is itself alienating. Some feminists would argue that the masculine interest in dominating nature leads to the pursuit of rational efficiency to the point of irrationality.[38] This would be a good example of the concept of "rational in means, mad in ends," exemplified by Captain Ahab in Melville's *Moby Dick*.

Ecotheology: The Search for God in Nature

In examining the early Christian and medieval periods, we saw how Christian theology—at least its mainstream component—tended to devalue nature. So powerful was this thrust that the noted historian Lynn White, Jr., was able to blame it for the modern ecological crisis. The postmedieval Protestant Reformation did little to change things. Indeed, with its emphasis on the direct relation of the individual soul to God and its stress on scripture rather than philosophy, it reinforced the earlier conviction that this world was of little relative worth compared to the things of heaven. Yet most of the critics of modern technological society call for a spiritual reorientation as a prerequisite for the economic and political changes they postulate are required to save the human race and its civilization.

Sociologist Richard L. Means has called for a new theology of nature to back up Christian ethics in a technological society.[39] In fact, efforts to create such theology, usually referred to as "ecotheology," have been underway in both Protestant and Catholic theological circles since the sudden spurt of interest in environmental issues in the United States (and eventually elsewhere) in the 1960s and 1970s.[40] How successful these efforts are as theology, however, is a disputed matter. They begin with the weight of theological tradition against them. As a contemporary Catholic theologian, John Carmody, summed it up in 1983: "By and large, Roman Catholic and Eastern

Orthodox theologians have paid ecology little heed. Some Protestant theologians have done yeoman work... but much mainline Protestant theology still rumbles along with little advertance to ecology, and almost all fundamentalist theology greets environmental issues with crashing silence."[41]

What is the theological and philosophical background for the new theological interest in the value of nature? Carmody noted that many of the interested Protestant theologians had been influenced by the "process theology" of Alfred North Whitehead.[42] But even many process theologians, such as Protestant John Cobb, remain within a somewhat pessimistic tradition, and hold the view that our limits are related to our imperfectability. Cobb says, "Perhaps the deepest level of our response to the awareness of limits is the recognition that we cannot free ourselves from guilt."[43] Carmody, like many other contemporary Christian ecotheologians, has been strongly influenced by other spiritual traditions. Not only should we take cues from other sacramental religions, he holds, but we have much to learn from other kinds of religions, such as those of the American Indians. He also writes that "the omnipresence of... transcendental revelation, forces us to take Asian religious experience more seriously than classical Christian theology did. For example, Hindu, Buddhist, and Taoist philosophies all offer rich reflections on nature that Christian theologians have yet to appropriate."[44]

Yet for all his attempts to create a new ecotheology, Carmody cannot entirely escape a basic anthropocentrism. He cannot accept the biocentric egalitarianism of some ecological thinkers. "I do not mean, of course, that trees and human beings are equal," he writes. "The proper relationship between nature and humanity admits of nature's serving many human needs." Yet at the same time he wishes, somehow, to allow nature some rights. Somehow it can serve people without also losing its integrity. He asks, rhetorically, "For whom are we to preserve nature?" He replies, "For God, for nature's own sake, and for the human generations to come." "Preserving nature," he says, "is a key part of a far-reaching vote for life."[45] Thus what we have on balance is a chastened anthropomorphism, but an anthropomorphism nevertheless.

The attempt of Carmody to create a new theology of nature is part of a major shift in attitude among at least a strong minority of contemporary Christian theologians.[46] This despite the fact that, as one scholar puts it, "We cannot... expect to find any simple biblical theology of nature for today. There are, in the Bible itself and in subsequent Christian thought, a diversity of views articulated in a succession of historical contexts."[47] This new interest in ecotheology spans several denominations, at least among the more traditional "mainstream" churches.[48] Even conservative evangelicals closely tied to literal interpretation of biblical texts have played a role.[49]

One area on which special attention has been lavished by some denominations is the land itself. Several years ago Catholics presented a lengthy statement called "Strangers and Guests: Toward Community in the Heartland—A Regional Catholic Bishops' Statement on Land Issues." A principal author of the document, John Hart, writes that "our stewardship of the earth means, then, that we must both work with the earth and let the earth live undisturbed. In both instances we are part of the earth's rhythm and harmony." "We must also be conscious when we work with the earth," he continues, "that we are striving to meet human *needs* and not human *wants*. Human needs are basic and finite. Human wants are superfluous and infinite."[50]

By no means is this new theological appreciation of humanity's place in nature without its critics. During the first Reagan administration not only did the Department of the Interior simply go out of its way to undo previous political and legal victories for environmental causes,[51] it did so as a matter of principle. In part this was due to the theological convictions of Secretary of the Interior James Watt, a sincere believer in the traditional view that humanity has a God-given mandate to exploit and use the earth, which has no intrinsic value.[52] Watt's beliefs were part of a more general approach to "Christian politics" on his part, related to the resurgence of the "religious Right" in the politics of the era. But environmentalism has been attacked from the "religious Left" as well, by those who have argued that it can get in the way of social justice.[53]

One can say that, in a Christian context, ecotheology is the attempt to find the basis for a new religious approach, one that is radically different from what has been the historic mainstream of a complex tradition. As one reviewer of Carmody's work put it, his problem "is *not* that Christianity and ecology are incompatible, but that reactivating an atrophied or neglected tradition is hard, lonely work."[54] Yet there is no question that a strong movement exists in many churches in America and elsewhere to tackle the new issues involved. A major conference was held, partly for symbolic reasons, in Assisi, Italy, in the fall of 1986 under the auspices of the World Wildlife Fund International in which the world's five major religions participated.[55] Despite its venue, Catholic participation was less than wholehearted for various reasons. But in March 1988 Pope John Paul II issued a major social encyclical, "Sollicitudo Rei Socialis," which among other things stated: "Among today's positive signs we must also mention a greater realization of the limits of available resources and of the need to respect the integrity and the cycles of nature and to take them into account while planning for development, rather than sacrificing them to certain demagogic ideas about the latter. Today this is called ecological concern."[56]

While the pope had not gone nearly as far as theologians such as Carmody and Hart might have wished, the document was revolutionary in being the first time in history that the Vatican had taken ecological concerns really seriously, at the level of both philosophy and social doctrine. What had caused the new papal attitude? Speculation exists that it was in part due to the report, "A Modern Approach to the Protection of the Environment," published in November 1987 by the pope's "scientific senate," the Pontifical Academy of Sciences.[57] Within weeks the Catholic bishops of the Philippines issued a pastoral letter, "What Is Happening to Our Beautiful Land?" in which they deplored that "our country is in peril. All living systems on land and in the seas around us are being ruthlessly exploited. The damage to date is extensive and, sad to say, often irreversible." They went on to add that "the task of preserving and healing is a daunting one given human greed and the relentless drive of our plunder economy." The Catholic church had been, like other religious bodies, "slow to respond to the ecological crisis," and they asked "for a Filipino theology of creation which would be sensitive to our unique living world, our diverse cultures and our religious heritage."[58] The pope eventually went far to meet the challenge of ecology. In his World Day of Peace letter issued on January 1, 1990, "Peace with God the Creator, Peace with All of Creation," he held that for Christians, "responsibility within creation and their duty toward nature and the Creator are an essential part of their faith," and "the ecological crisis is a moral crisis."[59] Cardinal Roger Etchegaray, archbishop of Marseilles and president of the Pontifical Commission for Justice and Peace, argued that the document put concern about ecology on the same level with calls for social and economic justice and said that, to the pope, "ecology is a moral imperative, not simply a fashion or a political movement."[60] If so, revolutionary changes in Catholic social teaching lie ahead.

In their 1991 meeting the American Catholic bishops issued a statement on the environment for the first time, with the warning that "care for the earth" should not lead to "choosing between people and the planet"; in any event the statement was overshadowed by proclamations on other issues.[61] Shortly afterward, Archbishop Renato Martino, Vatican ambassador to the United Nations, called for radical new thinking to keep "a habitable blue planet" from turning into a "scorched planet."[62] But neither the bishops nor the ambassador addressed the issue of population control, which many would consider a blind spot negating any official Catholic adherence to ecological reform.

Several Catholic theologians have gone beyond the pope and the bishops in trying to forge a theology that takes natural creation seriously. Perhaps the best known has been the Dominican priest Matthew Fox, author of *Original*

Blessing and founder of the Institute in Culture and Creation Spirituality. Central to "creation spirituality" is what Fox calls "panentheism." Panentheism differs from pantheism in that the latter affirms "that 'everything is God and God is everything.' Panentheism, on the other hand . . . slips in the little Greek word *en* and thus means 'God is in everything and everything is in God.' "[63] This new attitude toward the natural world requires a new way of conceptualizing Christ. The new core of Christianity "must be a vital and living cosmology, a cosmological mysticism, a Cosmic Christ."[64] In this cosmology, Christ is identified with the earth, and what our society through its technology is doing to the earth is another form of crucifixion: "The killing of Mother Earth in our time is *the number one ethical, moral, spiritual, and human issue of our planet.*"[65]

Of course, there is widespread opposition to his ideas among conservative Catholics. In late 1988, Fox was silenced for a year at the direction of the Vatican. Since his return to public life he has remained a highly controversial figure.[66]

Equally if not more creative in forging a new theology of creation focusing on ecological concerns is the work of the Catholic theologian, Father Thomas Berry, of the Passionist order.[67] Until recently most of his ideas were found in occasional papers released through the Riverdale Center for Religious Research, which he founded. But in 1988 his *The Dream of the Earth* was published as the first volume in the Sierra Club Nature and Natural Philosophy Library.[68] Berry is nothing if not outspoken. He has said such things as, "I say democracy is a conspiracy against the natural world, and our U.S. Constitution against the North American continent," and "We need a constitution for the North American continent, not only for the humans."[69]

Berry is fully ecumenical in outlook. His earliest interest was in Asian religions, especially Buddhism.[70] He freely acknowledges they have taught him much about how humans and nature should relate.[71] He was also strongly influenced by Teilhard de Chardin, and has been president of the American Teilhard Association, though he realizes that his master was flawed. "Teilhard meant a lot to me in three ways," he tells an interviewer. "He was the first person who told the story of the universe as a psychic-spiritual as well as a physical-material thing. He indicated how the human story identifies with the universe story. And he moved the religious issue from its redemption preoccupations to creation concerns." But, Berry goes on, "What he did not have was an ecology. He had no sensitivity to the natural world. He wanted to conquer and control the function of things. He was arguing for genetic engineering."[72]

Given his preoccupation with natural creation, Berry downplays interest in ordinary social and economic issues. Indeed, he seems to believe they distract from the really important physical ones. He contends we must change our way of life: "I think we can have a better standard of living, but a different standard of living."[73] War is dangerous, but it is not the greatest threat to human survival: "The immediate danger is not *possible* nuclear war but *actual* industrial plundering."[74] This takes place because we are under the spell of the myth of progress. Berry, in language similar to that of Lynn White, Jr., to whom he refers, is critical of what has been the primarily negative orientation of Western Christian theologians toward the natural world. Religion should rather "perceive the natural world as the primary revelation of the divine, as primary scripture, as the primary mode of numinous presence. Christian religion would cease its antagonism toward the earth and discover its sacred quality."[75] For guides to how the good society, informed by ecological consciousness, would operate, Berry turns to familiar figures among ecological critics of modernity, including E. F. Schumacher, Amory Lovins, and the Club of Rome.[76] But these individuals, groups, and movements cannot do the job ultimately: "All fail in judging the present and outlining the future because none of them are able to present their data consistently within a functional cosmology."[77]

Developing and popularizing a new cosmology, a "new story" of the Earth that takes into account its passage through time from the first instant of creation to the present is at the heart of Berry's efforts. Every civilization, he argues, depends for its existence and proper functioning on a commonly accepted version of its place in the universe: "It is all a question of story. We are in trouble just now because we do not have a good story. We are in-between stories. The Old Story, the account of how the world came to be and how we fit into it, is not functioning properly. We have not learned the New Story."[78] The Old Story, of course, is the traditional cosmology of the Christian West. Berry regards it as no longer adequate.[79] We got along well with the Old Story in the past, but the Old Story had many problems. The weakness of the Old Story was especially striking in its cosmology.[80] There must be simply one story, which Berry and his followers, such as physicist Brian Swimme,[81] are trying to create. According to Berry, showing the influence of Teilhard, "The Story of the Universe is the story of the emergence of a galactic system in which each new level of being emerges through the urgency of self-transcendence."[82]

Where does Berry's cosmology come from? It comes from nature first of all, nature as apprehended directly.[83] Yet Berry's vision of what this universe means is largely shaped by science, in the Thomistic tradition: "Thomas Aquinas had his Aristotle; Thomas Berry has his Newton, Darwin, Lyell, and Einstein." Now this science provides, unwittingly perhaps, its own Story of the Universe. In this vision, "Humans and yeast are kin. They organize themselves chemically and biologically in nearly indistinguishable patterns of intelligent activity. They speak the same genetic language. And all things whether living or not are descendents of the supernova explosion."[84] But humans, for Berry as for Chardin, are a special part of the universe that is aware of itself.[85] This new vision of the Story requires a new outlook on values as such.[86] This New Story is all-important, Berry would argue, because it is something that all can share in common.

Berry consistently berates humans for their lack of attention to the planet.[87] But when it comes to positive suggestions, he seems not simply to want to let earth alone but, perhaps to make this possible, for humans to transform their culture by going back to its origins and starting over again. Heidegger, we saw, wanted to start philosophy over again by going back before Socrates, to rescue being from its analyzers and simply contemplate it directly. Berry, too, wants to go back to the beginning. It is a biological rather than an intellectual contemplation he seems to be asking for, but perhaps it has more in common with Heidegger's solution than might appear at first blush: "It appears necessary that we go beyond our cultural coding, to our genetic coding, to ask for guidance," he writes. "We need to go to the earth, as the source whence we came, and ask for its guidance."[88] Like Heidegger, Berry asks us to seek direct knowledge of reality from the soil that bore us.

Central to the theological problems raised by Berry's new vision is its relation to Christian revelation as traditionally conceived. What Berry is really saying is that the universe itself is a revelation. Christianity, and the other major religious traditions of mankind as well—Hinduism, Buddhism, Taoism, Native American religions[89]—are simply attempts by humans to grasp part of this revelation. None of them is complete alone, yet all call attention to an essential part of the truth. What the New Story provides is a context in which the essential truth and unity of all of them can be fully appreciated.[90] These different traditions each have their own integrity. They cannot be merged in a common tradition. Seen in this light, traditional Christianity loses much of its privileged position. It is the universe itself that tells us what we must do in order to be saved.

Despite this, Berry continues to insist that, although our "biblical . . . and theological traditions have provided only minimal resistance to the degradation of the earth," there are powerful resources in them that can be used to aid in the renewal process. But, he writes, "We are in a situation beyond anything ever experienced before in human or earth history," and our traditional ways are not enough. "Our primary resource in this situation, it seems to me," he goes on, "is not our biblical revelation but the revelation granted to us through the universe, the planet earth, the living world about us, and the course of our own human development." In this way of viewing things, the earth is primary, man is secondary: "The earth and its well-being are our basic referent as regards reality and value. The earth can exist without us. We cannot exist without the earth."[91]

Much debate exists among theologians as to whether or not Berry's ideas are compatible with Catholic theology, and if so in what way.[92] But there is another way of attacking Berry which usually neither he nor his defenders face outright. He seems to lack interest in ordinary questions of social justice.[93] Leaving aside Berry's own personal distaste for social reformism, is it in fact impossible to marry his vision with a quest for social justice? Theoretically it would seem not impossible, since many of the depredations against nature in the contemporary world also involve depredations against human beings.

What is true in theory seems verified in reality. There is at least one figure who has been inspired by Berry's New Story in his own quest for a better world in the here and now. Sean McDonagh has been an Irish missionary priest working with poor tribal peoples in the Philippines, seeing their land destroyed about them by outside economic forces. He asks, "How is a follower of Jesus to respond to the rampant destruction and poisoning of the natural world, which, if the current rate continues or increases, will threaten all life on Earth?"[94] McDonagh's book, *To Care for the Earth*, contains vivid descriptions of the ills of the areas in the Philippines where he works. He refers to environmental degradation in his native Ireland as well. To put these problems in perspective, he retells at length the New Story, with direct reference to the works of Berry, Brian Swimme, Matthew Fox and to the Gaia hypothesis of James Lovelock. He is, in addition, impressed by what can be learned from the peoples among whom he works: "Tribal peoples are sensitive to spirit presence."[95] Even at the theological level, ecotheology and the more traditional theology of social justice can be combined, just as they can be combined in practical and political action against forces menacing both environmental survival and social justice. In years to come, we will see how fruitful these ecotheological initiatives will be.

Chapter Fourteen

PEOPLE ARE NOT ABOVE NATURE: THE CHALLENGE OF DEEP ECOLOGY

We have seen bitter arguments over whether or not nature is indeed imperiled. Technology has been presented as its potential savior or as its nemesis. But critics on both sides have shared the same basic assumption: that the outcome must be of benefit to humans. They judge technology on the basis of whether or not it contributes to human well-being. But what if their basic assumption is wrong? What if human beings are not separate from and superior to nature? If they are not, the whole argument takes on a radical new perspective. In the last analysis, humans *are* a part of nature. The only hope for salvation may be to adopt a system of values held by those such as John Muir, a system that is today called "deep ecology."[1]

Arne Naess

"Deep ecologists" have much in common with earlier ecological thinkers such as Thoreau and Leopold. But they go even further by looking upon humans as merely another part of nature, and perhaps not even the most important part. The term "deep ecology" was coined by the Norwegian philosopher and mountaineer Arne Naess in a 1983 article, "The Shallow and Deep, Long-Range Ecology Movements—A Summary."[2] Deep ecology, or what Naess now prefers to call "the New Philosophy of Nature,"[3] distinguishes itself from "shallow" ecology, or simple mainstream environmentalism, in that it is not simply concerned with what modern technology is doing to the human race but what it is doing to nature as a whole. Their concern is not only with man but with the whole biosphere.[4] This sentiment is largely a replay of the argument over the purposes of conservation waged in America in the early twentieth century between the supporters of Pinchot and Muir— that we must accept "biospherical egalitarianism" rather than anthropocentrism. Naess argues that "the *equal right to live and blossom is an intuitively clear* and obvious value axiom."[5]

Naess makes clear that the tenets of deep ecology "are not derived from ecology by logic or induction. Ecological knowledge and the life style of the ecological field-worker have *suggested, inspired, and fortified* the perspectives of the Deep Ecology movement," but its real roots lie elsewhere. What he calls ecophilosophy (ecosophy) goes beyond ecology. His work on ecosophy "developed out of his work on Spinoza and Gandhi and his relationship with the mountains of Norway." What is basically involved is less a matter of propositions than of general attitude. Even Naess "suggests that biocentric equality as an intuition is true in principle, although in the process of living, all species use each other as food, shelter, etc."[6] He says that "there is a basic intuition in deep ecology that we have no right to destroy other living beings without sufficient reason."[7] His new "ecosophy," as he puts it, "involves a shift from science to wisdom."[8] Deep ecology has more than a surface resemblance to various Eastern attitudes toward nature, and the *koan*like phrase for deep ecology, suggested by Naess himself, is "simple in means, rich in ends."[9]

The phrase "deep ecology" has caught on and is widely used among philosophers who deal with the question of the human relationship with nature. But the basic concept, especially its denial of the primacy of human values, has been a subject of much controversy. Many of its critics are themselves concerned with environmental degradation and agree with the general thrust of the movement. But they reject its basic biocentric egalitarianism. One argues that asking humans not to disrupt the biosphere places them in a class apart from other species and is itself inherently anthropocentric.[10] Naess defends the movement by, in effect, granting the point but arguing that not all deep ecologists are opposed to *all* human activity directed against other natural creatures.[11]

Gary Snyder

Prominent among the supporters of deep ecology has been the poet Gary Snyder.[12] Snyder was born and raised on a small dairy farm in western Washington State, in the middle of lumbering country. As a child he began to camp out alone at night in the woods.[13] Later he spent time doing fire-lookout duty in the Cascade Mountains, free—almost compelled—to contemplate nature. His father had been an organizer for the Industrial Workers of the World, the notorious I.W.W. According to one critic, Snyder found merit in the I.W.W. motto, "Forming a new society within the shell of the old," with its emphasis on constructive rather than destructive social change.[14] Snyder graduated from Reed College in Portland, Oregon, in 1951, studying the Far Eastern culture that had fascinated him ever since he had seen examples of its art in Seattle museums as a boy. In time he went to Japan, where he spent many years studying Zen Buddhism and eventually marrying a Japanese woman.

Snyder's poetry and ideas have always reflected a reverence for the wilderness, particularly that of the American West, combined with themes from oriental thought. One critic has written of Snyder that "the Western world with its dualism and antimonies he has made alien to himself. His knowledge of Zen Buddhism is not that of a dilettante, but in so far as this is possible for an occidental, of an adept."[15] Another holds that "the fundamental—if disguised—anthropocentric humanism of the Romantics has been replaced in . . . Snyder by a broader, biocentric view in which external nature is valued for its own sake and man is seen as one co-equal partner in the process of the whole."[16] Snyder's background gives special bite to his reverence for wildness. Noting that in the Western Hemisphere "we are blessed with a bit of remaining wilderness," he hopes that the "temples of our hemisphere will be some of the planet's remaining wilderness areas."[17]

Snyder has always insisted that his political, economic, and religious views and his poetry are inseparable: "You would not think a poet would get involved in these things. But the voice that speaks to me as a poet, what Westerners have called the Muse, is the voice of nature herself, whom the ancient poets called the great goddess, the Magna Mater. I regard that voice as a very real entity."[18]

Snyder has labored long as a poet, both in Japan and in the United States, and not without recognition. In 1975 he was given the Pulitzer Prize for his book, *Turtle Island*, which consists not only of some of his finest poems but also important essays and statements linking Zen Buddhism with the Native American tradition and with deep ecology. Snyder's success as a poet has not interfered with a career devoted, in large measure, to writing and speaking on issues relating to nature and technology. Indeed, it has enhanced it, as he clearly regards his mission as a poet to be a spokesman for neglected nature. For him, "Wildness is not just the 'preservation of the world,' it *is* the world."[19]

Central to his views is his allegiance to Buddhism.[20] He rejoices that "the Buddhists teach respect for all life, and for wild systems," and relates this attitude to ecology, since "man's life is totally dependent on an interpenetrating network of wild systems."[21] But Snyder, unlike many Western admirers of oriental thought, is no innocent. He is aware that not all has been well with Buddhism over the centuries: "Historically, Buddhist philosophers have failed to analyze out the degree to which ignorance and suffering are caused or encouraged by social factors, considering fear-and-desire to be given facts of the human condition." This is dangerous, since "this can be the death of Buddhism, because it is death to any meaningful function of compassion. Wisdom without compassion feels no pain."[22]

Snyder is eloquent in drawing sociopolitical conclusions from his basic perspective, but he knows better than to attack growth head on. He seeks to move civilization from quantitative to qualitative growth. Yet his rejection of modern civilization is as basic as it is complete, and is ultimately based on religious grounds: "What it comes down to simply is this: If what the Hindus, the Buddhists, the Shoshone, the Hopi, the Christians are suggesting is true, then all of industrial/technological civilization is really on the wrong track, because its drive and energy are purely mechanical and self-serving— *real* values are someplace else."[23] Snyder contends that our technology conditions how we live, and that by changing aspects of it we can change our whole lives: "A decentralized energy technology could set us free. It's only the prevailing economic and government policies that block us from exploring that further," he has said.[24] Note how he still remains conscious of the political dimension of dealing with technology.

An important part of Snyder's version of the deep ecology credo is his passion for localism. "Bioregionalism," he writes, "is the entry of place into the dialectic of history."[25] He starts from a special feeling for the continent as a whole. The introduction to *Turtle Island* begins: "Turtle Island—the old/new name for the continent, based on many creation myths of the people who have been living here for millennia, and reapplied by some of them to 'North America' in recent years."[26] When asked if he had a special attachment to California, he replied: "My sense of place is the whole West Coast. No, not the whole West Coast, Northern California, Oregon, and Washington are where I feel most at home."[27] Snyder practices what he preaches. Living in his chosen habitat, the Sierra Nevada foothills, he has been active in trying to make his deep ecology philosophy count in the real world. He and others who agree with his ideas have joined with longtime local residents to elect a planning commission concerned with controlling growth, and are involved with the U.S. Forest Service in the management of Tahoe National Forest.[28] His basic allegiance remains antianthropocentric. In the struggle over environmental philosophy, he sides with those whose values reflect an awareness of the whole of nature—the position of deep ecologists.

Ernest Callenbach

If Snyder is in effect the poet laureate of deep ecology, a special role belongs
to Ernest Callenbach. He coined the word "Ecotopia" to describe what a
society based on the ideology of deep ecology would look like in practice.[29]
He did so in a book called *Ecotopia: The Notebooks and Reports of William
Weston*.[30] Originally published by Banyan Tree Books in Berkeley in 1975,
it soon became an underground best seller, then a mass-market paperback,
and was translated into many foreign languages. His term has been picked
up by many others who do not necessarily accept the deep ecology ideology;
writers such as Joel Garreau use it to describe the area in which Callenbach's
utopia is set—the same Pacific Northwest region that Gary Snyder calls
home.[31]

Ecotopia is a classic utopia in form, in which the first visitor from the
United States, Weston, writing in 1989, describes the society that seceded
from the nation in 1980 through, paradoxically, nuclear blackmail. The
Ecotopians use technology, but sparingly. Their first consideration in using
it is how well it fits in with their philosophy.[32] "The basic philosophy of the
Ecotopians tends to be patterned after the American Indian,"[33] and Callenbach's Ecotopians have much in common with Native Americans, or at least
a common view of them. They have returned to hunting on a major scale;
deer are now numerous near the cities, and "game is said to be a source of
considerable meat in the Ecotopian diet; it is prized for its 'spiritual' qualities."
Weston is clearly impressed by the Ecotopian attitude toward the Indian:
"Many Ecotopians sentimental about Indians," he tells his diary, "and there's
some sense in which they envy the Indians their lost natural place in the
American wilderness." One custom linked to primitive peoples that the Ecotopians have embraced, on a trial basis, is "war games," ritual battles that
end as soon as someone is hurt, designed to test the courage of the
participants.[34]

The emphasis on war games—which take the place of all organized sports
in Ecotopia[35]—is somewhat curious given the high position women have in
Ecotopian life. Indeed, as Callenbach details in a later "prequel" to *Ecotopia*,
entitled *Ecotopia Emerging*,[36] the original impetus for secession, led by the
Survivalist party, stemmed from the problems of a woman, Lou Swift, who
invented a solar cell the dominant economic interests were determined to
suppress. Indeed, the whole movement was led primarily by women. In
Ecotopia, "women . . . have totally escaped the dependent roles they still tend
to play with us," Weston writes in his notebook. They "hold responsible jobs,
receive equal pay, and of course they control the Survivalist Party." The head

of state is a woman. Not only that, but child-rearing is equally shared by the sexes. It is interesting that, reflecting implicitly the fact that the American environmental movement has been primarily a white upperclass endeavor, blacks in Ecotopian society have largely chosen a form of self-segregation. Just as whites have their own local communities, blacks have entities generically called Soul City. Says Weston, "Life within the black territories . . . has more holdovers from pre-Independence days than Ecotopia as a whole." For Weston, "this admission that the races cannot live in harmony is surely one of the most disheartening developments in all of Ecotopia."[37]

Central to Ecotopian philosophy and policy is a basic feature we have already seen in deep-ecology thinking: a fundamental aversion to large numbers of human beings as such. Population has actually dropped since independence, thanks largely to government promotion of birth control—all female controlled, it is noted. But "some radical Survivalist Party thinkers believe that a proper population size would be the number of Indians who inhabited the territory before the Spaniards and Americans came—something less than a million for the whole country, living entirely in thinly scattered bands."[38] One striking feature of Ecotopian philosophy is its unwillingness to accept the idea that humans, though they are an integral part of nature, are part of an evolutionary upsurge that must be not only maintained but manipulated. The Ecotopians are not interested in anything like eugenics or genetic engineering. Unlike the Americans of Weston's day, they "are more willing to live with the biological constitutions we now possess."[39]

Other Movements and Movers

Callenbach has gone farther than anyone else in trying to spell out what an ideal society would look like from the viewpoint of deep ecology, though others have pushed particular aspects of it, without necessarily accepting the total philosophy. Sessions and Devall argue for "bioregionalism" in *Deep Ecology*,[40] and journalist Kirkpatrick Sale pushes the same idea in *Dwellers in the Land*[41] without being totally committed to the whole deep ecology cause. Bioregionalism has as its "underlying premise the idea that the environmental crisis begins at home and revolves around individuals' perceptions of their place in the world."[42] The term "bioregionalism" was coined in 1976 by Peter Berg, who ran a bioregional clearinghouse in San Francisco called the Planet Drum Foundation. Berg argues that "all the planetwide pollution problems originate in some bioregion. . . . It's the responsibility of the people who live there to eliminate the source of them."[43]

Closely related to bioregionalism is the concept of "reinhabitation." This idea, as developed by Sessions, involves accepting a local region as the basis of life,[44] and can stand or fall aside from deep ecology as a whole. Its relationship to Gary Snyder's ideas about really inhabiting a locality are obvious.

Not strictly part of deep ecology as a movement, but related to it, is the scientific hypothesis of Gaia as formulated by the British chemist and freelance scientist, James Lovelock.[45] According to Lovelock and his followers, the earth did not so much create life as the other way around. The existence of life made atmosphere possible in the first place.[46] The idea of a really living earth has had tremendous appeal to all who put nature above society and sometimes above humanity itself, as many deep ecologists do. But the idea has also gained adherents among such scientists as Lovelock's frequent collaborator, University of Massachusetts biologist Lynn Margulis.[47]

Initially the Lovelock-Margulis theory met with some rejection even among those concerned with ecological problems, in part, paradoxically, because Lovelock seemed to contend that Gaia was so strong that it could even overcome human insults to the environment.[48] But recently, ironically, the Gaia concept has been gaining favor among more conventional scientists as it has been modified in important respects. At a recent Gaia conference sponsored by the American Geophysical Union, Lovelock yielded basic points to his opponents, in effect retreating from the "strong" Gaia hypothesis, and conceding that the paper of James Kirchner of Berkeley "was a clear-cut demolition of Gaia." Kirchner had argued that the strong versions were untestable, while "the weaker forms, in which life merely influences the environment are so obviously correct that they do not merit status as hypotheses." When Lovelock tried to argue that Kirchner's points were not new but had been in effect long incorporated in his theories, Kirchner replied that it was one thing to admit this orally, another to put it in print, saying "you've changed, but your old ideas are still out there."[49] Given the way things work in the environmental community, however, to judge by its past adherence to earlier theories of the earth as a living being, it may be a long time before the newly chastened version of Gaia catches up with the old one. In the meantime, the broader use of the term continues to serve as a synonym for a holistic ecological view of the world.[50]

A major source of controversy surrounding the deep ecology movement has been its attitude toward human population. Indeed, the conflict over population has long troubled the environmental movement as a whole, beginning with a classic confrontation between Paul Ehrlich and Barry Commoner and resurfacing in the late 1980s.[51] A major impetus for its coming to the fore again was an article in *Earth First!: The Radical Environmental Journal* by Miss Ann Thropy on May 1, 1987. This article—the pseudonym is obvious—argued that AIDS might be a blessing in disguise because it would reduce the world's population. A follow-up article in the same publication on December 22, 1987, by Daniel Keith Conner, "Is Aids the Answer to an Environmentalist's Prayer?" asked, among other things, "What defense could be worse for us or better for Gaia than AIDS?"[52]

Earth First! was founded in 1980 by defectors from mainline environmental organizations who believed that groups such as the Sierra Club and the Wilderness Society were not militant enough. Its primary leader was David Foreman, a former Washington lobbyist for the Wilderness Society. As he later wrote, "The people who started Earth First! decided there was a need for a radical wing that would make the Sierra Club look moderate."[53] The emphasis from the first was on direct activism, legal or otherwise, in defense of the environment, rather than on lobbying or lawsuits. Foreman has written that originally Earth First! was begun by "a half-dozen people camping under a full moon in the Mexican desert, dreaming of a wilderness preserve and scheming 'how to put a monkey-wrench in the gears of the machinery destroying natural diversity.' "[54] The monkeywrench was not simply a metaphor. Foreman is the author of a self-published book, *Ecodefense: A Field Guide to Monkeywrenching*, which sold out its first edition of five thousand copies in 1985. Earth Firsters! pioneered such tactics of "ecotage" as nailing spikes into trees so that they would destroy lumbering equipment, even though the tactic might also injure or kill lumbermen, even though they claimed the trees were marked. They soon equivocated, however, on whether or not such action was justified,[55] and in time California and Oregon activists rejected tree spiking to avoid frightening and antagonizing lumber workers.[56] But the impulse to strike out directly at those who despoil the wilderness still remains their major reason for being.

Earth First! grew rapidly. There were in 1987 some seventy-two chapters of Earth First! in twenty-four states, most of them in the West, and by 1989 it was estimated that the group had about ten thousand members and many times as many sympathizers in the more conventional environmental organizations.[57] Much of the inspiration for their direct-action philosophy comes from the ideas of the late Edward Abbey, as set forth in his novel, *The Monkey Wrench Gang*.[58] It has been argued that his concern over population growth has led to an attitude toward immigration that was in effect racist, holding that "our" culture, a "product of northern European civilization," was threatened with Latinization.[59] Abbey himself was noted for preferring other natural creatures to humans—he once said he "would rather kill a *man* than a snake"[60]—and many deep ecologists would sympathize.

The radical philosophy and rapid growth of Earth First! soon attracted the attention of the FBI, a development encouraged by the timber companies and others who were the movement's special targets. The FBI accordingly infiltrated the organization, seeking above all to find a way to indict Foreman, its leader.[61] Foreman was among five people arrested in May 1989 and scheduled to go on trial in federal court in April 1991. The government, using the testimony of an Earth First! member turned informant, claimed that it would prove that the defendants were practicing to sabotage power lines for nuclear facilities in three states. Between his arrest and trial, Foreman produced a new book, *Confessions of an Eco-Warrior*.[62] In August 1991 he and his codefendants pleaded guilty to a conspiracy to destroy a ski lift on a mountain sacred to the Hopis, in hopes of securing a plea bargain.[63] Four of the defendants received prison terms for vandalism. Foreman's own sentencing was delayed for five years, and he may be allowed to plead guilty to a lesser offense.[64]

The sword hanging over Foreman's head may well be designed to cow him. But even before the trial, he publicly broke with Earth First! over what he now considers its countercultural extremism.[65] A cultural conservative of the old rural Southern mode, he has even been saying some nice things about the Sierra Club. But as an old-fashioned patriot, he remains committed to a deep-ecology-inspired defense of the land against human intrusion, whatever the cost. In any case, monkeywrenching, even after the trial, is not dead; to think so is "absurd," says Foreman. Says one activist, "We're still here and we're still fighting and the FBI isn't going to stop us." Soon after the sentencing, a group calling itself Hunt Saboteurs threatened to disrupt a bighorn sheep hunt in Arizona.[66] Deep ecology is not simply an abstract philosophy, but retains an activist fringe.

Chapter Fifteen

REFORM OR REVOLUTION?

By the late twentieth century, rising concern over humanity's relationship to nature and to technology had reached the point where major movements for reform at the practical level had arisen throughout the industrialized world. They were influenced and conditioned by many of the ideologies we have examined. But environmentalism was becoming a new ideological perspective in itself.[1] Most of the reform movements began as ad hoc responses to perceived real world problems and were reformist in nature. There is much irony in the fact that the contemporary environmental movement, designed not only to maintain a natural environment in which human life could flourish but also to protect nature itself against human encroachment, arose in the United States. North American civilization was a civilization of people who came to the continent from elsewhere. It had none of the natural piety toward the land that might have arisen—as it did—in the Native American population. As the Canadian political philosopher George Grant put it, "None of us can be called autochthonous, because in all there is some consciousness of making the land our own." The gods of the Rockies "cannot manifest themselves to us as ours. They are the gods of another race, and we cannot know them because of who we are and what we did."[2] This has been especially true south of the Canadian border. The stress on using nature as a means to material enrichment has been a major factor in creating the American national character.[3]

Despite this, by the mid-twentieth century the old dogmas of the absolute priority of economic growth were being questioned on both sides of the border.[4] This process had started in the United States, as we have seen, with the conservation movement of the late nineteenth and early twentieth centuries. By the late twentieth century, conservation had mutated into environmentalism. As one historian puts it: "Conservation gave way to environment after World War II amid a rising interest in the quality of life beyond efficiency in production. The two tendencies often came into conflict as resources long thought of as important for their material commodities came to be prized for their aesthetic and amenity uses."[5]

Indeed, there was a fundamental difference in motivation. The conservation movement was still basically anthropocentric. It used a utilitarian rationale for saving nature for the sake of human material needs. The environmental movement, however, increasingly had a nonutilitarian conception of people's relationship to nature.[6] There were differences not simply in ideology but also in the social composition of the movements. The conservation movement was based primarily on a scientific leadership, while the environmental movement was far more popular in nature.[7] Between the heydays of the two movements there had been other attempts to react against modern urban industrial civilization, but they were of little practical importance. In the late 1920s there was some interest in going "back to the land," an idea pushed by intellectuals such as Ralph Borsodi in *This Ugly Civilization*.[8] Later, Romantic groups such as the Nashville Agrarians played a role in rejecting urban civilization. But neither had serious political importance. The real turning point came when in the 1930s the political dynamics underlying the struggle over the American government's relationship toward nature were completely turned upside down. Historian Stephen Fox underlines the point, writing that "when Franklin Roosevelt became President the organized conservation movement was controlled by members of the opposition party."[9] But after the New Deal, "most conservationists were Democrats."[10]

While this political revolution was going on, the changes in philosophical emphasis from conservation to ecology and environment were beginning to take place, to come to full fruition in later decades. This new concern with the earth itself and the need for *all* Americans to work for its preservation was highlighted in President Jimmy Carter's farewell address.[11] But Carter, after all, had been soundly defeated, and even though polls indicated most Americans shared his basic environmental concerns, his successor, Ronald Reagan, went out of his way to sabotage the environmental efforts of his predecessor.[12] Reagan's appointment of the ideologically antienvironmental, born-again Christian James Watt as Secretary of the Interior helped the environmental organizations flourish. But their increased membership did not compensate for the many changes in governmental policy during Reagan's eight years in office.

During the 1970s and 1980s the environmental movement flourished to the degree that it became a major player in politics in the United States and elsewhere. This was despite—or perhaps because of—the fact that no clear-cut philosophy or ideology had emerged in the various organizations in-volved.[13] Reform environmentalism sets itself out as a practical program for changing the current posture of our modern industrial world toward nature. But it does not really endorse a total revolution in ideas and institutions, although there is in it more than a hint of utopia as a distant star to guide present incremental action. A similar and often allied force is the appropriate-or alternative-technology movement, which calls for a basic change in our present total commitment to large- scale, nature-defying technology for its own sake and sets forth step-by-step alternatives to existing technologies. Yet, like environmentalism, it judges what is good or bad by a standard that is, in relation to the present system, revolutionary in the extreme.[14]

The appropriate-technology movement stems largely from the work of a few seminal thinkers such as the late German-born British economist E. F. Schumacher.[15] It advocates the use of small- scale, decentralized technologies that ideally can be maintained by renewable energy sources, rather than large-scale, centralized technologies that require scarce fossil fuels or nuclear power. Related schools of thought go by such names as "soft" as opposed to "hard" technology, as espoused particularly in the area of energy technology by physicist Amory Lovins. Lovins argues that nuclear technology, for example, requires a centralized state and economy.[16] More recently, with movements toward a "New Age" receiving much publicity, the term "postmodern" tech-nology has arisen.[17] But what all these groups have in common is the con-viction that some technologies are good and some are bad for human freedom, for dignity, and even for survival. All hold that standards can be found and applied to test which are which, and that the best interest of humanity requires that the good technologies be allowed and the bad rejected by society as a whole. Because this implies judgment, a movement called "technology as-sessment" has arisen to do the job, which in turn has led to the creation of the Office of Technology Assessment to advise the U.S. Congress about the potential of various technologies.[18]

Many would argue that the task of technology assessment is impossible, since it requires knowledge about the future that we can never possess.[19] Others contend that technology assessment always tends to be negative about change. What would we have done about the automobile, some ask, had we foreseen its consequences in deaths, pollution, and suburbanization? This query begs the whole question about the ultimate social value of motor transportation. As already noted, the father of contemporary technology as-sessment is E. F. Schumacher,[20] and many of the attacks on appropriate technology are directed on him and his contention in *Small Is Beautiful* that smaller is (usually) better.[21] Thus technophile Samuel Florman, noting that Schumacher has written that "only in small social groups . . . is it possible for people to live worthwhile lives" cites to the contrary the lives portrayed in such American classics as *Winesburg, Ohio; Spoon River Anthology,* and *Main Street.*[22] Some from the Left especially argue that to concentrate on technology ignores the essentially political basis of oppression. Thus Alexander Cockburn and James Ridgeway write that "appropriate technology is a bogus issue. Indeed, it is one guise for a dangerous and reactionary politics, implying that the mere introduction of new sorts of machines will change the political economy for the better."[23] But this of course begs the question, simply denying the central contention of the appropriate-technology school that *technology* itself is the independent variable that determines political and economic systems.

Many proponents of appropriate technology would be guarded in their endorsement of the computer and highly dubious about the use of genetic engineering. But while such technologies have been threatening to change the structure of human society and even the structure of the human body itself, progress in the technologies of simple destruction has been proceeding apace. Both supporters of technology assessment and reform environmentalists can agree that this presents a danger that makes other consequences of the conquest of nature through technology pale in comparison. The invention and spread of the ICBM and nuclear weapons poses an obvious continuing threat to the continued existence of human civilization. Not only would another major war, waged with these weapons, be an environmental disaster beyond measure, but some have argued that even a minor exchange of nuclear explosions might create a "nuclear winter" that could threaten the survival of the human race, though this position is increasingly open to question.[24]

Thus, the causes of world peace and planetary survival are in essence the same, although different leaders and movements may pursue them separately. But there is one major exception. The special circumstances of post-World War II Central Europe gave birth to a movement that combined rejection of nuclear war with rejection of much of modern industrial civilization, a movement with a truly revolutionary attitude and program for societal regeneration. Strongest in former West Germany, but with expressions elsewhere including the United States, the Green movement seeks to create a new politics based on respect for nature and control over technology.

The Green Movement

The Green movement's objective is the reassertion of the primacy of politics, of the open and conscious control by the political order of humanity's relationship with nature and technology. Its general perspective was prefigured by philosopher of science Stephen Toulmin when he wrote that "postmodern" science had led us to be attracted to two conflicting philosophies, which he called the white (Epic) and the green: "The green philosophy is a contemporary counterpart of Stoicism. It has roots, most typically, in the theories of ecology and in the practices of 'natural living.' "[25]

That the Green movement as a real-world political force should have arisen in former West Germany was no accident.[26] Several factors were at work. Some were cultural, including the long-standing German infatuation with nature and a less powerful, but still real, distrust of many aspects of modern technology. But there were specific political factors as well. Germany is at the center of Europe. It was considered to be the probable major theater of action in any future war between the "superpowers" that did not immediately become a war of ICBMs but would be fought instead, at least initially, with tactical nuclear weapons. The general distrust of nuclear power and nuclear weapons explicit in the basic Green attitude took on special meaning in these circumstances. Though the Social Democrats were opposed to stationing nuclear weapons in Germany for their own political reasons, the Greens could attack the idea in terms of a broader perspective. In addition, the West German political system provided for proportional representation, which encouraged groups with special programs to participate in politics because they could obtain parliamentary strength in proportion to their adherents.

The basic groundwork for the Green movement in Germany was laid by the mainstream ecological literature of modern industrial society.[27] In understanding the German Greens it is important to keep in mind that in the period after World War II, West German intellectuals started out from a position much more critical of modern civilization than did their American counterparts. Despite the fact that the Communist party was de facto outlawed by the Bonn constitution, many Germans sympathized with much of the Marxist critique of modernity. The East German expellees Rudoph Bahro and Wolfgang Harich "were at the forefront of an effort to create an updated, ecological version of Marxist critical theory."[28] One commentator says that to understand the emergence of the Greens in Germany one has to take into account the thesis of American political scientist Ronald Inglehart about the existence of a "silent revolution" in values in Western Europe in the 1970s in which nonmaterial values were coming to the fore,[29] and the "vacuum that has existed on the left of the SPD since the 1950s."[30]

Before the entry of the Greens into parliament, ecology was hardly considered a real political topic. *The Global 2000 Report* commissioned by President Carter had sold three times as many copies in smaller West Germany than it had in the United States, an index of German interest in ecological problems. Even so, a discussion of the book in the Bundestag was attended by only one-tenth of the members. It was this kind of indifference to ecological issues by mainstream parties that led to the formation of the Greens.[31] The Greens first entered parliament in 1983, twenty-seven deputies strong. They were the first new party to be elected in thirty years, and, regardless of ideological origins, insisted on being seated between the right-wing Christian Democratic Union (CDU) and the left-wing Socialist Party of Germany (SPD).

Almost since its inception, the Green party has been plagued by a series of ideological rifts that have been echoed in disputes over tactics. The basic ideological conflicts in Germany's Green movement are reflected, though not simply, in the contest between the "Fundis" and the "Realos"—between those who wish to maintain the utmost purity of Green principles and those who are concerned with immediate political action to attain specific goals.[32] This division is further complicated by the antihierarchical orientation of the party, which has led it to proclaim that elected members be rotated in office so that the party not be permanently dominated by any one group. In addition, although the party is committed to a feminist point of view, women still do not have quite the power within it which they think they deserve.

Where does the spiritual orientation of the Greens come from? Many, if not most, of the Greens interviewed by an American research team considered themselves Christians, though few were involved with institutionalized religion.[33] The American Green theorist, Charlene Spretnak, has written that in interviewing German Greens they often spoke of the movement's spiritual orientation, but were hesitant to be specific, fearful of the way the Nazis had manipulated religion.[34] However, Marxist-oriented Greens sought to squelch talk of nature, which had been prevalent in the West German Greens' first campaign for seats in the European parliament in 1979. Spretnak concluded that "the spiritual dimension of Green politics is unlikely to come out of West Germany, though it provides motivation for many West German Greens."[35] Nevertheless, many West German Greens have become fascinated by the holistic sense of nature found in the Native American, pre-Christian European, Taoist, and Buddhist traditions.[36] What this boils down to is a generally romantic attitude toward nature, with deep roots in the German past. Many have condemned this attitude as such, particularly when found in alliance, however parlous, with Marxism. Thus the guardian of Roman Catholic orthodoxy, German-born Cardinal Joseph Ratzinger, in an interview with the Italian journal *Il Sabato* said that "it is my impression that Marxism is today in crisis, and that even its more refined version, that combined with romantic elements, has lost attraction and force. What is coming to the fore is rather a vision, and it is in the line of ecology and the Greens." He went on to condemn what he regarded as the new synthesis by saying that it is expressed in "a somewhat antitechnical, somewhat antirational concept of man as united to nature. It is a concept that has an antihumanist element."[37]

We have already seen that, given the drift of German intellectual life in the post-World War II period, Marxism had the inside track. Just about the only prominent Green who escaped this line was Petra Kelly, who had spent crucial years of her life in the United States, even working in the antiwar campaign of Eugene McCarthy. Since her fluent English enabled her to be a major interpreter of the Greens to the English-speaking world, this anomaly tended to skew outside perceptions of the movement.[38] It is clear that there is a major, perhaps by now a predominant, trend in the West German Green movement that is deeply indebted to Marxist analysis. This is true even when, in the name of "ecosocialism," much of the essence of the mainstream Marxist position regarding nature is rejected. Werner Hulsberg leans on Engels's *Dialectics of Nature* to argue that "the myth was hereby laid to rest according to which Marx had been an admirer of technology and had seen it as 'the foundation of the present and the motor force of the future.' "[39] But as we saw earlier in our discussion of Marxism, Hulsberg is far from the mainstream of Marxist thought.

A major role in the rise of Marxist sentiment among the Greens was played, paradoxically, by the East German exile, Rudolf Bahro. Bahro was forced out of East Germany because of his politics, above all his controversial book, *The Alternative in East Europe*. This book, interestingly, was influenced by Teilhard de Chardin's *The Phenomenon of Man*. Of Teilhard's book, Bahro said, "It is a thoroughly materialist work, and I share its thesis that man has an ultimate goal and an ultimate purpose. Man cannot be reduced to the realities of a particular moment, for there are other, transcendental realities that form part of the historical process."[40] This materialist transcendentalism is the key to Bahro's ideas, to why he became a Green, and why, in time, he left the party for political agitation beyond party politics.

Bahro, despite his affection for his Marxist background, eventually repudiated most of the post-Enlightenment West: "It is no accident that European civilization first led to capitalism and to the aggressive policies bound up with it. It is all historically conditioned, and I wonder whether man's whole development since the Renaissance has, in this respect, not been an aberration." Marx was wrong, he says, in believing that progress was inevitable. Look at biological evolution: "A species can die, the evolutionary process can take a wrong turn." Modern industrial civilization is such a wrong turn: "Our industrial system is incompatible with the preservation of our natural environment." Marx "was still very much a man of the bourgeois period . . . in that he laid stress on objectivity in the making of things." Not only did human evolution begin "to go wrong with the English industrial revolution," but, like Heidegger, Bahro holds that the real mistake was in the beginning: "The European culture of the past two or three thousand years, which announced its birth in the *Iliad*, has been extremist in its most inner disposition, modeling itself on individual competition and the Olympia principle of 'more, higher, faster, better.' " These principles have "in the recent period led to capitalism, they were its preconditions."[41] This is indeed a long way from Marx and his adulation of the bourgeoisie for having played a "progressive" role in the taming of the Earth.

Bahro calls for the reorganization of society that recognizes the proper relation of technology to nature. But most of the particulars of how society should be reconstructed are not new: "They have been described by utopian socialists or communists for hundreds of years."[42] In time his messianism would cause Bahro to leave even the German Greens. The reason was not simply that the Fundos, to whom he adhered, were apparently losing out to the Realos, but it was a despair with politics itself. In Bahro's public utterances he "calls for a new spiritual awakening. He talks of the 'politics of salvation,' of the 'need to correct the last ten thousand years of evolution,' of 'a change so deep one must speak of a break with basic European patterns of behavior.' "[43] In his statement of resignation, he berated the Greens for not being radical enough: "The Greens have identified themselves . . . with the industrial system and its political administration. Nowhere do they want to get out."[44]

The Greens enjoyed their greatest initial success in Germany, but the movement has appeared in other countries as well.[45] In Belgium they were able to elect members to parliament two years before the German party did, and Ireland elected its first Green in 1989.[46] The Green party took 7 percent of the vote in New Zealand in 1990.[47] The first Green major elected in Eastern Europe won in Cracow in the same year.[48] There is a fledgling Green movement in Italy as well; its low status has been dramatized by the fact that it had to bring in outsiders, such as the American activist Jeremy Rifkin, to bolster its appeal.[49] But in the European parliamentary elections of 1989, the party took five seats with 6.2 percent of the vote.[50] In Sweden several years ago, the Green Environment party emerged as a major political force, the first new party to gain seats in the Swedish parliament since 1917;[51] but they virtually disappeared in the 1991 elections, which saw a turn to the right that ousted the long-standing Social Democratic government. Australia does not have a real Green party as such, but environmental issues have come to the fore in the labor movement, which has carried on strikes to protect the environment, so-called Green bans.[52] Former Australian Communist party head Jack Mundey has been a leader in trying to move that group in a Green direction.[53]

France has been an especially difficult country for the Greens. While there are many Green supporters, France does not have proportional representation. The fact that most elections go to a second ballot can give organized minority interests some power. But after initial Green successes, the French parliament changed the law in 1976 to make it necessary to have more votes in the first round to participate in the second—designed to hurt Green chances.[54] Because of the majoritarian traditions of the French Revolution, the French government has long been notorious for riding roughshod over dissidence and has few mechanisms for citizen input into decisions. Thus the only challenge to genetic engineering in France came from West German representatives in the European parliament.[55] But despite these obstacles, *Les Vertes* ran strong in the French municipal elections in 1989.[56] In the elections for the European parliament later that year, they did even better than the German Greens.[57]

Even more trying for the Greens is the situation in Great Britain, in which elections are biased against minority parties at every stage. Even though they received a larger share of the vote than in any other European country, in the European parliamentary elections they won no seats at all.[58] But Green ideas have increasing resonance even there. Before her downfall, Conservative Prime Minister Margaret Thatcher was surprising many by becoming a spokesperson for various environmental causes.[59] Even in Mexico, one of the most environmentally stressed nations on earth, a Green party has arisen, though it is handicapped by a problem all such parties outside the developed world face: the priority of immediate poverty. "I have to worry about my kids. I can't worry about the trees," as one voter told its leader.[60]

Given the increasing success of the Green cause in politics in foreign countries, it suffered a noticeable shock in late 1990. In the first all-German elections after unification, the Greens were virtually wiped out. They received less than the 5 percent of the necessary vote and won no seats whatsoever in the west, though a few won in the east in alliance with other groups. They blamed their loss on internal struggles between Realos and Fundis, on their issues being stolen by other parties such as the Social Democrats, and above all on their opposition to German unity. "We fell under the wheels of German unification, of German euphoria," said party spokesman Hans-Christian Strobele after the loss.[61] Are the Greens finished in Germany? One political observer, Walter Kaltefleiter, director of the Institute of Political Science in Kiel, said after the election that "all parties in Germany now have strong environmental positions, and the public still considers the environment perhaps the most important issue in the country." But, he went on, "The Greens lost their monopoly on this issue. It's still too early to write them off, but their period of influence is over."[62] Perhaps. But the euphoria of unity is fast wearing off. Ironically, this is due in part to the terrible environmental destruction caused in East Germany under communism, making the economic costs of reunification much more than the taxpayer in the west expected. The great postunification victory of Chancellor Helmut Kohl is being repudiated in various local elections that have caused his party to lose control of the upper house of parliament. In these key elections, a-not-yet-dead Green party has played a role.[63]

The American Greens

The United States has a long tradition of interest in the environment; it has an even longer tradition of citizen involvement in politics. Why has it not developed a Green movement like that of Germany? One reason of course is that the American electoral system is radically different. It abjures proportional representation and shares the British system of winner-take-all politics. As a result, groups seeking to influence governmental action have almost always found it more fruitful to act as pressure groups. Thus organizations such as the Sierra Club and the Wilderness Society have primarily acted as lobbyists before legislative bodies and executive agencies. Sometimes they have joined with primarily legally oriented groups such as the Natural Resources Defense Council in an attempt to use the courts to further their aims. There has been for several years a political action committee, the League of Conservation Voters, which endorses and supports individual candidates. But there has never been a serious environmental party as such.

The late 1980s, however, saw an attempt to launch a full-fledged Green political movement.[64] As might be expected, it has been and is subject to some of the same ideological rifts as Germany's movement. Indeed, some of the foremost American sympathizers and interpreters of the German Greens have played major roles. The "deep" versus "shallow" ecology conflict has also been an important factor. So too has the issue of the role to be played by Marxist analysis. In early 1984 news of the Greens' victories in West Germany and the publication of *Green Politics* by Capra and Spretnak led to a flood of letters to the American organizations mentioned in the book.[65] One result was that a conference of sixty members of ecology, peace, and community-action groups met in Minneapolis in August 1984 to form an organization called the Committees of Correspondence. Their task was to unite various existing movements to work jointly toward "values such as self-government, love of the land, community responsibility, respect for diversity, and sustainability."[66] In July 1987 over fifteen hundred people met in Amherst, Massachusetts, to thrash out the directions the new movement should take.

Ideological rifts have been present from the outset. Murray Bookchin, whom we have already met, runs the Institute for Social Ecology at Goddard College in Vermont, has published *Green Perspectives Newsletter*, and is the major leader of the "social ecology" faction. The deep ecology group has many spokespersons, including the spiritual wing of the movement represented primarily by figures such as Charlene Spretnak.[67] Brian Tokar, author of the leading American introduction to Green politics, *The Green Alternative: Creating an Ecological Future*,[68] holds that the differences are in large part regional, with social ecology holding more appeal in New England with its pastoral, socialized setting, and deep ecology being embraced by the more wilderness-oriented West.[69]

Bookchin, who talks of a Left Green Network within the American Green movement,[70] attacked deep ecology at the 1987 conference, calling it "vague, formless, [and] often self-contradictory," a "black hole of half-digested, ill-formed and half-baked ideas," and an "ideological toxic dump."[71] He harshly condemns the ideas of David Foreman[72] and followers of deep ecology such as Kirkpatrick Sale.[73] Bookchin says Edward Abbey "beats them all" and attacks him for saying that the United States "is a product of northern European civilization" and that we must not allow "our country" to "become Latinized in whole or in part."[74] But while he discounts any idea that he wishes to create the same "turf wars" in the American Green movement that Germany has seen,[75] he says "social ecologists and Left Greens have to be concerned"[76] about people such as Charlene Spretnak who stress the spiritual aspect.[77] Bookchin takes particular umbrage at her writing that "it is when we turn to the issue of spiritual matters that we are faced with a huge hole in Green politics: *What is sustainable religion?*"[78] Thus we see that there has been the same desire on the part of some American leftists to co-opt the Green movement that we saw in Germany and to create a "Green socialism."[79]

Spretnak, for her part, insists on the integral spirituality she says must pervade the Green movement down to the lowest level: "To be successful, the expression of the spiritual dimension of Green politics must present some rather complex ideas in very simple and commonsense terms *without watering down the power inherent in spiritual impulses.*" Her concept of spirituality encompasses elements of all the existing religions, combining what she feels is their movement toward greater reverence for nature in an ecumenical vision of what might be: "Sustainable religion in the Green vision of society entails the 1200 'primary religious bodies' in our country emphasizing four areas *that are already contained in their traditions*: spiritual development through inner growth, ecological wisdom, gender equality, and social responsibility." We do not need to create a new religion, but to use the old ones in a better way: "We need to encourage a shift, which will not be a small one or an easy one, toward sustainability and toward deeply meaningful religion that does not separate itself from Nature, from our bodies, and from women."[80]

The American Green movement is now in disarray, however. This is not primarily the result of disputes over deep ecology, Green socialism, or spirituality; most of these issues do not really touch the concerns of members or groups at the local level. Despite its success in gathering numbers of activists together annually, it has yet to demonstrate real political seriousness. It remains plagued by an inability to rally racial minorities to its cause, despite the wishes of some of its members and the fact that that environmental degradation bears most heavily on them.[81] More important, though it continues to adopt platforms,[82] it consistently holds back from the decision to enter the political arena on a national scale.[83] But even more basic in the final analysis, it is succumbing to a disease that has infected its prototype in Germany and that has crippled many antiestablishment social movements in the United States. Proclaiming self-expression as a supreme virtue, it prefers to talk rather than to act. So marked was the personal posturing and infighting at the 1990 gathering in the Colorado Rockies that political science professor John Rensenbrink, a revered figure who was as much as possible the leader of the movement, announced that he was leaving the organization.[84] Many who sympathize with its aims are now giving up the organized Green movement as a hopelessly lost cause. But The Greens, as they now choose to call themselves, still continue to be active throughout the nation, and now have official ballot access in Alaska and California.

The Struggle Continues

Whatever the future holds for the Green movement as such, there is no question that the United States, and the international community generally, is becoming increasingly conscious of the practical dangers posed to human health and survival by technologically conditioned assaults on nature. The problems of air and water pollution, including the destruction of forests by acid rain,[85] are already the subject of international negotiations and treaties.[86] So too has population growth, though ideologically and politically delicate, long been a matter of concern.[87] The increasing desertification of the earth,[88] the export to poor nations of the toxic waste of the rich,[89] and the apparently increasing shortages of critical resources such as key minerals and arable land are more and more the subjects of discussion. Quite often these dangers invite and produce conflicts between the poor nations who will be their immediate victims and the richer nations that will be hurt in the long run. A prime example is the destruction by man of the tropical forests of the Amazon basin.[90] Despite these conflicts, the United Nations was able in 1982 to agree on a World Charter for Nature, which was adopted by a vote of 111 to 1. The only negative vote was conspicuously cast by the United States.[91]

One archetypical environmental problem that has reached not only the consciousness of experts but also the public in general—even being featured in a cover story in *Time* magazine[92]—is the "Greenhouse effect." People have begun to pay attention to the long-standing warning of many scientists that the pollution of the Earth's atmosphere would cause a permanent global rise in temperatures. The consequences, while varied, would on the whole be negative for the human race. In any event, they could force radical and costly readjustments in agriculture, housing, and living generally.[93] Swedish chemist and physicist Svante Arrhenius had first sounded the alarm in 1896. But it was not until the summer drought of 1988 that American policymakers began to take seriously the need for measures to try to mitigate the effects of unchecked industrialization, since much of the damage could not be reversed.[94] Concern had already been voiced—in the most anthropocentric terms possible—about the rising epidemic of skin cancer caused by the depletion of the ozone layer.[95] At the same time, the long-standing disquiet over the continued destruction of animal and plant species—a problem both in anthropocentric and deep ecological terms—was reinforced by the pending impact of expected changes in temperature.[96]

The major questions remain. There is now a general worldwide understanding of the practical problems of uncontrolled technology and the resultant stress on nature, increasingly causing prompt practical actions to deal with them. But can our relation to technology and nature be made to contribute to our ultimate well-being without a revolution in ideas and institutions that will require us to repudiate most of our long-cherished and deeply ingrained cultural habits? Can such a revolution take place? Do we really want it to occur? Despite the technological determinists, the choice remains ours alone.

Chapter Sixteen

WHERE DOES HUMANITY GO FROM HERE?

We should by now have gained several insights from our survey. The drive to conquer nature through technology is virtually universal. All cultures create material technologies to make their lives easier. The modern Western world has been the leader in this drive, especially since the industrial revolution, as the poorer nations of the world have been striving to catch up. Some, such as Japan, may now actually surpass the West in technological prowess. What is true of technology itself—the desire to have more material power over the world—is also true of attitudes toward nature. While Western civilizations have been spearheading the intellectual justification of conquest of nature, most peoples advocate it. Certain religious and philosophical systems of the Eastern world may deny the supremacy of humanity over nature in theory. But when the opportunity arises, peoples such as the Indians, the Chinese, and the Japanese are usually just as willing as Europeans or Americans to despoil the Earth.

However, technology is not a monolith. Some technologies have greater impact on the natural environment and the fabric of society than others. All have as their objective the modification of the natural world to fit human needs and wishes. Some do less violence to the rhythms of nature than others. Some create less pollution than others. Some threaten human freedom less than others. Some, such as, nuclear weapons threaten human existence more than others.

But technology is not an irresistible force. Technology as such cannot be rejected by human cultures, though they can and do choose which technologies to embrace and which to resist. It is thus quite possible for individual societies to choose to embrace particular technologies in accordance with their vision of the good life. But no society *as a whole* makes choices about technologies. Different technologies have different effects on different groups within societies. Every society known to history has been ruled by groups that are in some conflict with other groups; thus the technologies chosen and stressed by societies reflect who is in power. In every society in the world today, the strong use technology as a means to control the weak. Negative side effects of industrial technologies such as air and water pollution are most powerfully felt by the poor within developed nations. Internationally, it is the poorer nations that bear the greatest burden of environmental degradation. Although it is often argued that concern with the environment is a special preserve of the elite, the evidence is strong that it is shared by the poor as well, both domestically and internationally;[1] but the poor just don't have access to the means of expressing and disseminating this concern.

Given the almost universal willingness of modern society to use technology to exploit nature, can anything be done to prevent this process from wrecking the natural fabric upon which the human race depends for survival?

Saving the Planet

What can we do to prevent the changes we are making in our environment from destroying the physical basis of human civilization? First, we must recognize that the free market by itself cannot save us. Although the free market is necessary as a system for making the ordinary decisions about the allocation of resources and incentives in economic life, economic activity determined by the market often has side effects—"externalities," in the economist's jargon—that market theory cannot adequately resolve. As economist Herman Daly has stated: "When vital issues (e.g., the capacity of the earth to support life) have to be classed as externalities, it is time to restructure basic concepts." He further contends that "all conclusions in economic theory about the social efficiency of pure competition and the free market are explicitly premised on the absence of externalities. The undeniable importance of externalities in today's world is therefore a serious challenge to the relevance of these conclusions." Among the externalities not dealt with by conventional market theory are acid rain, the pollution of the oceans, the "greenhouse effect," the destruction of the ozone layer, and the rupture of community life at both local and national levels. Ironically, "the market depends on the community to regenerate moral capital, just as it depends on the biosphere to regenerate natural capital."[2] The market as it currently operates is destroying both the social and physical bases of its existence, hence it must be supplemented by other social forces for the economy and the planet itself to survive.

The private economy is itself important, of course. Enlightened business organizations have a role to play. The economy and the environment are not necessarily in conflict, and "the answers to economic and environmental problems are often the same."[3] But what is needed is regular collaboration between business and governmental agencies. In Europe and Japan such cooperation already takes place, often in a creative way. In Japan, for instance, though most research and development money comes from private sources, government agencies channel it into areas of greatest national need. There are some success stories about American business acting on its own; the 3M Corporation, for example, has long saved money by dealing with its waste products in an environmentally benign manner. Yet in many other cases, American business has joined forces with government in dealing with environmental problems. These are the exceptions, but increasingly leaders of the private sector recognize that business cannot do it alone: "Instead of getting government off our backs, we need it to pick us up off the ground."[4]

The fact remains that, much as many Americans still may think of government as an enemy, it is the political order through which authoritative social decisions have always been made by human beings. But government at what level? The popular slogan, "Think globally, act locally," is only a half-truth. The restoration of a national environment and the harnessing of technology can be accomplished at the national level, but other global problems of environmental protection, such as driftnet regulation and ozone layer protection, can be dealt with only at the international level.[5] The greenhouse effect, though it will affect different areas of the planet in different ways, does not respect national boundaries or punish polluting nations proportionately: the destruction of the ozone layer will increase the risk of cancer throughout the globe.[6] Ocean pollution, too, is a global phenomenon. Dealing with these crises will require action that transcends national boundaries. The fact that so many of these problems are caused by or exacerbated by powerful international corporations makes international action imperative.

This is not to suggest world government in a structural sense, but coordinated action by the sovereign states that compose the world community. As former Soviet foreign minister Edvard Shevardnadze told the United Nations in August 1989, "The biosphere recognizes no division into blocs, alliances, or systems. . . . All of us . . . need an international program to manage the risks involved in economic activities and to shift to alternative technologies that spare both man and nature."[7] International conferences have been held since 1988 to examine the greenhouse effect, though concerted action has been another matter. All conferences have faced a perennial problem, whether they seek to mitigate global warming or the destruction of the ozone layer; the United States, a major contributor to both conditions, has been dragging its feet, maintaining that current data is inconclusive and more study is needed.[8] Since it contributes far more greenhouse gases to the earth's atmosphere than any other nation and is considered a global leader, its failure to cooperate effectively dooms meaningful international action.[9]

Environmental issues have, of course, received increasing international attention. The July 1991 conference of the Group of Seven, the leading industrial powers, stressed in its closing declaration the "formidable environmental challenges" the international community faces.[10] When nineteen Latin American presidents met at the First Ibero-American Summit in Mexico that same month, they were greeted by a proposal of the Group of 100, consisting of leading Latin American intellectuals, for a "Latin American Ecological alliance."[11] Some argue that since "global action by the whole community of nations is the only hope for mitigating disaster," traditional notions of sovereignty are now outmoded.[12] That sovereignty is not what it used to be is illustrated by the European Economic Community, whose member nations have, by treaty, relinquished much of their historic powers. But the Persian Gulf war and its aftermath went even further, as have the European attempts, however abortive, at mediation of the 1991–92 Yugoslav civil war.[13] In both cases, intervention in the affairs of hitherto sovereign states took place with their reluctant consent but without prior agreement on their part. It is important to note that such intervention met with the approval of the international community.

There are increasing pressures for "outsiders" to intervene in sovereign states to protect the basic human rights of their citizens. In decades to come, we can be sure that similar demands will arise for intervention to protect the health and survival of human beings throughout the world. The pressure on Brazil to mitigate its destruction of the Amazon basin, though it does not involve direct physical intervention, does in some sense impugn its sovereignty, as many Brazilians claim, and is are a herald of things to come.

Other precedents, albeit cloudy ones, already exist for intervention by persuasion rather than force. In these cases, domestic policies are altered largely through financial pressures. For decades, developing nations have been under pressure to keep population growth in check in order to qualify for greater amounts of foreign aid. The new trade relations between the United States and Mexico involve commitments—however unrealistic—on the part of Mexico to mend its ways environmentally.[14] Programs exist whereby nations agree to protect some of their wilderness areas in return for mitigation of their international debts.[15] For many years the United States has made the existence of a free market economy a major qualification for eligibility of American aid to developing nations. In 1991 the rich nations of the developed world in effect made the former Soviet Union the target of similar requirements. However desirable the ends of such intervention may or may not be, the nations under pressure often argue that their sovereignty is being violated, and they are quite right. But ideological barriers to such pressures still exist even at the international level. For example, attempts to use environmental concerns to bar foreign imports of tuna caught by Mexican boats were recently rebuffed by the General Agreement on Tariffs and Trade (GATT) panel as an improper interference with international commerce, an illustration of how deeply entrenched free-market values remain.[16]

If some dangers must be dealt with primarily at the international level, many can and must be handled by government at the national and local levels. Unless one assumes that complete free trade is a desirable way of ordering the international economy—an increasingly dominant belief, though by no means a completely justifiable one[17]—national tariffs will play a major role in influencing the relationship between nature, technology, and community. Ironically, much of the dispute over international trade between the United States and Japan and the European Economic Community in recent years has been characterized by the United States' insistence that other nations abandon traditional farming for the hypertechnologized and ultimately unsustainable American system of agribusiness. Here again it appears that American policy is acting against the long- run interests of the biosphere.

Since only individual nations can levy taxes, tax policy can greatly influence energy policy. Solar power, for example, can and should be subsidized, as it briefly was in the United States under the Carter administration. Alternatively, our reliance on oil will increase, as it irresponsibly has been allowed to do under the Reagan and Bush administrations, with the latter even advocating environmentally dangerous drilling in wilderness areas. Subsidies and regulation also play a role in influencing consumers and industry. Energy-efficient homes can and should be mandated, as in Sweden,[18] or else they are left to the vagaries of the market as they unfortunately are in the United States.

New, less-polluting technologies can be encouraged or discouraged by government policies. France, for example, instituted a Mission for Clean Technologies in 1979 to provide grants to spur their development.[19] The United States established the Environmental Protection Agency (EPA) to monitor the impact of economic activity on the environment, though for over a decade it has been mired in a political struggle with Detroit about how far the automobile industry must go to increase fuel efficiency; thus auto emissions are still a growing problem in most of the nation. Despite the still unsolved problem of how safely to dispose of nuclear wastes, the nuclear power industry now claims to be less polluting than fossil fuels.[20] Policies toward nuclear energy differ from nation to nation. France has long stressed the nuclear option, gets most of its electrical energy from nuclear power, and intends to go on doing so. Sweden, too, has long been heavily reliant on nuclear energy; yet as a result of the Chernobyl accident in the former USSR, it has decided to phase out nuclear power completely by the next century.[21] American policy is complicated by its vast investment in military nuclear technology; some calculate that the costs of cleaning up all of the radioactive waste produced by America's military nuclear activity will exceed the costs of producing it in the first place.[22] Ironically, the military itself may get a new lease on life in managing the cleanup.[23]

Clearly, various nations contribute to local and global pollution in different ways and to different degrees. Western Europe, out of economic necessity, has been very active in reducing pollution through more efficient waste controls and emission standards. The former Communist bloc, as we have seen, has already destroyed many natural processes within its borders, perhaps irreparably. Nations such as the United States and Japan have sought to control pollution within their own boundaries. But both, especially Japan, have shown little interest in preventing its spread in the developing nations in which international corporations with branches in the Third World operate. The effective control of pollution on a global scale therefore depends largely on actions taken by individual nations.

A major factor in creating the greenhouse crisis has been the loss of agricultural land to cities, and above all to deforestation. Through action or inaction, sovereign nations control the extent to which cities will spread and farmland disappear; but farmland, though useful for the purpose of absorbing carbon dioxide, does so far less than forests. Individual nations also control the extent to which their forests are cut down, whether to create farmland for local or for export crops. The Amazon is being cut down in large measure so that cows can graze and be bred to provide hamburgers for Americans. Much of the Southeast Asian forests—and much of the virgin forest of the American Northwest—have been clear-cut to provide wood for the voracious appetite of Japan. Cognizant of the central role of trees in the biosphere, Americans as divergent as the TreePeople in Los Angeles, the American Forestry Association, and even President Bush have advocated a massive program of planting trees to correct the imbalance in the atmosphere.[24] Others argue that it already impossible to correct the damage.[25] Protection of wilderness is also a national responsibility. When one includes Alaska, most of the world's remaining wilderness is in the United States. The United States has pioneered in wilderness protection, and the American concept of national parks has spread to other nations. But in the United States, as elsewhere, wilderness and what it represents is in danger.[26]

Perhaps the greatest single danger to maintaining the habitability of the planet is population growth.[27] Different nations have differing attitudes toward population. Most developed nations still consider growth desirable *for themselves*, while many developing nations seek to restrict their population growth, with the encouragement, approval, and assistance of the developed. As free trade becomes more of a world ideal, it is obvious that when people cannot migrate freely, jobs will. Insofar as living standards in the poorer nations are upgraded, the pressures of population on resources will increase as well. It is simply impossible for the whole world eventually to consume as much per capita as Americans now do. In the long run, not only must world population growth level off to virtually zero, but so must growth in consumption of resources per person. The major areas of population growth are in the poorer nations of the Third World. No matter what pressures or inducements are offered, they will not cease their growth until their living standards rise.[28] The only just solution is an international adjustment over a period of time to equalize living standards throughout the globe. This means that developing nations, as they seek to grow economically, must be encouraged to use environmentally appropriate technologies rather than the wasteful ones in use in the developed countries. But it also means that the developed nations must learn to live far more frugally than they do at present.

Though national energy, land-use, and transportation policies can set the framework, such frugality cannot be primarily mandated from above. Most of it must stem from initiatives at the local level. Even if individual citizens become more conscious of the impact of their life-styles upon the environment, changing them will require some governmental encouragement, incentives, and action. Recycling may require subsidization for collection and reuse of materials, which usually involves taxation policies. Creation of energy-efficient homes and cities can involve changes in building codes. In many nations, such as the United States, these matters are in large measure under the jurisdiction of state, provincial, county, or municipal authorities.

Changing Directions

If all of the above steps are necessary to prevent ecological disaster, will the world muster the will to take action? "We don't respond to processes. We respond to events," one science writer notes,[29] citing the fact that cold-blooded animals such as frogs are averse to a sharp rise in heat but can blithely boil to death in water in which the temperature is slowly raised. The many processes by which the environment is being degraded are not perceived by the average human as "events." Even if we are aware of them, we often do not see them as the result of our personal choice and therefore subject to our control. In addition, even when they are, those groups that seek to reverse these trends are perceived—at times even by themselves—as acting in a selfish fashion. At the same time, those groups and individuals who profit by environmental degradation have no qualms about acting in their own interest, often claiming that what they do is necessary for progress, and therefore is for the good of all. Thus an American labor union that fears that free trade with Mexico will hurt its members and raises the unrelated but real issue that Mexican environmental laws are almost never enforced is seen by many as acting in bad faith; at the same time, manufacturers seeking to take advantage of what they see as cheap Mexican labor can claim to have the financial interest of the American consumer in mind and be viewed as doing something for the sake of the larger community. Many businesses, as we have seen, have found that environmentally responsible production processes are profitable, but most will still be tempted to cut corners to remain competitive with others who are less responsible, if there are no laws requiring responsible conduct.

How should we deal with this problem of obscured or mixed motivations? In any complex political process, such as that in the United States, interest-group alliances determine results. As a general principle, those who seek to preserve natural processes must accept the fact that they will have to take advantage of others who will be their allies in particular struggles, yet who may themselves struggle for different, more narrow, even in some cases more selfish ends. What counts is the result. A golf course saved from housing or industrial development is still open space, although the golf club's members may not care about such things. And when a labor union prevents a corporation from using a computer system because it eliminates members' jobs, this can still be a victory for human dignity as such.

In time, one hopes, support for environmental protection and restoration will not come primarily from labor unions, businessmen, large-lot real estate developers, and others motivated by less than pure devotion to the cause of environmentalism. But until that day, their support remains essential and will always play a role in the political dynamics of saving the planet and creating a truly human and humane society on it.

Increasingly, symptoms of environmental degradation such as pollution and hot summers are becoming obvious to all. Furthermore, most of the heedless economic growth that underlies current environmental crises has taken place in the name of progress. But allied with the concept of progress has been faith in the ability of science to solve all problems. It is now the scientific community that is sounding the alarm, and in an organized fashion.[30] One can only hope that the symptoms become obvious before recovery is impossible.

Are there realistic grounds for such hope? Political and economic leaders can be persuaded by argument, and many are coming to realize that simply giving large-scale technology complete freedom to flourish in the name of progress is not the centuries-old presumed panacea for all human need. But leaders cannot act completely on their own; they must reflect the basic views of the society as a whole. The revolution in environmental consciousness has to date taken place largely in the minds of the relatively affluent and better educated. Opponents of the environmental movement who have effectively charged it with being "elitist" have been on solid ground. But it is noteworthy that most radical changes in society have originated among such elites.

However, in recent decades the people at the bottom of the social hierarchy have become aggrieved at a situation that increasingly causes a deterioration in the quality of their own lives as well. While much deforestation stems from the demand of the rich nations for forest products, much also stems from the population pressure of poor farmers seeking land. The poor are becoming increasingly politicized, aware that in a world ruled by the rich their neighborhoods are targeted for waste disposal and as environmental hazard sites. On a larger scale, developed nations continue to regulate industrial pollution at home while ignoring its increase in Southeast Asia, Africa, or Latin America. In such a world, the rich nations will continue to seek to diminish pollution domestically while winking at its increase in Southeast Asia, Africa, or Latin America. But more and more unions in the rich nations and social movements in the poor ones protest against dangerous working conditions and the rape of their local environment by multinational corporations. In October 1991 a conference in Washington, D.C., brought together hundreds of leaders of black, Hispanic, Native American, and Asian groups from throughout the United States in the First National People of Color Leadership Summit on the Environment. The theme of the conference—supported by mainstream groups such as the Sierra Club—was "environmental racism," which contends that areas signaled out for toxic dumps, incinerators, and other hazardous sites, were primarily those inhabited by racial minorities.[31] Prominent among the groups represented was the Southwest Organizing Project, run by Hispanic residents of Albuquerque, which for years has protested against contamination of ground water in minority communities and radioactive waste dumps on Indian reservations.[32] Earlier that same month, a conference of the Student Environmental Action Coalition, the largest student-run political organization on American campuses, called for a new agenda for the American environmental movement that would join issues of race, class, and injustice with traditional preservationist concerns.[33]

Such pressures from below have yet to reach a point where they constitute a threat to the status quo. In many cases, especially in developing nations, the leaders are an elite cut off from their people, and they are still quite willing and able to exploit them and the local earth. But throughout the world, women, racial minorities, consumer groups, and ordinary citizens are beginning to question an order of things that more and more seems to make their aspirations for a better life untenable. As time goes on, toxic dumps will no longer be simply the burden of the poorest in society.[34] More and more, the poor may revolt against bearing the first, most immediate consequences of overexploitation of the planet's ecosystem.

But they can never triumph unaided. The more affluent groups within societies must be willing to yield to the needs of the disadvantaged. The wealthier nations must be more willing to share with the less- developed ones. This does not necessarily translate into foreign aid in the conventional sense. Creation of trading relationships that do not destroy the natural resources or human capital of the poor nations for the sake of the rich would be much more effective. This is not a matter of altruism. If the planetary ecosystem collapses, it will play no favorites.

But before enlightened self-interest takes hold, we must radically reevaluate what our self-interest really is. Trite or foolish as it may sound, the world requires a spiritual revolution—a revolution in the outlook on life, a change of heart, a *metanoia*. Almost everyone throughout the civilized world pays lip service to the environment today, but not all are realizing the extent of the change in attitudes and behavior necessary to save the planet.

Human beings—especially in the developed world—require a basic change in values. What is precluding us from dealing with our environmental problems in a sufficiently radical way? It is the simplistic concept of progress that assumes bigger is better, and more is preferable to less. We need not give up the idea of progress itself; faith in progress is fundamental to the Western culture that has now irrevocably conquered the world. But it is possible to progress in the things that have sustained humanity through the ages—art and music and human relations—without using up more and more of the earth's scarce resources. While we are giving up "bigger" and "more," we must also give up the related value we place on aggression to get them. That nature is "red in tooth and claw" has been accepted, especially by the West, since the era of Darwin. What makes humanity unique is that so much of its aggression takes place not only against other species or its own fellows, but, through its technology, against the natural world. Although the adherents of ecofeminism may often overstate their case, they are surely right about one thing: unless we end the male—or at least the "macho"—control of our societies, we will continue to seek to dominate nature rather than cooperate with it, and we will in time destroy ourselves in the process.

Contrary to what many may believe, human beings have great resources for good that are untapped in modern societies. People are not naturally selfish.[35] Societies throughout the world—and the world as a whole—must rediscover the importance of community. Unfortunately, economic development has long warred against community development.[36] But it is noteworthy that many societies in the Far East, especially Japan and the "Little Dragons" such as Singapore and South Korea, have retained a sense of community based on their Confucian roots, and this success has not hurt them in contemporary economic competition. Now it is time for Americans and many Europeans to give up their radical individualism as well, because it has done much to fuel the current environmental crisis, the dehumanization of work, and the increasing social instability. When this belief was combined with an emphasis on the economic realm in the early modern period, the stage was set for an all- out race to see who could use the resources of the Earth faster and more completely. Downplaying economic growth will also give new value to the community. Today, even within the citadel of classic liberal individualism, the United States, many liberal political theorists are calling for a revival of community for the sake of the common good.[37]

Finally, we must rethink our place in nature.[38] Ideological disputes can continue between those who think we must value nature for anthropomorphic reasons literally to save our skins, those who believe we must value it as a God-given trust,[39] and those who follow "deep ecology" and believe humans are simply another part of nature. But only by recognizing our basic kinship with the animal kingdom and the earth and sea can we avoid using our technological powers to destroy nature—and ourselves in the process.

NOTES

1. To Serve Man or to Serve Nature?

1. Lewis Mumford, *The Myth of the Machine: Technics and Human Development* (New York: Harcourt, Brace and World, 1966), 48–58.
2. Johan Huizinga, *Homo Ludens: A Study in the Play-Element in Human Culture* (London: Temple Smith, 1970).
3. Carolyn Merchant, *The Death of Nature: Women, Ecology, and the Scientific Revolution* (San Francisco: Harper and Row, 1979).
4. See Robert Ardrey, *African Genesis* (New York: Atheneum, 1961).
5. Erich Neumann, *The Great Mother: An Analysis of the Archtype* (Princeton: Princeton University Press, 1955).
6. On women in early culture, see Elise Boulding, *The Underside of History: A View of Woman through Time* (Boulder, Colo.: Westview Press, 1976).
7. Discussion of early technology from the *homo faber* point of view is found in V. Gordon Childe, *Man Makes Himself* (New York: New American Library, 1951; first published 1964); and his *What Happened in History* (Baltimore: Penguin Books, 1964; first published 1942).
8. Murray Bookchin, *The Ecology of Freedom: The Emergence and Dissolution of Hierarchy* (Palo Alto, Calif.: Cheshire Books, 1982), 101.
9. J. Donald Hughes, *Ecology in Ancient Civilizations* (Albuquerque: University of New Mexico Press, 1975), 32.
10. Ibid., 38.
11. See Robert W. Tucker, *The Marx-Engels Reader*, 2d ed. (New York: W. W. Norton, 1972), 653–658. See also Karl A. Wittfogel, *Oriental Despotism* (New Haven: Yale University Press, 1957).
12. N. K. Sanders, ed., *The Epic of Gilgamesh* (Harmondsworth, Eng.: Penguin Books, 1972), 70–84.
13. Hughes, op. cit., 38.
14. Frances Macdonald Cornford, *Before and After Socrates* (Cambridge: University Press, 1960), 42.
15. For a list of definitions millennia later, see Arthur O. Lovejoy, Gilbert Chinard, George

226

Boas, and Ronald S. Crane, A *Documentary History of Primitivism and Related Ideas*, vol. 1 (Baltimore: Johns Hopkins University Press, 1935), 447–456.

16. Werner Jaeger, *Paideia: The Ideals of Greek Culture*, vol 1, trans. by Gilbert Highet (New York: Oxford University Press, 1945), 168, 182, 323.
17. R. G. Collingwood, *The Idea of Nature* (Oxford: Clarendon Press, 1945), 8.
18. On this idea today, see J. E. Lovelock, *Gaia: A New Look at Life on Earth* (New York: Oxford University Press, 1980).
19. Frederick Copleston, S.J., *A History of Philosophy*, vol. 1, *Greece and Rome* (Westminster, Md.: Newman Press, 1950), 320.
20. Jaeger, op. cit., vol. 2, 183; vol. 3, 27–29.
21. J. Donald Hughes, "Ecology in Ancient Greece," *Inquiry* 18:116 (1975). For dissent, see John Rodman, "The Other Side of Ecology in Ancient Greece: Comments on Hughes," *Inquiry* 19:108–112 (1976).
22. Hughes, *Ecology*, op. cit., 48.
23. Ibid., 51.
24. Vincent Scully, *The Earth, The Temple, and the Gods* (New Haven: Yale University Press, 1962), 3.
25. Henry Rushton Fairclough, *Love of Nature among the Greeks and Romans* (New York: Longmans, Green, 1930), passim.
26. Clarence J. Glacken, *Traces on the Rhodian Shore: Nature and Culture in Western Thought from Ancient Times to the End of the Eighteenth Century* (Berkeley: University of California Press, 1967), 118.
27. On Greek farming generally, see G. E. Fussell, "Farming Systems in the Classical Era," *Technology and Culture* 21:16–44 (1961). On their destructive aspects, see Marshall Massey, "Carrying Capacity and the Greek Dark Ages," *Co-Evolution Quarterly* 40:29–35 (Winter 1983).
28. A modern argues that if Plato's ideas about government had been followed, such destruction might have been avoided. See Robert Webking, "Plato's Response to Environmentalism," *The Intercollegiate Review* 20:27–35 (Winter 1984).
29. Thorlieff Boman, *Hebrew Thought Compared with Greek* (London: SCM Press, 1960), 123–183.
30. Cornford, op. cit., 63. The world-soul, incidentally, was regarded as feminine, as the word *physis* is in Greek; see Merchant, op. cit., xix, 10.
31. Jaeger, op. cit., vol. 3, 211.
32. *Politics*, I, 8. On teleological ideas among the Greeks, see Glacken, op. cit., 35–51.
33. Hughes, *Ecology*, op. cit., 65; also Hughes, "Ecology in Ancient Greece," 122–123.
34. See J. Donald Hughes, "The Environmental Ethics of the Pythagoreans," in Robert C. Schultz and J. Donald Hughes, eds., *Ecological Consciousness: Essays from the Earthday X Colloquium*, University of Denver, April 21–24, 1980 (Washington, D.C.: University Press of America, 1980), 149–164.
35. Cornford, op. cit., 30.
36. "In Greek thinking... the very notion of 'humanity' tends to be associated with at least a minimal development of technology"; Mulford Q. Sibley, *Nature and Civilization: Some Implications for Politics* (Itaska, Ill.: F. E. Peacock, 1977), 168.
37. M. I. Finley, "Technical Innovations and Economic Progress in the Ancient World," *Economic History Review*, 18:236 (1965).
38. Leo Strauss, *Thoughts on Machiavelli* (Seattle: University of Washington Press, 1969), 298.
39. On the meaning of *techne*, see Carl Mitcham, "Philosophy and the History of Technology,"

in George Bugliarello and Dean B. Doner, eds., *The History and Philosophy of Technology*, introduction by Melvin Kranzberg (Urbana: University of Illinois Press, 1973), 172–184.

40. S. T. Lowry, "The Classical Greek Theory of Natural Resource Economics," *Land Economics* 41:204 (1966).

41. *Politics*, X, 10.

42. Witold Rybczynski, *Taming the Tiger: The Struggle to Control Technology* (New York: Penguin Books, 1985), 167–168.

43. This was true even in utopian speculation. See Frank E. Manuel and Fritzie P. Manuel, *Utopian Thought in the Western World* (Cambridge: Belknap Press of Harvard University Press, 1979), 64–104.

44. Collingwood, op. cit., 8.

45. Fairclough, op. cit., 181.

46. Sir Archibald Geikie, K.C.B., *The Love of Nature among the Greeks and the Romans: During the Later Decades of the Republic and the First Century of the Empire* (London: John Murray, 1912), 283–297, 322–348.

47. On the Roman city, see Lewis Mumford, *The City in History: Its Origins, Its Transformations, and Its Prospects* (New York: Harcourt, Brace and World, 1961), 205–242.

48. Hughes, *Ecology*, op. cit., 89.

49. Ibid.

50. On Roman ideas about nature, see Glacken, op. cit., 51–79.

51. Lovejoy et al., op. cit., 252–253.

52. John Passmore, *Man's Responsibility for Nature: Ecological Problems and Western Traditions* (New York: Charles Scribner's Sons, 1974), 14.

53. Quoted in John Passmore, *The Perfectability of Man* (London: Duckworth, 1970), 54.

54. Quoted in Passmore, *Man's Responsibility*, op. cit., 14.

55. Merchant, op. cit., 23.

56. See Glacken, op. cit., 122–149.

57. Hughes, *Ecology*, op. cit., 109.

58. Ibid., 126.

59. Collingwood, op. cit., 8.

60. Boman, op. cit.

61. For example, Passmore, op. cit., 10.

62. Jeanne Kay, "Concepts of Nature in the Hebrew Bible," *Environmental Ethics* 10:312, 325 (1988).

63. Frederick Elder, *Crisis in Eden: A Religious Study of Man and Environment* (Nashville: Abingdon Press, 1970), 84.

64. Kay, op. cit., 314.

65. See Bernhard W. Anderson, "Creation in the Noachic Covenant," in Philip N. Joranson and Ken Butigan, eds., *Cry of the Environment: Rebuilding the Christian Creation Tradition* (Santa Fe, N.M.: Bear and Co., 1984), 45–61.

66. On the theme of Israel, God, and the land, see H. Paul Santmire, *The Travail of Nature: The Ambiguous Ecological Promise of Christian Theology* (Philadelphia: Fortress Press, 1985), 190–192. It has been argued that in fact the Jewish tradition and care about nature are basically opposed. See Steven S. Schwartzschild, "The Unnatural Jew," *Environmental Ethics* 6:347–362 (1984); and the reply by Jeanne Kay, "Comments on 'The Unnatural Jew,' " ibid., 7:189–191, 19–85. On the whole question of Jewish environmental ethics, see also David Ehrenfeld and Philip J. Bentley, "Judaism and the Practice of Stewardship," *Judaism* 34:301–311 (1985); and Jonathan Helfand, "The Earth Is the Lord's:

Judaism and Environmental Ethics," in Eugene C. Hargrove, ed., *Religion and Environmental Crisis* (Athens: University of Georgia Press, 1986), 38–39.

67. One scholar comments that "by and large, their biblical perspective was anthropocentric. The land was a wonderful gift from God, and so it should have been used well, but the land had few rights over against its human stewards"; John Carmody, *Ecology and Religion: Toward a New Christian Theology of Nature* (New York: Paulist Press, 1983), 88–89. This is, however, an overstatement.
68. Kay, op. cit., 316–317,319, 321.
69. Richard H. Hiers, "Ecology, Biblical Theology, and Methodology: Biblical Perpsectives on the Environment," *Zygon: A Journal of Religion and Science*, 43–49 (1984).
70. Kay, op. cit., 313, 317–318.
71. Glacken, op. cit., 156.
72. Elder, op. cit., 90.
73. Hiers, op. cit., 48, 50.
74. Passmore, op. cit., 12.
75. W. Lee Humphreys, "Pitfalls and Promises of Biblical Texts as a Basis for a Theology of Nature," in Glenn C. Stone, ed., *A New Ethic for a New Earth* (Andover, Conn.: Faith-Man-Nature Group, 1971), 99.
76. Theodore Olson, *Millennialism, Utopianism, and Progress* (Toronto: University of Toronto Press, 1982), fn. p. 25.
77. Kay, op. cit., 322.
78. Glacken, op. cit., 160.
79. Kay, op. cit., 326.
80. John Black, *The Dominion of Man: The Search for Ecological Responsibility* (Edinburgh: Edinburgh University Press, 1970), 46–47.
81. Santmire, op. cit., 194, 198.
82. On this theme, see Helen A. Kenik, "Toward a Biblical Basis for Christian Theology," in Matthew Fox, O.P., *Western Spirituality: Historical Roots, Ecumenical Routes* (Santa Fe, N.M.: Bear and Co., 1981), 27–75.
83. Boman, op. cit., 156.
84. James Barr, "Man and Nature: The Ecological Controversy and the Old Testament," in David and Ellen Spring, eds., *Ecology and Religion in History* (New York: Harper Torchbooks, 1974), 66, 67, 68.
85. John Maquarrie, "Creation and Environment," in Spring and Spring, op. cit., 36.
86. James Barr, "Man and Nature: The Ecological Controversy and the Old Testament," in ibid., 51.
87. William E. Fudpucker, "Through Christian Technology to Technological Christianity," in ibid., 53.
88. "Technique and the Opening Chapters of Genesis," in Carl Mitcham and Jim Grote, eds., *Theology and Technology: Essays in Christian Analysis and Exegesis* (Lanham, Md.: University Press of America, 1984), 125.
89. Ibid., 176. See also his "The Relationship between Man and Creation in the Bible," ibid., 139–155.
90. Black, op. cit., 46, 52–55.
91. Passmore, op. cit., 29, 33, 37.
92. Carmody, op. cit., 91.
93. William Kuhns, *Environmental Man* (New York: Harper and Row, 1969), 28.
94. Black, op. cit., 121.
95. Santmire, op. cit., 66.

96. D. S. Wallace-Hadrill, *The Greek Patristic View of Nature* (Manchester: Manchester University Press, 1976), 80, 90.
97. He is referred to as a Greek father because he wrote in that language. On him, see Glacken, op. cit., 189–196.
98. Wallace-Hadrill, op. cit., 93, 103.
99. Ibid., 114. On Origen generally, see Santmire, op. cit., 49–53; and Glacken, op. cit., 183–186.
100. Santmire, op. cit., 40.
101. On Iranaeus, see ibid., 35–44.
102. Wallace-Hadrill, op. cit., 120, 129.
103. See Matthew Fox, O.P., *Original Blessing* (Santa Fe, N.M.: Bear and Co., 1983).
104. "St. Augustine's Theology of the Biophysical World," in Schultz and Hughes, op. cit., 100.
105. Santmire, *The Travail*, op. cit., 61, 70. Richard H. Hiers accepts this view, op. cit., 46, as does Glacken, op. cit., 199.
106. Mitcham in Bugliarello and Doner, op. cit., 186.
107. Carmody, op. cit., 103.
108. See David J. Herlihy, "Attitudes toward the Environment in Medieval Society," in Lester J. Bilsky, *Historical Ecology: Essays on Environment and Social Change* (Port Washington, N.Y.: Kennikat Press, 1980), 104.

2. *Religion, Nature, and Technology*

1. George Ovitt, *The Restoration of Perfection: Labor and Technology in Medieval Culture* (New Brunswick: Rutgers University Press, 1987), 55.
2. Melvin Kranzberg, in Kranzberg, ed., *Ethics in an Age of Pervasive Technology* (Boulder, Colo.: Westview Press, 1980), 61.
3. Robert I. Faricy, S.J., "Christ and Nature," in F/N/M Papers, no. 1, *Christians and the Good Earth* (Alexandria, Va.: Faith-Man-Nature Group, n.d.), 76.
4. George D. Economou, *The Goddess Natura in Medieval Literature* (Cambridge: Harvard University Press, 1972), 3.
5. Alfred Biese, *The Development of the Feeling for Nature in the Middle Ages and Modern Times* (New York: Burt Franklin, 1964), 68; first published 1905.
6. See M. D. Chenu, O.P., *Nature, Man, and Society in the Twelfth Century: Essays on New Theological Perspectives in the Latin West*, selected and trans. by Jerome Taylor and Lester K. Little (Chicago: University of Chicago Press, 1968), 99–145.
7. Saint Denis the Aeropagite, Athenian convert of Saint Paul and revered authority in the Middle Ages.
8. Chenu, op. cit., 126.
9. Glacken, op. cit., 240–248.
10. On the condemnations and their context, see ibid., 248–250.
11. On Aquinas's philosophy generally, see Frederick Copleston, S.J., *A History of Philosophy*, vol. 2, *Medieval Philosophy, Augustine to Scotus* (Westminster, Md.: Newman Press, 1952), 302–434; for his views on creation and politics, see M. D. Chenu, O.P., "Body and Body Politics in the Creation Spirituality of Thomas Aquinas," in Fox, ed., *Western Spirituality*, op. cit., 193–214.
12. On his political ideas, see Thomas Gilby, O.P., *The Political Thought of St. Thomas Aquinas* (Chicago: University of Chicago Press, 1958).
13. Quoted in Chenu, *Nature*, op. cit., 11.

14. Carmody, op. cit., 230.

15. Glacken, op. cit., 230.

16. See Ovitt, op. cit., 84.

17. Ibid., 236; Santmire, op. cit., 91–93.

18. See Edward Armstrong, *Saint Francis, Nature Mystic: The Derivation and Significance of the Nature Stories in the Franciscan Legend* (Berkeley: University of California Press, 1973), 104n. In the Middle Ages, animals could be the object of legal actions and excommunicated by the church under Canon Law.

19. Santmire, op. cit., 94.

20. Robert A. Nisbet, *The Social Philosophers: Community and Conflict in Western Thought* (New York: Thomas A. Crowell, 1973), 327, 333.

21. Christopher Dawson, *Progress and Religion: An Historical Enquiry* (New York: Sheed and Ward, 1938), 178–179.

22. Lynn White, Jr., *Machina Ex Deo: Essays in the Dynamism of Western Culture* (Cambridge: MIT Press, 1968), 94.

23. Richard J. Woods, "Environment as Spiritual Horizon: The Legacy of Celtic Monasticism," in Joranson, op. cit., 62–84.

24. Armstrong, op. cit., 35–41, 146–147.

25. Armstrong defines a nature mystic as "a person of Christian faith who, through the apprehension of the beauty, goodness, and glory of God revealed in Creation, is uplifted to an ineffable experience"; ibid., 17.

26. White, *Machina Ex Deo*, op. cit., 91, 93.

27. Armstrong, op. cit., 30.

28. Santmire, op. cit., 116–117. For other views on Francis and his life, see Glacken, op. cit., 204–216; and Paul Wiegund, "Escape from the Birdbath: A Reinterpretation of St. Francis as a Model for the Ecological Movement," in Joranson, op. cit., 148–157.

29. See Matthew Fox, O.P., "Creation-Centered Spirituality from Hildegard of Bingen to Julian of Norwich: 300 Years of an Ecological Spirituality in the West," ibid., 85–106; and Fox, "Meister Eckhart on the Fourfold Path of a Creation-Centered Spiritual Journey," in Fox, *Western Spirituality*, op. cit., 215–258.

30. Lynn White, Jr., *Medieval Religion and Technology: Collected Essays* (Berkeley: University of California Press, 1978), 145, 146.

31. Eugene C. Hargrove, *Foundations of Environmental Ethics* (Englewood Cliffs, N.J.: Prentice Hall, 1989), 201.

32. Ovitt, op. cit., 87.

33. Quoted in Paul T. Durbin, "Philosophy of Technology," in Paul T. Durbin, ed., *A Guide to the Culture of Science, Technology, and Medicine* (New York: Free Press, 1984), 283.

34. Kuhns, op. cit., 129.

35. On medieval technology, see Terry S. Reynolds, "Medieval Roots of the Industrial Revolution," *Scientific American* 251:122–130 (1984).

36. *Evolution and Christian Hope: Man's Concept of the Future from the Early Church Fathers to Teilhard de Chardin* (Garden City, N.Y.: Anchor Books, 1966), 121–142.

37. *Medieval Religion*, op. cit., passim.

38. Arnold Pacey, *The Maze of Ingenuity: Ideas and Idealism in the Development of Technology* (New York: Holmes and Meier, 1975), 57.

39. Rybczynski, op. cit., 12.

40. White, *Medieval Religion*, op. cit., 238.

41. One index of the changing attitude toward work was the altered status of Saint Joseph. Though he was originally a figure of popular disdain in the West, his cult was taken up

in part due to Eastern influences, and by 1399 his feast was adopted by the Franciscans, then the Dominicans, and by the whole Catholic church in 1621; ibid., 185.

42. Jacques Le Goff, *Time, Work, and Culture in the Middle Ages*, trans. by Arthur Goldhammer (Chicago: University of Chicago Press, 1980), 77.

43. Carl Mitcham, "The Religious and Political Origins of Modern Technology," in Paul T. Durbin and Friedrich Rapp, eds., *Philosophy and Technology* (Dordrecht, Neth.: D. Reidel, 1983), 272.

44. *The Rule of St. Benedict*, in Latin and English, ed. and trans. by Abbot Justin McCann (Westminster, Md.: Newman Press, 1952), 13.

45. For example, White, *Machina Ex Deo*, op. cit., 65.

46. Ovitt, op. cit., 106, 137–138, 150.

47. Ibid., 119.

48. Ibid., 17.

49. See the whole of "Merchant's Time and Church's Time in the Middle Ages," in Le Goff, op. cit., 29–42, 62.

50. Ibid., 63, 67, 70.

51. Quoted in ibid., 50, 51.

52. See Dawson, op. cit., 179–184.

53. White goes so far as to say that "man was no longer the focus of the visible universe. In this sense Copernicus is a corollary to St. Francis"; *Medieval Religion*, op. cit., 39.

54. Ibid., 126.

55. Ibid., 90, 126, 130.

56. See David S. Landes, *Revolution in Time: Clocks and the Making of the Modern World* (Cambridge: Belknap Press of Harvard University Press, 1983), 59–66.

57. Lewis Mumford, *Technics and Civilization* (New York: Harcourt, Brace, 1934), 13–14.

58. White, *Machina Ex Deo*, op. cit., 219.

59. Jean Gimpel, *The Medieval Machine: The Industrial Revolution of the Middle Ages* (New York: Penguin Books, 1977), 147.

60. Ibid., 127–129.

61. Mumford, *Technics*, op. cit., 107–150; see also Carlo Cipolla, *Clocks and Culture, 1300–1700* (New York: Walker, 1967), 158–174.

62. On windmills, see Gimpel, op. cit., 144; also "The Advent and Triumph of the Windmill," in Marc Bloch, *Land and Work in Medieval Europe: Selected Papers*, trans. by J. E. Anderson (Berkeley: University of California Press, 1967), 136–168.

63. White, *Medieval Religion*, op. cit., 144.

64. Ibid. But see also Rodney Hilton and P. H. Sawyer, "Technical Determinism: The Stirrup and the Plough," *Past and Present* 24:90–100 (1963).

65. Melvin Kranzberg and Joseph Gies, *By the Sweat of Thy Brow: Work in the Western World* (New York: G. P. Putnam's Sons, 1975), 60.

66. For example Bloch, op. cit., passim, 176.

67. Mumford, *Technics*, op. cit., 67.

68. Kranzberg and Gies, op. cit., 77–78. On mining generally, see Gimpel, op. cit., 59–74.

69. See Glacken, op. cit., 468–470.

70. Gimpel, op. cit., 78.

71. Ibid., 82.

72. See Herlihy, op. cit., 108–109.

73. Carlo M. Cipolla, *Before the Industrial Revolution: European Society and Economy, 1000–1700* (New York: W. W. Norton, 1976), 110.

74. On medieval forestry, see Gimpel, op. cit., 75–81; and Glacken, op. cit., 318–341.

75. On the diffusion of technology, see Cipolla, op. cit., 174–181.
76. Ibid., 134. On medieval cities generally, see Mumford, *The City*, op. cit., 253–343.
77. On guilds, see Kranzberg and Gies, op. cit., 65–74.
78. On working conditions, see Cipolla, op. cit., 63–95; and Gimpel, op. cit., 93–113.
79. Reay Tannahill, *Sex in History* (New York: Stein and Day, 1981), 147.
80. Ovitt, op. cit., 180.
81. Ibid. On the position of women in the Middle Ages, see Tannahill, op. cit., 131–161, and Amaury de Reincourt, *Sex and Power in History* (New York: Delta Books, 1975), 137–160, 207–231.
82. Tannahill, op. cit., 144.
83. Ovitt, op. cit., 195, 202.
84. On plagues and population, see Cipolla, op. cit., 147–157.
85. See William H. McNeill, *The Rise of the West: A History of the Human Community* (Chicago: University of Chicago Press, 1963), 456; and his *The Pursuit of Power: Technology, Armed Force, and Society since A.D.1000* (Chicago: University of Chicago Press, 1982), 64.
86. Martin van Creveld, *Technology and War: From 2000 B.C. to the Present* (New York: Free Press, 1989), 20.
87. White, *Medieval Religion*, op. cit., 29, 280.
88. On medieval military technology, see Gimpel, op. cit., 231–236; and McNeill, *The Pursuit of Power*, op. cit., 63–116.
89. Cipolla, op. cit., 103.

3. *Technology Flourishes*

1. Merchant, op. cit., 27.
2. Collingwood, op. cit., 103.
3. On the full significance of Galileo's ideas, see Mumford, *The Myth*, op. cit., 51–76.
4. On Descartes, see Fritjof Capra, *The Turning Point: Science, Society, and the Rising Culture* (New York: Bantam Books, 1983), 101–108.
5. On Bacon, see James Collins, *A History of Modern European Phiosophy* (Milwaukee: Bruce Publishing, 1954), 51–73; Manuel and Manuel, op. cit., 243–260; Mumford, *The Myth*, op. cit., 105–129; and for criticism of Mumford's views, see Olson, op. cit., 273–274.
6. Thomas A. Spragens, Jr., *The Irony of Liberal Reason* (Chicago: University of Chicago Press, 1981), 56.
7. Merchant, op. cit., 169.
8. Arnold Pacey, *The Culture of Technology* (Cambridge: MIT Press, 1983), 114–115.
9. Carl Mitcham and Jim Grote, "Aspects of Christian Exegesis: Hermeneutics, the Theological Virtues, and Technology," in Mitcham and Grote, op. cit., 33.
10. See, for example, Merchant, op. cit., 168.
11. Brian Easley, *Witch Hunting, Magic and the New Philosophy: An Introduction to Debates of the Scientific Revolution, 1450–1750* (Atlantic Highlands: Humanities Press, 1980), 247–248.
12. Bookchin, op. cit., 286. See also Langdon Winner, *Autonomous Technology: Technics-out-of-Control as a Theme in Political Thought* (Cambridge: MIT Press, 1977), 135–139.
13. From a different perspective, French historian Jules Michelet held that witchcraft was in fact the origin of much modern science. See Le Goff, op. cit., 21.
14. On witchcraft generally, see Charles Williams, *Witchcraft* (Cleveland: Meridian Books,

1959); on witchcraft in England, see Keith Thomas, *Religion and the Decline of Magic* (New York: Charles Scribner's Sons, 1971).

15. This is part of the thesis of Merchant, op. cit., passim, and of much of Easley, op. cit., passim.
16. Easley, op. cit., 533.
17. Santmire, op. cit., 127.
18. Ibid., 124, 125.
19. Ibid., 128, 129.
20. Ibid., 122, 131.
21. George Grant, *Technology and Empire: Perspectives on North America* (Toronto: House of Anansi, 1969), 24.
22. Santmire, op. cit., 133.
23. Paulo Rossi, *Philosophy, Technology, and the Arts in the Early Modern Era*, trans. by Salvator Attanasio, ed. by Benjamin Nelson (New York: Harper and Row, 1970), 32–33.
24. Ibid., 16–17.
25. Glacken, op. cit., 464.
26. Ibid.
27. Rossi, op. cit., 15.
28. Ibid., 100.
29. Cipolla, op. cit., 227.
30. William Leiss, *The Domination of Nature* (New York: George Braziller, 1972), 27.
31. John U. Nef, *The Conquest of the Material World* (Cleveland: World Publishing, 1967), 4.
32. Ibid., 39. On mining in this era, see Mumford, *Technics*, op. cit., 60–77.
33. Mumford, *Technics*, op. cit., 75.
34. Glacken, op. cit., 468–469.
35. Mumford, *Technics*, op. cit., 70.
36. Leiss, op. cit., 79.
37. Capra, op. cit., 66
38. Quoted in Easley, op. cit., 171, 219.
39. Ibid., 214.
40. Quoted in ibid., 139.
41. Quoted in ibid., 214.
42. Glacken, op. cit., 423. See also Merchant, op. cit., 248.
43. On Spinoza's philosophy generally, see Collins, op. cit., 199–251; on Leibniz, ibid., 252–310.
44. See George Sessions, "Western Process Metaphysics (Heraclitus, Whitehead, and Spinoza)," in Bill Devall and George Sessions, *Deep Ecology* (Salt Lake City: Peregrine Smith Books, 1985), 236–242.
45. Frederick L. Baumer, *Modern European Thought: Continuity and Change in Ideas, 1600–1950* (New York: Macmillan, 1977), 90.
46. James Collins, *Spinoza on Nature*, foreword by George Kimball Plochman (Carbondale: Southern Illinois University Press, 1984), 66.
47. But Spinoza did in practice believe that animals had no rights; see Sessions in Devall and Sessions, op. cit., 240.
48. Arthur C. Lovejoy, *The Great Chain of Being: A Study in the History of an Idea* (Cambridge: Harvard University Press, 1936), 188.
49. Glacken, op. cit., 506.
50. Ibid., 508.

51. Ibid., 377, 477.
52. On Kant's philosophy generally, see Collins, *History*, op. cit., 455–543.
53. Passmore, op. cit., 80.
54. On Hobbes's ideas, see Leo Strauss, *Natural Right and History* (Chicago: University of Chicago Press, 1973), 166–202; and Collins, op. cit., 101–137.
55. "Hobbes pointed out that the highest certainty was to be found in purely artificial things, things people put together and can also take apart"; Winner, op. cit., 279.
56. For instance, William Ophuls, *Ecology and the Politics of Scarcity: Prologue to a Political Theory of the Steady State* (San Francisco: W. H. Freeman, 1977), 142–166; and Ophuls, "Leviathan or Oblivion?" in Herman E. Daly, ed., *Toward a Steady-State Economy* (San Francisco: W. H. Freeman, 1973), 215–230. For criticism of his position, see Robert Webking, "Liberalism and the Environment," *Political Science Reviewer* 11:223–249 (1981); and Robert W. Hoffert, "The Scarcity of Politics: Ophuls and Western Political Thought," *Environmental Ethics* 8:5–32 (1986).
57. On their ideas, see Marie Louise Berneri, *Journey through Utopia*, foreword by George Woodcock (New York: Schocken Books, 1950), 143–173; Manuel and Manuel, op. cit., 332–361; Kenneth Rexforth, *Communalism: From Its Origins to the Twentieth Century* (New York: Seabury Press, 1974), 133–154; and the classic of Edward Bernstein, *Cromwell and Communism: Socialism and Democracy in the Great English Revolution* (New York: Schocken Books, 1963; first published 1895).
58. Easley, op. cit., 224.
59. Ibid., 230.
60. James Farr, "Technology in the Digger Utopia," in Arthur L. Kallenberg, J. Donald Moon, and Donald R. Sabia, Jr., eds., *Dissent and Affirmation: Essays in Honor of Mulford Q. Sibley* (Bowling Green, Ohio: Bowling Green University Popular Press, 1983), 129.
61. On Locke's ideas, see Collins, op. cit., 311–365; Strauss, op. cit., 202–251; and Robert Goldwin, "John Locke," in Leo Strauss and Joseph Cropsey, eds., *History of Political Philosophy* (Chicago: Rand McNally, 1963), 433–468.
62. For a classic dissent from this view see Richard Cox, *Locke on War and Peace* (Oxford: Clarendon Press, 1960).
63. Ophuls, *Ecology*, op. cit., 154, 223.
64. Ibid., 144. See also Ophuls, "Leviathan or Oblivion?" op. cit., 221–222.
65. C. B. Macpherson, "Democratic Theory: Ontology and Technology," in David Spitz, ed., *Political Theory and Social Change* (New York: Atherton Press, 1967), 211.
66. *Second Treatise*, sec. 42, as rendered by Goldwin in Strauss and Cropsey, op. cit., 468.
67. Strauss, *Natural Right*, op. cit., 245.
68. Goldwin in Strauss and Cropsey, op. cit., 450.
69. *Second Treatise*, sec. 37.
70. Strauss, *Natural Right*, op. cit., 241.
71. *Second Treatise*, sec. 42.
72. Strauss, *Natural Right*, op. cit., 258, 250–251.
73. This reading of Locke represents agreement with the consensus of such otherwise opposed interpreters as the conservative Leo Strauss and the quasi-Marxist C. B. Macpherson. For criticism of Macpherson's thesis in "Democratic Theory," see Hwa Yol Jung, *The Crisis of Political Understanding: A Phenomenological Perspective in the Conduct of Political Inquiry* (Pittsburgh: Duquesne University Press, 1979), 139–140. For an exposition of the thesis that Locke's endorsements of dominion over nature are found primarily in the *Second Treatise* and not so much in his thought as a whole, see Kathleen M. Squadrito, "Locke's View of Dominion," *Environmental Ethics* 1:255–262 (1979).

4. Technology Triumphs

1. Cipolla, op. cit., 273.
2. William Irwin Thompson, At the Edge of History (New York: Harper and Row, 1971), 93.
3. Pacey, op. cit., 206.
4. Anthony F. C. Wallace, The Social Context of Innovation: Bureaucrats, Families, and Heroes in the Early Industrial Revolution as Foreseen in Bacon's New Atlantis (Princeton: Princeton University Press, 1982), 5, 8, 151.
5. This was especially true in the nascent, and very important, iron industry; ibid., 67, 101, and passim.
6. Baumer, op. cit., 148.
7. Nef, Conquest, op. cit., 219.
8. John U. Nef, War and Human Progress: An Essay on the Rise of Industrial Civilization (Cambridge: Harvard University Press, 1950), 23.
9. Nef, Conquest, op. cit., 325.
10. Rybczynski, op. cit., 10.
11. Mumford, Technics, op. cit., 108. Authorities differ on when Hero lived; dates given range from 200 BC. to 300 BC.
12. On attempts to create a usable steam engine, see ibid., 158–160.
13. Science and Common Sense (New Haven: Yale University Press, 1951), 39.
14. David S. Landes, The Unbound Prometheus: Technological Change and Industrial Development in Western Europe from 1750 to the Present (Cambridge: Cambridge University Press, 1969), 61.
15. P. Hans Sun, "Notes on How to Begin to Think about Technology in a Theological Way," in Mitcham and Grote, op. cit., 174–175.
16. Nef makes an elaborate argument that the usage dates from a series of lectures by Arnold Toynbee in the 1880s in which he coined the term "industrial revolution" and used dates for it that originated not in economic but in political considerations, beginning it with the accession to the throne of George III in 1760 and ending it with the Reform Act of 1832; War, op. cit., 274–275.
17. Landes, Unbound Prometheus, op. cit., 1.
18. Cipolla, op. cit., 222.
19. Ibid., 276.
20. Wallace, op. cit., 153.
21. Landes, Unbound Prometheus, op. cit., 61, 66–67, 127, 129.
22. Ibid., 23. Nef, Conquest, op. cit., 230–234.
23. Kranzberg and Gies, op. cit., 111.
24. For Smith and his ideas, see Robert L. Heilbroner, The Worldly Philosophers: The Lives, Times, and Ideas of the Great Economic Thinkers (New York: Simon and Schuster, 1953), 33–66.
25. On Ure see Pacey, op. cit., 268–270.
26. Kranzberg and Gies, op. cit., 95, 96–97, 191.
27. Quoted in Pacey, op. cit., 211, 212.
28. The classic arguments of R. H. Tawney and Max Weber for the Protestant origins of capitalism are supported by Landes, Unbound Prometheus, op. cit., 23–24, while Nef is highly skeptical, op. cit., 221–224. See also Amintore Fanfani, Catholicism, Protestantism, and Capitalism (London: Sheed and Ward, 1949).

29. Nisbet, op. cit., 352.
30. Karl Polanyi, *The Great Transformation* (New York: Rinehart, 1944), 139–140.
    31. On the Luddites, see Kranzberg and Gies, op. cit., 102–103; Rybczynski, op. cit., 37–47; and Afrian J. Randall, "The Philosophy of Luddism: The Case of the West of England Woollen Workers, ca. 1790–1809," *Technology and Culture* 27:1–17 (1986).
32. Mumford, *Technics*, op. cit.,86, 89.
33. See *Technics*, ibid., 92. For a technical discussion, see McNeill, *The Pursuit of Power*, op. cit., 128–129.
34. McNeill, *The Pursuit of Power*, op. cit., 248.
35. Nef, *War*, op. cit., 29.
36. Quoted in ibid., 28.
37. See McNeill, *The Pursuit of Power*, op. cit., 173.
38. Ibid., 278. For a larger perspective, see ibid., 269–294.
39. Nef, *War*, op. cit., 65–66.
40. Ibid., 377.
41. Ibid.
42. Quoted in Easley, op. cit., 28.
43. Thomas, op. cit., 166, 167.
44. Ibid., 22.
45. On the Ordinance, see Glacken, op. cit., 491–494.
46. On Buffon, see ibid., 663–681.
47. Quoted in ibid., 666.
48. Thomas, op. cit., 193–194, 197.
49. Ibid., 276. The word in the form "conservancy" was first used in the later Middle Ages with reference to the preservation of the river Thames.
50. On the concept, see Max Oelschlaeger, *The Idea of Wilderness: From Prehistory to the Age of Ecology* (New Haven: Yale University Press, 1991).
51. Thomas, op. cit., 260. On wilderness in English taste see ibid., 254–269.
52. ibid., 268.
53. Chester L. Cooper, "Growth in America: An Introduction," in Cooper, ed., *Growth in America* (Westport, Conn.: Greenwood Press, 1976), 21.
54. On Burke's ideas, see Russell Kirk, *The Conservative Mind: From Burke to Santayana* (Chicago: Henry Regnery, 1953), 11–61; and Peter J. Stanlis, *Edmund Burke and the Natural Law* (Ann Arbor: University of Michigan Press, 1958).
55. On Malthus's views, see Glacken, op. cit., 637–654.
56. On Morris, see E. P. Thompson, *William Morris: Romantic to Revolutionary* (New York: Pantheon Books, 1976), and Rybczynski, op. cit., 139–146.
57. Krishnan Kumar, *Utopia and Anti-Utopia in Modern Times* (London: Basil Blackwell, 1987), 456n. *News from Nowhere* was originally written as a response to Bellamy's *Looking Backward*.
58. T. Jackson Lears, *No Place of Grace: Antimodernism and the Transformation of American Culture, 1880–1920* (New York: Pantheon Books, 1981), passim.
59. On Kant, see Collins, op. cit., 455–543.
60. Because of his importance, the literature on Darwin and Darwinism is vast and of course largely laudatory. For an interesting critical view, see Jeremy Rifkin, *Algeny* (New York: Viking Press, 1983).
61. Baumer, op. cit., 343–344.
62. See Loren Eisely, *Darwin's Century* (New York: Doubleday, 1958).
63. Floyd W. Matson, *The Broken Image: Man, Science, and Society* (Garden City, N.Y.:

Anchor Books, 1966), 27. As the scientist-philosopher Loren Eisely describes Darwin's theory, "It was a great blow to man. Earlier man had seen his world displaced from the center of space; he had seen the Empyrian heaven vanish to be replaced by a void filled only with the wandering dust of worlds. . . . Finally, now he was to be taught that his trail ran backward until in some lost era it faded into the nightworld of the beast"; *The Firmament of Time* (New York: Atheneum, 1960), 141

64. Kranzberg, in introduction to Bugliarello and Doner, op. cit., xxiii.

65. Leiss, op. cit., 82.

66. Krishnan Kumar, *Prophecy and Progress: The Sociology of Industrial and Post-Industrial Society* (Harmondsworth, Eng.: Penguin Books, 1978), 29. On Saint Simon and his ideas, see ibid., 29–44; Manuel and Manuel, op. cit., 581–614; on those of his disciples ibid., 615–640.

67. Winner, op. cit., 140, 141, 174.

68. Kumar, op. cit., 43.

69. Ibid., 37.

70. Manuel and Manuel, op. cit., 654. On Fourier, see ibid., 641–675; and Fourier, *Design for Utopia: Selected Writings of Charles Fourier*, introduction by Charles Gide, foreword by Frank E. Manuel, trans. by Julian Franklin (New York: Schocken Books, 1971).

71. Lee Cameron McDonald, *Western Political Theory: The Modern Age* (New York: Harcourt, Brace and World, 1962), 296.

72. *Socialism Utopian and Scientific*, in Tucker, op. cit., 690.

73. On Kropotkin's ideas, see Manuel and Manuel, op. cit., 739–742; and Nisbet, op. cit., 372–382.

74. Speaking of the anarchist tradition, Nisbet writes: "Joined as it is with one variant or another of the utopian-ecological community, there is nothing surprising in the fact that such pluralism . . . is the closest thing we have to a genuine ideological alternative to Western society as it is presently constituted"; op. cit., 432.

75. Ibid., 448.

5. Americans Conquer Their Continent

1. See Eric Walgren, *The Vikings and America* (New York: Thames and Hudson, 1986).

2. Charles L. Sanford, *The Quest for Paradise: Europe and the American Moral Imagination* (Urbana: University of Illinois Press, 1961), 39.

3. David E. Shi, *The Simple Life: Plain Living and High Thinking in American Culture* (New York: Oxford University Press, 1985), 8.

4. Sydney E. Ahlstrom, "Reflections on Religion, Nature, and the Exploitative Mentality," in Cooper, ed., op. cit., 19.

5. See Sidney E. Mead, *The Nation with the Soul of a Church* (New York: Harper and Row, 1975).

6. John K. Roth, *American Dreams: Meditations on Life in the United States* (San Francisco: Chandler and Sharp, 1976), 4.

7. "The association of the New World with unlimited riches is a commonplace in the history of ideas, but until one realizes how immediate, coarse, and brutal was the response of European greed to the prospect of boundless wealth, one cannot understand how quickly the radiant image became crossed with streaks of night"; Howard Mumford Jones, *O Strange New World. American Culture: The Formative Years* (New York: Viking Press, 1964), 40–41.

8. Michael Kammen, *People of Paradox: An Inquiry Concerning the Origins of American Civilization* (New York: Vintage Books, 1973), 7.

9. Joseph Dorfman, *The Economic Mind in American Civilization, 1606–1825*, vol. 1 (New York: Viking Press, 1946), x.

10. Jones, op. cit., 352.

11. Kammen, op. cit., 17.

12. Roderick Nash, *Wilderness and the American Mind*, rev. ed. (New Haven: Yale University Press, 1973), xi. He goes on to add that "our attitude toward the wilderness is far older and more complex than we usually assume. The main component of that attitude is fear and hatred"; ibid., xii. See also Frederick Turner, *Beyond Geography: The Western Spirit against the Wilderness* (New York: Viking Press, 1980). For the argument that exploitation of the continent started at the outset, see Kirkpatrick Sale, *The Conquest of Paradise: Christopher Columbus and the Columbian Legacy* (New York: Alfred A. Knopf, 1990).

13. Nash, op. cit., 36.

14. Hans Huth, *Nature and the American: Three Centuries of Changing Attitudes* (Berkeley: University of California Press, 1957), 7–8. On Edwards's thought generally, see Vernon L. Parrington, *Main Currents in American Thought: An Interpretation of American Literature from the Beginnings to 1920*, vol. 1, *1620–1800* (New York: Harcourt, Brace, 1927), 148–163.

15. Joseph M. Petulla, *American Environmental History: The Exploitation and Conservation of Natural Resources* (San Francisco: Boyd and Fraser, 1977), 47.

16. Huth, op. cit., 9–10.

17. For aspects of this developing criticism of the traditional state of the Indians, see Ronald L. Meek, *Social Science and the Ignoble Savage* (Cambridge: Cambridge University Press, 1976).

18. Quoted in Daniel J. Boorstin, *The Americans*, vol. 1, *The Colonial Experience* (New York: Vintage Books, 1958), 58.

19. J. Donald Hughes, *American Indian Ecology*, preface by Jamake Highwater (El Paso: Texas Western Press, 1983), 11, 17.

20. See J. Baird Callicott, "Traditional American Indian and Western European Attitudes toward Nature: An Overview," *Environmental Ethics* 4: 310 (1982), and Calvin Martin, *Keepers of the Game: Indian Animal Relationship and the Fur Trade* (Berkeley: University of California Press, 1978). For a traditional Indian view, see John G. Neihardt, ed., *Black Elk Speaks* (Lincoln: University of Nebraska Press, 1932), and for sympathy with that view see Fred Fertig, "Child of Nature, the American Indian as Ecologist," *Sierra Club Bulletin* 55:4–7 (1976).

21. Daniel A. Guthrie, "Primitive Man's Relationship to Nature," *BioScience* 21:722 (1971).

22. Kammen, op. cit., 150.

23. Boorstin, op. cit., 260.

24. Petulla, op. cit., 29, 31.

25. Walt Anderson, *A Place of Power: The American Episode in Human Evolution* (Santa Monica, Calif.: Goodyear, 1976), 106.

26. Earl Finbar Murphy, *Governing Nature* (Chicago: Quadrangle Books, 1967), 5–6.

27. Dorfman, op. cit.

28. Roger Burlingame, *Machines That Built America* (New York: Harcourt, Brace, 1953), 14.

29. Quoted in Elting S. Morison, *From Know-How to Nowhere: The Development of American Technology* (New York: Basic Books, 1974), 22. On American lack of skill in this period, see ibid., 16–39; and Carole Shammas, "How Efficient Was Early America?" *Journal of Social History* 14:3–24 (1980).

30. John W. Oliver, *History of American Technology* (New York: Ronald Press, 1956), 3, 10, 59.

31. Brooke Hindle, *Technology in Early America: Needs and Opportunities for Study* (Chapel Hill: University of North Carolina Press, 1966), 5.

32. Oliver, op. cit., 63.

33. Stuart Bruchey, *The Roots of American Economic Growth, 1607–1861: An Essay in Social Causation* (New York: Harper and Row, 1965), 59.

34. Quoted in Anderson, *A Place of Power: The American Episode in Human Evolution*, op. cit., 114.

35. On the ideology of the Revolution generally, see Clinton Rossiter, *The Political Thought of the American Revolution* (New York: Harcourt, Brace and World, 1963); J.C.A. Pocock, *The Machiavellian Moment: Florentine Political Thought and the Atlantic Republican Tradition* (Princeton: Princeton University Press, 1955); Bernard Bailyn, *The Ideological Origins of the American Revolution* (Cambridge: Harvard University Press, 1967); and Gordon S. Wood, *The Creation of the American Republic, 1776–1787* (Chapel Hill: University of North Carolina Press, 1969).

36. Bailyn, op. cit., 25. "As the political crisis with England deepened, the American republican theorists used such ancient analogies to contrast the pervasive luxury, immorality, and avarice of the British ruling class with an idealized image of an agrarian American republic in the classic tradition"; David E. Shi, *In Search of the Simple Life: American Voices, Past and Present* (Salt Lake City: Gibbs M. Smith, 1986), 79.

37. Wood, op. cit., 58.

38. Shi, *The Simple Life*, op. cit., 56.

39. Wood, op. cit., 110.

40. Shi, *In Search*, op. cit., 85.

41. Wood, op. cit., 118.

42. Quoted in ibid., 97.

43. Wood, op. cit., 571.

44. Shi, *The Simple Life*, op. cit., 77.

45. Wood, op. cit., 606.

46. Benjamin R. Barber, "The Compromised Republic: Public Purposelessness in America," in Robert Horwitz, ed., *The Moral Foundations of the American Republic* (Charlottesville: University Press of Virginia, 1977), 23.

47. John P. Diggins, *The Lost Soul of American Politics: Virtue, Self-Interest, and the Foundations of Liberalism* (New York: Basic Books, 1984), 15, 119.

48. John F. Kasson, *Civilizing the Machine: Technology and Republican Values in America, 1776–1900* (New York: Grossman Publishers, 1976), 7, 10.

49. "Franklin's *Autobiography*, which told the story of the greatest urban intellectual of the period, is a record of civic devotion but not a contentious or carefully crafted celebration of urban culture"; Morton and Lucia White, *The Intellectual Versus the City: From Thomas Jefferson to Frank Lloyd Wright* (New York: Mentor Books, 1964), 18.

50. Sanford, op. cit., 119.

51. Parrington, op. cit., 172.

52. Quoted in Kasson, op. cit., 38.

53. *Notes on the State of Virginia*, ed. and with an introduction by William Peden (Chapel Hill: University of North Carolina Press, 1955), 165.

54. Kasson, op. cit., 19. Another critic notes that "while Jefferson loved nature, it was not as a primitivist. His home was at Monticello, not next to the Natural Bridge. Beyond an attachment to certain practices of American Indians, he found no merit in following the

course of people untouched by civilization. On the contrary, his faith lay in progress through technology, individual freedom, and the stages of development expected even of the Indians"; Charles A. Miller, *Jefferson and Nature: An Interpretation* (Baltimore: Johns Hopkins University Press, 1988), 252.

55. Kasson, op. cit., 25.
56. Quoted in White and White, op. cit., 29. It should be noted that Jefferson distinguished between workshops and machines and in Leo Marx's view did not see technology as an independent agent in society; *The Machine in the Garden: Technology and the Pastoral Ideal in America* (New York: Oxford University Press, 1964), 149.
57. Kasson, op. cit., 35. The significant parts of the report are excerpted in Kenneth Dolbeare, *Directions in American Political Thought* (New York: John Wiley, 1969), 151–165. On Hamilton's ideas, see Parrington, op. cit., 292–307; and David W. Minar, *Ideas and Politics: The American Experience* (Homewood, Ill.: The Dorsey Press, 1964), 151–173. On the early enthusiasm for industrialism see Dorfman, op. cit., 246–257. On Hamilton's views on this question, see ibid., 404–417.
58. Quoted in Kasson, op. cit., 32. On Coxe, see ibid., 28–32, and Marx, op. cit., 155–169.
59. Quoted in Dorfman, op. cit., 255.
60. Burlingame, op. cit., 8, 9.
61. On Jarvis's career, see Morison, op. cit., 40–71.
62. Burlingame, op. cit., 9.
63. Anderson, op. cit., 63. On American industrialization prior to the Civil War, see Brooke Hindle and Steven Lubar, *Engines of Change:. The American Industrial Revolution, 1790–1860* (Washington, D.C.: Smithsonian Institution Press, 1986).
64. Petulla, op. cit., 73.
65. Anderson, op. cit., 79–82.
66. Quoted in Roderick Nash, *Wilderness and the American Mind*, 3d ed. (New Haven: Yale University Press, 1982), 56.
67. Ibid., 65.
68. An expression of this is found in Vernon L. Parrington, *Main Currents in American Thought: An Interpretation of American Literature from the Beginnings to 1920*, vol. 2, *1800–1860: The Romantic Revolution in America* (New York: Harcourt, Brace, 1927), 212–237.
69. Huth, op. cit., 35.
70. Cecelia Tichi, *New World, New Earth: Environmental Reform in American Literature from the Puritans through Whitman* (New Haven: Yale University Press, 1979), 171.
71. One critic speaks of "Cooper's portrayal of the West as a wilderness upon which civilization must be imposed"; Bernard Rosenthal, *City of Nature: Journeys to Nature in the Age of American Romanticism* (Newark: University of Delaware Press, 1980), 40, 96.
72. On Emerson, see Parrington, op. cit., 386–399; White and White, op. cit., 35–45; Ralph Henry Gabriel, *The Course of American Democratic Thought*, 2d ed. (New York: Ronald Press, 1940), 40–48; and Don M. Wolfe, *The Image of Man in America*, 2d ed. (New York: Thomas Y. Crowell, 1970), 70–83.
73. A. J. Beitzinger, *A History of American Political Thought* (New York: Dodd, Mead, 1972), 202.
74. He also believed that the universe was "dynamic, organic, and diversitarian"; Robert H. Walker, "The Universe When He Moults Is Sickly," in Cooper, ed., op. cit., 178.
75. Gabriel, op. cit., 44.
76. Quoted in White and White, op. cit., 40.
77. Quoted in Gabriel, op. cit., 45.

78. Quoted in White and White, op. cit., 39.
79. Shi, *The Simple Life*, op. cit., 130.
80. White and White, op. cit., 210–211.
81. Marx, op. cit., 230, 233.
82. Kasson, op. cit., 117, 123.
83. Marx, op. cit., 234–235.
84. Kasson, op. cit., 125.
85. Rybczynski, op. cit., 60.
86. Quoted in Kasson, op. cit., 130.
87. Joyce Carol Oates, "The Mysterious Mr. Thoreau," *New York Times Book Review* 93:31 (May 1, 1988).
88. Ibid. On Thoreau's ideas, especially those relating to nature and technology, see Parrington, op. cit., 400–413; Gabriel, op. cit., 48–53; and Rybczynski, op. cit., 135–139.
89. Gabriel, op. cit., 49; Nash, op. cit., 86.
90. Quoted in Tichi, op. cit., 164.
91. Quoted in Roth, op. cit., 91.
92. Quoted in Tichi, op. cit., 161.
93. Quoted in Donald E. Worster, *Nature's Economy: The Roots of Ecology* (Garden City, N.Y.: Anchor Books, 1979), 74, 85.
94. Ibid., 103.
95. *Walden, or Life in the Woods* (Boston: Houghton Mifflin, 1906), 242, 244.
96. Rosenthal, op. cit., 88.
97. Tichi, op. cit., 165. Shi agrees; op. cit., 173.
98. Rybczynski writes that "his Yankee practicality precluded a retreat into authentic primitiveness, which was as inconvenient then as it is now. Like a magpie, he scavanged the detritus of the very society of which he was so critical"; op. cit., 137.
99. Quoted in Rosenthal, op. cit., 101.
100. Quotes from Tichi, op. cit., 167.
101. Quotes from ibid., 168.
102. On Bancroft's *History*, see ibid., 188–205.
103. Ibid., 189.
104. See Rosenthal, op. cit., 62–63.
105. Tichi, op. cit., 192–194.
106. Ibid., 197.
107. "Clearly," Tichi writes, "as he imagines the primeval scene, he loathes it"; ibid., 198, 199, 253.
108. Mark Sagoff, *The Economy of the Earth: Philosophy, Law, and the Environment* (New York: Cambridge University Press, 1988), 135. On Cole, see Hargrove, op. cit., 97–98, and Douglas G. Adams, "Environmental Concerns and Ironic Views of American Expansionism Portrayed in Thomas Cole's Paintings," in Joranson and Butigan, op. cit., 296–305.
109. Kammen, op. cit., 257.
110. Ibid., 258.
111. Shi, op. cit., 110.
112. Daniel J. Boorstin, *The Americans*, vol. 2, *The National Experience* (New York: Vintage Books, 1965), 240.
113. Huth, op. cit., 135.
114. Quoted in Nash, op. cit., 101.
115. Quoted in ibid., 41.

116. Quoted in Arthur A. Ekrich, Jr., *Man and Nature in America* (New York: Columbia University Press, 1963), 40, 211.

117. Quoted in Nash, op. cit., 101.

118. Boorstin, op. cit., vol. 2, 273.

119. "The three Republican presidents who followed Jefferson in the White House virtually completed the identification of Hamiltonianism and Jeffersonianism"; Bruchey, op. cit., 122, 135.

120. On public land policies in this era, see Petulla, op. cit., 72–88, and John T. Schlebecker, *Whereby We Thrive: A History of American Farming, 1607–1972* (Ames: Iowa State University Press, 1975), 57–70.

121. Petulla, op. cit., 111, 128.

122. On farm marketing in this period, see Schlebecker, op. cit., 71–85, 130–137.

123. John A. Kouwenhoven, *The Arts in Modern American Civilization* (New York: W. W. Norton, 1967), 41.

124. On this period, see Boorstin, *The Americans: The Colonial Experience*, op. cit., 243–265.

125. Oliver, op. cit., 104–112, 145.

126. Quoted in ibid., 144.

127. Hugo A. Meier, "Technology and Democracy, 1800–1860," *Mississippi Valley Historical Review* 43:618–619 (1957).

128. Howard P. Segal, *Technological Utopianism in American Culture* (Chicago: University of Chicago Press, 1985), 81.

129. Kouwenhoven, op. cit., 63.

130. Meier, op. cit., 633–634.

131. Tichi, op. cit., 163, 164.

132. Kasson, op. cit., 240.

133. Quoted in Meier, op. cit., 626.

134. Quoted in Dorfman, op. cit., 484.

135. Sanford, op. cit., 164.

136. Marx, op. cit., 208.

137. Boorstin, *The Americans*, op. cit., 22.

138. Ibid., 24, 25.

139. On Whitney, see Merritt Roe Smith, "Eli Whitney and the American System of Manufacturing," in Carroll W. Pursell, Jr., *Technology in America: A History of Individuals and Ideas* (Cambridge: MIT Press, 1981), 45–61. Strong dissent to the usual view comes from Robert L. Woodbury, "The Legend of Eli Whitney and Interchangeable Parts," *Technology and Culture* 1:235–253 (1960).

140. Kouwenhoven, op. cit., 36.

141. Burlingame, op. cit., 32.

142. Boorstin, op. cit., vol. 2, 32.

143. Quoted in ibid., 30.

144. Ibid., 33.

145. Ibid.

146. Boorstin, op. cit., vol. 2, 28. On the Lowell project, see Thomas Kiernan, *The Road to Colossus: A Celebration of American Ingenuity* (New York: William Morrow, 1985), 68–73; Kasson, op. cit., 64–106; and Shirley Kolack, "The Impact of Technology on Democratic Values: The Case of Lowell, Massachusetts," *Bulletin of Science, Technology & Society* 8:405–410 (1988).

147. Kasson, op. cit., 69.

148. Ibid., 78.

149. Ibid., 93, 104–105.
150. On agricultural technology during this era, see Petulla, op. cit., 96–101; and Schlebecker, op. cit., 97–109.
151. Boorstin, op. cit., vol. 2, 97–98, 101–102; see also Petulla, op. cit., 102–125.
152. Oliver, op. cit., 180.
153. Boorstin, op. cit., vol. 2, 103–107. On railroads, see also Petulla, op. cit., 125–128. On transportation and agriculture, see Schlebecker, op. cit., 86–97.
154. On mining during this period, see Petulla, op. cit., 149–167.
155. "For many Americans it soon came to mean that wealth was the sole, and in the new republican society, the proper means of distinguishing one person from another"; Gordon S. Wood in Bernard Bailyn, David Bryon Davis, David Herbert Donald, John L. Thomas, Robert H. Wiebe, and Gordon Wood, *The Great Republic: A History of the American People* (Lexington, Mass.: D. C. Heath, 1977), 397.
156. Boorstin, op. cit., vol. 2, 123.
157. Bruchey, op. cit., 206.
158. According to one historian, "Even excluding these oppressed minorities [Indians, Negro slaves, and freed Negroes], one finds many indications that economic inequality increased substantially from 1820 to 1860"; Davis in Bailyn et al., op. cit., 458.
159. Ibid., 459.
160. "The overwhelming majority of unskilled workers remained unskilled workers. It is true that in the 1850s many of the sons of unskilled workers moved into semi-skilled factory jobs. But this generational advance was almost always limited to the next rung on the ladder"; ibid., 460.
161. Ibid., 461.
162. Norman Ware, *The Industrial Worker, 1840–1860: The Reaction of American Industrial Society to the Advance of the Industrial Revolution* (Chicago: Quadrangle Books, 1964), xvi.
163. Davis in Bailyn et al., op. cit.
164. Ibid., 462.
165. Quoted in Bruchey, op. cit., 166.
166. On American technology on the eve of the Civil War, see Morison, op. cit., 87–98.

6. Americans Exploit Their Continent

1. See Paul F. Boller, Jr., *American Thought in Transition: The Impact of Evolutionary Naturalism, 1865–1900* (Chicago: Rand McNally, 1965), 1–46.
2. Donald Worster, *American Environmentalism: The Formative Period, 1860–1915* (New York: John Wiley, 1973), 13.
3. Donald Worster, *Rivers of Empire: Water, Aridity, and the Growth of the American West* (New York: Pantheon Books, 1985), 113.
4. Quoted in Worster, *American Environmentalism*, op. cit., 21.
5. Petulla, op. cit., 221.
6. Huth, op. cit., 168.
7. Nash, op. cit., 168.
8. Ibid., 112.
9. David M. Noble, *The Eternal Adam and the New World Garden: The Central Myth of the American Novel since 1830* (New York: Grosset and Dunlap, 1971), 5.
10. Ibid., 7.
11. Worster, *Rivers*, op. cit., 50.

12. Ibid., 4. On the politics of water and land distribution in the West until the present day, see also Wallace Stegner, *The American West as Living Space* (Ann Arbor: University of Michigan Press, 1987); and Marc Reisner, *Cadillac Desert: The American West and Its Disappearing Water* (New York: Viking Press, 1986).

13. Henry Nash Smith, *Virginland: The American West as Symbol and Myth* (Cambridge: Harvard University Press, 1950), 191.

14. Ibid., 205–206.

15. Shi, op. cit., 201.

16. Petulla, op. cit., 131.

17. Worster, op. cit., 218.

18. On Powell, see Gabriel, op. cit., 178–181; and Worster, op. cit., 132–143.

19. Quoted in Gabriel, op. cit., 180.

20. Quoted in ibid.

21. Henry Nash Smith, op. cit., 186–200.

22. Worster, *American Environmentalism*, op. cit., 58.

23. Worster, *Rivers*, op. cit., 134.

24. Quoted in Anderson, op. cit., 148–149.

25. Worster, *Rivers*, op. cit., 154.

26. See Samuel P. Hays, *Conservation and the Gospel of Efficiency: The Progressive Conservationists* (Cambridge: Harvard University Press, 1959). See also Petulla, op. cit., 218–219.

27. On early conservationists, see Ekrich, op. cit., 81–99.

28. Huth, op. cit., 176. For another view, see John Rodman, "Four Forms of Ecological Consciousness Reconsidered," in Tom Attig and Donald Scherer, eds., *Ethics and the Environment* (Englewood Cliffs, N.J.: Prentice Hall, 1983), 82–92.

29. Attributed by Stephen Fox, "Conservation: Past, Present, and Future," in Neil Evernden, ed., *The Paradox of Environmentalism* (Toronto: York University Press, 1984), 22.

30. Samuel Hays, "The Limits-to-Growth Issue: A Historical Perspective," in Cooper, ed., op. cit., 118.

31. Devall and Sessions, op. cit., 47.

32. "Compared to Thoreau, who cringed at the excess of wildness and idealized the half-cultivated, Muir was wild indeed"; Nash, op. cit., 127. The literature on Muir is vast and growing. See Michael Cohen, *The Pathless Way: John Muir and American Wilderness* (Madison: University of Wisconsin Press, 1984); Stephen Fox, *John Muir and His Legacy: The American Conservation Movement* (Boston: Little, Brown, 1981); Frederick Turner, *Rediscovering America: John Muir in His Time and Ours* (San Francisco: Sierra Club Books, 1988); and Roderick Fraser Nash, *The Rights of Nature: A History of Environmental Ethics* (Madison: University of Wisconsin Press, 1989), 33–54. On his conflict with Pinchot, see Nash, *Wilderness*, op. cit., 129–140.

33. See Michael P. Cohen, *The History of the Sierra Club* (San Francisco: Sierra Club Books, 1989).

34. Shi, op. cit., 196.

35. Huth, op. cit., 151.

36. Worster, op. cit., 185.

37. Quoted in ibid.

38. See Fox, op. cit., 139–147.

39. Quoted in Richard Hofstadter, *The American Political Tradition and the Men Who Made It* (New York: Vintage Books, 1954), 212. On Roosevelt's ideas generally, see ibid., 206–

237, and Hofstadter, *The Age of Reform: From Bryan to F.D.R.* (New York: Vintage Books, 1955), passim.

40. Quoted in Hofstadter, *American Political Tradition*, op. cit., 212.
41. On this era, see Petulla, op. cit., 267–268.
42. Ibid., 277; Huth, op. cit., 186.
43. On Leopold's career, see Fox, op. cit., 244–249; and Curt Meine, *Aldo Leopold: His Life and Work* (Madison: University of Wisconsin Press, 1987). For his ideas, see *A Sand County Almanac and Sketches Here and There*, illustrated by Charles W. Schwartz (New York: Oxford University Press, 1982; first published in 1949, a year after his death). For discussion of his ideas, see J. Baird Callicott, ed., *Companion to a Sand County Almanac: Interpretive and Critical Essays* (Madison: University of Wisconsin Press, 1987); Susan L. Flader, *Thinking Like a Mountain: Aldo Leopold and the Evolution of an Ecological Attitude toward Deer, Wolves, and Forests* (Columbia: University of Missouri Press, 1974; Lincoln: University of Nebraska Press, 1978); Nash, *Rights*, op. cit., 63–73; Ernest Partridge, "Are We Ready for an Ecological Morality?" *Environmental Ethics* 4:175–190 (1982); James D. Heffernan, "The Land Ethic: A Critical Appraisal," *Environmental Ethics* 4:235–247 (1982); and Jon D. Moline, "Aldo Leopold and the Moral Community," *Environmental Ethics* 8:99–120 (1986).
44. Clay Schoenfeld, "Aldo Leopold Remembered," *Audubon* 80:79 (May 1978).
45. On his development, see Flader, op. cit., passim; and Worster, *Nature's Economy*, op. cit., 284, 289.
46. Merchant, op. cit., 252.
47. *A Sand County Almanac*, op. cit., vii.
48. Ibid., viii.
49. Ibid., viii–ix.
50. Ibid., 200, 203.
51. Ibid., 207, 209.
52. Ibid., vii.
53. Huth, op. cit., 111.
54. Nash, *Wilderness*, op. cit., 108.
55. Huth, op. cit., 125.
56. Carl H. Moneyhon, "Environmental Crisis and American Politics," in Bilsky, op. cit., 142.
57. On the conservation policies of Taft and Wilson, see Petulla, op. cit., 278–285.
58. Kasson, op. cit., 142.
59. Ibid., 153. "Machine ornamentation represented an effort by engineers, manufacturers, and American society as a whole to assimilate the machine into republican civilization . . . [and] one even discovers items which make explicit the technologists' boast that machinery constituted the true republican art form"; ibid., 160. On the issue generally, see ibid., 139–180.
60. Ibid., 174.
61. Petulla, op. cit., 174.
62. Siegfried Giedion, *Mechanization Takes Command: A Contribution to Anonymous History* (New York: W. W. Norton, 1969), 38–39. On actual developments in this period, see Petulla, op. cit., 192–201; and Schlebecker, op. cit., 138–205.
63. On the significance of railroads in this era, see Petulla, op. cit., 182–184.
64. Bernard Weisberger, *The New Industrial Society* (New York: John Wiley, 1969), 13.
65. For discussion of business development during this period, see Thomas C. Cochran and William Miller, *The Age of Enterprise: A Social History of Industrial America* (New York:

Harper Torchbooks, 1961), 119–358. On the era generally, see Robert W. Wiebe, *The Search for Order, 1877–1920* (New York: Hill and Wang, 1967); and on the later years, see especially James Weinstein, *The Corporate Ideal in the Liberal State, 1900–1918* (Boston: Beacon Press, 1969).

66. See Daniel T. Rodgers, *The Work Ethic in Industrial America, 1850–1920* (Chicago: University of Chicago Press, 1978), and Herbert G. Gutman, *Work, Culture, and Society in Industrializing America: Essays in American Working-Class and Social History* (New York: Vintage Books, 1977).

67. On Adams's life and ideas see Boller, op. cit., 236–249; White and White, op. cit., 64–82; Wolfe, op. cit., 225–238; T. Jackson Lears, *No Place of Grace: Antimodernism and the Transformation of American Culture, 1880–1920* (New York: Pantheon Books, 1981), 262–297; and Cecelia Tichi, *Shifting Gears* (Chapel Hill: University of North Carolina Press, 1987), 137–168.

68. *The Education of Henry Adams*, introduction by James Truslow Adams (New York: Modern Library, 1931; first published, posthumously, 1918).

69. *Mont-Saint-Michel and Chartres*, introduction by Ralph Adams Cram (Boston: Houghton Mifflin, 1933; first published 1913).

70. Marx, op. cit., 345.

71. Quoted in ibid., 350.

72. See Winner, op. cit., 48–49.

73. Vernon Louis Parrington, *Main Currents in American Thought*, vol. 3, *1860–1920: The Beginnings of Critical Realism in America* (New York: Harcourt, Brace, 1930), 227. On Brooks Adams, see ibid., 227–236; Boller, op. cit., 232–235; Gabriel, op. cit., 243–246, 380–381; and Lears, op. cit., 132–136.

74. There was a wartime edition in 1943 with a long, generally favorable introduction by the noted Progressive historian Charles A. Beard, later reprinted in paperback (New York: Vintage Books, 1955).

75. Quoted in Beitzinger, op. cit., 428.

76. On Twain, see Parrington, op. cit., 86–101; and Kasson, op. cit., 202–215.

77. Parrington, op. cit., 87.

78. Quoted in ibid., 91.

79. Marx, op. cit., 324.

80. Quoted in Kasson, op. cit., 211, from the New York edition of 1889. It does not appear in the definitive edition edited by Bernard L. Stein and published by the University of California Press in 1983.

81. Twain's pessimism was not simply about the past, but also about the American present; ibid., 211.

82. Ibid., 203.

83. On Howells generally, see Parrington, op. cit., 241–253. On his utopianism, see Elisabeth Hansot, *Perfection and Progress: Two Modes of Utopian Thought* (Cambridge: MIT Press, 1974), 169–192; and Kasson, op. cit., 223–232.

84. White and White, op. cit., 102–122.

85. Tichi, *Shifting Gears*, op. cit., 54.

86. On the Shakers and Oneida, see Dolores Hayden, *Seven American Utopias: The Architecture of Communitarian Socialism, 1790–1975* (Cambridge: MIT Press, 1976), passim.

87. On George and his ideas, see Gabriel, op. cit., 208–215; Parrington, op. cit., 124–136; and Stephen J. Ross, "Political Economy for the Masses: Henry George," *Democracy: A Journal of Political Renewal and Social Change* 2:125–134 (July 1982).

88. Petulla, op. cit., 201–204.

89. See Hofstadter, *The Age of Reform*, op. cit., 3–130; Lawrence Goodwyn, *Democratic Promise: The Populist Movement in America* (New York: Oxford University Press, 1976); Norman Pollack, *The Populist Response to Industrial America* (Cambridge: Harvard University Press, 1962); Victor C. Ferkiss, "Populist Influences on American Fascism," *Western Political Quarterly* 10:350–373 (1957); Paul S. Holbo, "Wheat or What? Populism and American Fascism," *Western Political Quarterly* 14:727–736 (1961); Ferkiss, "Populism: Myth, Reality, Current Danger," *Western Political Quarterly* 14:737–740 (1961); and David Montgomery, "On Goodwyn's Populists," *Marxist Perspectives* 1:166–173 (1978).

90. Pollack, op. cit., especially takes this point of view.

91. Weisberger, op. cit., 121.

92. On Bryan, see Hofstadter, *American Political Tradition*, op. cit., 186–205.

93. See David W. Noble, *The Progressive Mind, 1890–1917* (Chicago: Rand McNally, 1970); Russell B. Nye, *Midwestern Progressive Politics: A Historical Study of Its Origin and Development, 1870–1950* (East Lansing: Michigan State University Press, 1951); William O'Neill, *The Progressive Years: America Comes of Age* (New York: Dodd, Mead, 1975); and Robert W. Wiebe, *Businessmen and Reform: A Study of the Progressive Movement* (Cambridge: Harvard University Press, 1967).

94. See Gabriel Kolko, *The Triumph of Conservatism: A Reinterpretation of American History, 1900–1916* (New York: Free Press of Glencoe, 1963), and Weinstein, op. cit.

95. David Dickson, *The New Politics of Science* (New York: Pantheon Books, 1984), 318.

96. Noble, op. cit., 41.

97. Bellamy, *Looking Backward* (Boston: Ticknor, 1888).

98. The literature on Bellamy is vast and conflicting. For an earlier interpretation which sees him as a noble reformer, see Parrington, op. cit., 302–319. But see also Gabriel, op. cit., 220–224; Hansot, op. cit., 113–144; Kasson, op. cit., 190–202; Manuel and Manuel, op. cit., 759–764; and Daphne Patai, ed., *Looking Backward, 1988–1888: Essays on Edward Bellamy* (Amherst: University of Massachusetts Press, 1988).

99. Cecelia Tichi, introduction to Penguin American Library edition (New York, 1982), 22.

100. On the movement, see Tichi, ibid., op. cit., 22–23; and Arthur Lipow, *Authoritarian Socialism in America: Edward Bellamy and the Nationalist Movement* (Berkeley: University of California Press, 1982).

101. The literature on American Marxism in the twentieth century is overwhelming. See, as introduction, Daniel Bell, *Marxist Socialism in the United States* (Princeton: Princeton University Press, 1967); Milton Cantor, *The Divided Left: American Radicalism 1900–1975* (New York: Hill and Wang, 1978); and James Weinstein, *Ambiguous Legacy: The Left in American Politics* (New York: Frederick Watts, 1975).

102. *The Theory of the Leisure Class: An Economic Study of Institutions*, introduction by C. Wright Mills (New York: Mentor Books, 1953).

103. On Veblen's thought, see John P. Diggins, *The Bard of Savagery: Thorstein Veblen and Modern Social Theory* (New York: Seabury Press, 1978); and Wolfe, op. cit., 295–301.

104. Veblen, *The Engineers and the Price System* (New York: B. W. Huebsch, 1921).

105. On Veblen's influence on technocracy, see Henry Elsner, Jr., *The Technocrats: Prophets of Automation* (Syracuse: Syracuse University Press, 1967), 17–135.

106. On the technocratic movement generally, see ibid., passim; William E. Akin, *Technocracy and the American Dream: The Technocratic Movement, 1900–1941* (Berkeley: University of California Press, 1977); and W. Warren Wagar, "The Steel Grey Savior," *Alternative Futures: The Journal of Utopian Studies* 2:38–54 (1979).

107. See Edwin T. Layton, Jr., *The Revolt of the Engineers: Social Responsibility and the American Engineering Profession* (Cleveland: Case Western Reserve University Press, 1971);

Peter Meiksins, "The 'Revolt of the Engineers' Reconsidered," *Technology and Culture* 29:219–246 (1988); and Layton, "American Ideologies of Science and Engineering," *Technology and Culture* 16:48–66 (1975).

108. On the ideology of business during the twenties, see Francis X. Sutton, Seymour E. Harris, Carl Kaysen, and James Tobin, *The American Business Creed* (Cambridge: Harvard University Press, 1956).

109. On the lack of ideology in the New Deal, see Hofstadter, op. cit., 315–352. On its relationship to earlier progressivism see Otis L. Graham, *An Encore for Reform: The Old Progressives and the New Deal* (New York: Oxford University Press, 1967), and Nye, op. cit., 236–270. See also R. Alan Lawson, *The Failure of Independent Liberalism, 1930–1941* (New York: Capricorn Books, 1971).

110. As Roosevelt confidant Robert E. Sherwood noted, the New Deal "was, in fact, as Roosevelt conceived it and conducted it, a revolution of the Right, rising up in its own defense"; quoted in Sidney Fine, *Laissez-Faire and the General Welfare State: A Study in Conflict in American Thought, 1865–1901* (Ann Arbor: University of Michigan Press, 1964), 282. See also Barton J. Bernstein, "The New Deal: The Conservative Achievements of Liberal Reform," in Bernstein, ed., *Towards a New Past: Dissenting Essays in American History* (New York: Vintage Books, 1969), 263–288.

111. See Ferkiss, "Populist Influences," op. cit.; and Alan Brinkley, *Voices of Protest: Huey Long, Father Coughlin, and the Great Depression* (New York: Knopf, 1982).

112. The manifesto has recently been reprinted: *I'll Take My Stand: The South and the Agrarian Tradition*, by twelve Southerners (Baton Rouge: Louisiana State University Press, 1977); original edition, New York: Harper and Bros., 1930). On the Nashville Agrarians, see George T. Nash, *The Conservative Intellectual Movement in America since 1945* (New York: Basic Books, 1979), 38–39, 57–58, 202–206; Robert M. Cruden, ed., *The Superfluous Men: Conservative Critics of American Culture, 1900–1923* (Austin: University of Texas Press, 1976), 161–208; and Ronald Lora, *Conservative Minds in America* (Chicago: Rand McNally, 1971), 107–123.

113. See Donald J. Bush, "Futurama: World's Fair as Utopia," *Alternative Futures: The Journal of Utopian Studies*, 2:3–20 (Fall 1979).

7. Marxist Socialism

1. On Hegel's relation to Marx on this subject, see Isaac D. Balbus, *Marxism and Domination: A Neo-Hegelian, Feminist, Psychoanalytic Theory of Sexual, Political, and Technological Liberation* (Princeton: Princeton University Press, 1982), 11–21.

2. Alfred Schmidt, *The Concept of Nature in Marx* (London: NLB, 1971), 17. His views on Marx are attacked by Marxist geographer Neil Smith in *Uneven Development: Nature, Capital, and the Production of Space* (London: Basil Blackwell, 1984). Smith says "the result of Schmidt's excellent philosophical pedantry is a vision of nature quite opposite to the spirit and practical intent of Marx's later work" (ibid., 23), and speaks of the "specific misreading of Marx that facilitated Schmidt's misconception of nature" (ibid., 25).

3. Schmidt, op. cit., 79.

4. And he means *man*. "Marx's image of the savage who wrestles with nature is not an expression so much of Enlightenment hubris as it is of Victorian arrogance. Woman, as Theodor Adorno and Max Horkheimer observed, has no stake in this conflict. It is strictly between man and nature"; Bookchin, op. cit., 10. See also N. Smith, op. cit., 24–25.

5. *Grundrisse*, cited in Schmidt, op. cit., 122.

6. *German Ideology.* From a translation by Schmidt of portions that do not appear in standard English versions, see ibid., 49.

7. Ibid., 95.

8. Tucker, op. cit., 344–345. On the implications of this for modern industry, see Mike Cooley, *Architect or Bee? The Human/Technology Relationship,* introduction by David Noble, comp. and ed. by Shirley Cooley (Boston: South End Press, 1982).

9. *Capital: A Critique of Political Economy,* vol. 1, *The Process of Capitalist Production,* trans. from the 3d German ed. by Samuel Moore and Edward Eveling, ed. by Friedrich Engels (New York: International Publishers, 1979), 86.

10. Schmidt, op. cit., 102.

11. A contemporary Marxist philosopher calls attention to this; see Howard L. Parsons, editor and compiler, *Marx and Engels on Ecology* (Westport, Conn.: Greenwood Press, 1977), 37, 41.

12. See ibid., passim. That Marx is not as anthropocentric and anti-nature as sometimes thought is argued in Donald C. Lee, "On the Marxian View of the Relationship between Man and Nature," *Environmental Ethics* 2:3–16 (1980). A contrary position is taken by Charles Tolman, "Karl Marx, Alienation, and the Mastery of Nature," *Environmental Ethics* 3:63–74 (1981), and Val Routley, "On Karl Marx as Environmental Hero," *Environmental Ethics* 3:237–244 (1981). Lee's reply is "Toward a Marxian Environmental Ethic," *Environmental Ethics* 4:339–343 (1982). For a critique of their arguments stating that Marx remained committed to human domination of nature, see John P. Clark, "Marx's Organic Body," *Environmental Ethics* 11:243–258 (1989).

13. Jürgen Habermas, *Knowledge and Human Interests,* trans. by Jeremy Shapiro (Boston: Beacon Press, 1971), 32–33.

14. Stanley Aronowitz, *The Crisis in Historical Materialism: Class, Politics, and Culture in Marxist Theory* (New York: Praeger, 1981), 47.

15. Schmidt, op. cit., 155.

16. *Manifesto,* in Tucker, op. cit., 477.

17. See David Mitrany, *Marx against the Peasant: A Study in Social Dogmatism* (Chapel Hill: University of North Carolina Press, 1951).

18. Friedrich Engels, *Dialectics of Nature,* trans. and ed. by Clemens Dutt, preface and notes by J.B.S. Haldane (New York: International Publishers, 1940), 13–14.

19. Tucker, op. cit., 184.

20. Engels, op. cit., 172.

21. *Crisis,* op. cit., 48.

22. Sanford A. Lakoff in Lakoff, ed., *Knowledge and Power: Essays on Science and Government* (New York: Free Press, 1966), 52. See also Mark Warren, "The Marx-Darwin Question: Implications for the Critical Aspects of Marx's Social Theory," *International Sociology* 2:251–269 (1987).

23. Quoted in Melvin J. Lasky, *Utopia and Revolution: On the Origins of a Metaphor, or Some Illustrations of the Problem of Political Temperament and Intellectual Climate and How Ideas, Ideals, and Ideologies Have Been Historically Related* (Chicago: University of Chicago Press, 1976), 486.

24. See Parsons, op. cit., 84; Balbus, op. cit., 126–128, 362–367.

25. Ophuls, *Ecology,* op. cit., 223.

26. Engels, op. cit., 290–291.

27. Nisbet, op. cit., 29.

28. Sibley, op. cit., 172.

29. See the thoughtful critique of Albert Borgmann, *Technology and the Character of Con-*

temporary Life: A Philosophical Inquiry (Chicago: University of Chicago Press, 1984), 81–85.

30. Langdon Winner, *The Whale and the Reactor: A Search for Limits in an Age of High Technology* (Chicago: University of Chicago Press, 1986), 57. See also Victor C. Ferkiss, *The Future of Technological Civilization* (New York: George Braziller, 1974), 67–71, 76–77.

31. Denis de Rougement, *The Future Is within Us*, trans. by Anthony J. C. Kerr (New York: Pergamon Press, 1983), 16.

32. For an interesting if flawed discussion, see Vernon Venable, *Human Nature: The Marxian View* (Cleveland: World Publishing, 1966), passim.

33. *Manifesto*, in Tucker, op. cit., 475–477.

34. See, for example, Peter Dale Scott, "Peace, Power, and Revolution: Peace Studies, Marxism, and the Academy," *Alternatives: A Journal of World Policy* 11:354–355 (Winter 1983–1984).

35. Bookchin, op. cit., 313.

36. This issue is discussed in Daniel Little, *The Scientific Marx* (Minneapolis: University of Minnesota Press, 1986), 49–51. Little's view is that *Capital* clearly does not embody technological determinism; ibid., 50. Melvin Kranzberg, however, holds that "the Marxists have been the chief proponents of a rigorous technological determinism"; introduction to Bugliarello, op. cit., xxiii.

37. Nisbet, op. cit., 356–357.

38. "On Authority," in Tucker, op. cit., 663 (emphasis added).

39. See his *Socialism* (New York: Saturday Review Press, 1972), and *The Twilight of Capitalism* (New York: Simon and Schuster, 1976).

40. See their *Monopoly Capitalism: An Essay on the American Social and Economic Order* (New York: Monthly Review Press, 1966). On recent American Marxism, see also Kenneth M. Dolbeare, Patricia Dolbeare, and Jane Hadley, *American Ideologies: The Competing Political Beliefs of the 1970s* (Chicago: Rand McNally, 1973), 217–254.

41. "The Chief Task of Our Day," in Robert C. Tucker, *The Lenin Anthology* (New York: W. W. Norton, 1975), 436.

42. "Immediate Tasks of the Soviet Government," ibid., 448–449.

43. See his remarks in ibid., op. cit., 345, 401, 402, 430.

44. *Literature and Revolution* (New York: International Publishers, 1925), 251–256.

45. See Kendall Bailes, "The Politics of Technology: Stalin and Technocratic Thinking among Soviet Engineers," *American Historical Review* 79:445–469 (1974).

46. For an early dissident view of ecological problems (among others) by a distinguished Soviet scientist (long banned in the Soviet Union), see Andrei P. Sakharov, *Progress, Coexistence, and Intellectual Freedom* (New York: W. W. Norton, 1968). For early surveys of the approach of Soviet society to ecological problems, see Marshall Goldman, *The Spoils of Progress* (Cambridge: MIT Press, 1972), and Philip R. Pryde, *Conservation in the Soviet Union* (Cambridge: Cambridge University Press, 1972). For a discussion within a Marxist context, see William Mandel, "The Soviet Ecology Movement," *Science and Society* 36:385–416 (1972).

47. Ilya Novik, *Society and Nature: Socio-Ecological Problems*, trans. by H. Campbell Creighton (Moscow: Progress Publishers, 1981), 24.

48. Peter Matthiessen, "The Blue Pearl of Siberia," *New York Review of Books* 33:37–47 (February 16, 1991), passim.

49. Alan Cooperman, "Nuke Debate Rekindled," *Albuquerque Journal*, October 13, 1991. For recent studies of Soviet policy, see Charles E. Ziegler, *Environmental Policy in the USSR* (Amherst: University of Massachusetts Press, 1987); Fred Singleton, ed., *Environ-*

mental Problems in the Soviet Union and Eastern Europe (Boulder, Colo.: Lynne Reinner Publishers, 1987); and Douglas R. Weiner, *Models of Nature: Ecology, Conservation, and Cultural Revolution in Soviet Russia* (Bloomington: Indiana University Press, 1988). See also Bill Keller, "Developers Turn Aral Sea into a Catastrophe," *New York Times*, December 20, 1988; Marjorie Sun, "Environmental Awakening in the Soviet Union," *Science* 241:1033–1035 (1988); Mary Battiata, "Eastern Europe Faces Vast Environmental Blight," *Washington Post*, March 20, 1990; Jeremy Cherfas, "East Germany Struggles to Clean Up Its Air and Water," *Science* 248:295–296 (1990); and Hilary F. French, "Restoring the East European and Soviet Environments," in Lester F. Brown et al., eds., *State of the World 1991: A Worldwatch Institute Report on Progress toward a Sustainable Society* (New York: W. W. Norton, 1991), 93–112.

50. For a history by a dissident scientist, see Zhores A. Medvedev, *Soviet Science* (New York: W. W. Norton, 1978).

51. Ophuls, *Ecology*, op. cit., 205.

52. Carl Mitcham, "Philosophy of Technology," in Durbin, op. cit., 300.

53. See Radovan Richta et al., *Civilization at the Crossroads: Social and Human Implications of the Scientific and Techological Revolution*, rev. ed. (Prague: International Arts and Science Press, 1969).

54. On the STR, see Mitcham, op. cit., 300–304; Victor C. Ferkiss, "Daniel Bell's Concept of Post-Industrial Society: Theory, Myth, and Ideology," *Political Science Reviewer* 9:87–100 (1979); Julian M. Cooper, "The Scientific and Technical Revolution in Soviet Theory," in Frederick J. Fleron, ed., *Technology and Communist Culture: The Socio-cultural Impact of Technology under Socialism* (New York: Praeger, 1977), 146–179; Eric P. Hoffman, "Soviet Views of 'The Scientific-Technological Revolution,' " *World Politics* 30:615–644 (1978), and "The Scientific Management of Society," *Problems of Communism* 26:59–67 (May–June 1977); Robin Laud, "Post-Industrial Society: East and West," *Survey* 21:1–17 (1975); and Jan F. Triska and Paul M. Cocks, *Political Development in Eastern Europe* (New York: Praeger, 1977), 54–72.

55. Quoted in E. Arab-Ogly, *In the Forecasters' Maze* (Moscow: Progress Publishers, 1975), 231. For a Soviet view of the STR, see P. N. Fedoseev, "The Social Significance of the Scientific and Technological Revolution," *International Social Science Journal* 27:152–162 (1975).

56. On the actual relationship between technology and politics in the former USSR, see S. Frederick Starr, "Technology and Freedom in the Soviet Union," in Lewis H. Lapham, ed., *High Technology and Human Freedom* (Washington, D.C.: Smithsonian Institution Press, 1985), 107–119.

57. For criticism from a Marxist perspective, see Smith, op. cit., 28–31. On the project itself, see Richard King, *The Party of Eros: Radical Social Thought and the Realm of Freedom* (Chapel Hill: University of North Carolina Press, 1972).

58. On Marcuse's concept of "repressive technology," see Balbus, op. cit., 234–257.

59. *One-Dimensional Man* (Boston: Beacon Press, 1969), xii. See also his *Eros and Civilization* (Boston: Beacon Press, 1955), *An Essay on Liberation* (Boston: Beacon Press, 1969), and "Some Implications of Modern Technology," *Studies in Philosophy and Social Science* 9:414–439 (1941). On Marcuse, see C. Fred Alford, *Science and the Revenge of Nature: Marcuse and Habermas* (Gainesville: University Presses of Florida, 1985); Peter Clecak, *Radical Paradoxes: Dilemmas of the American Left, 1945–1970* (New York: Harper Torchbooks, 1974), 175–229; Alisdair MacIntyre, *Herbert Marcuse: An Exposition and a Polemic* (New York: Viking, 1970); George Kateb, "The Political Thought of Herbert Marcuse,"

Commentary 49:57 (January 1970); and John Norr, "German Social Theory and the Hidden Face of Technology," *European Journal of Sociology* 15:323–336 (1974).

60. "Socialist Humanism?" in Erich Fromm, ed., *Socialist Humanism: An International Symposium* (Garden City, N.Y.: Anchor Books, 1966), 116.
61. Alford, op. cit., 4.
62. "Socialist Humanism," op. cit., 115.
63. Alford, op. cit., 6.
64. "Habermas . . . assumes that the basic structure of science is given by the objective character of human labor. . . . As long as nature does not change radically—and Habermas does not expect it to—the basic structure of science cannot change either"; Alford, op. cit., 169.
65. Ibid., 6.
66. Aronowitz, op. cit., 37.
67. Habermas, "A Reply to My Critics," in John B. Thompson, ed., *Habermas: Critical Essays* (Cambridge: MIT Press, 1982), 243–245.
68. *Knowledge and Human Interests*, trans. by Jeremy Shapiro (Boston: Beacon Press, 1971), 32–33.
69. "The Problem of Knowledge in Habermas," *Telos* 40:66 (Summer 1979).
70. On communication ethics, see his *Communication and the Evolution of Society*, trans. by Thomas McCarthy (Boston: Beacon Press, 1979), and *Legitimation Crisis*, trans. by Thomas McCarthy (Boston: Beacon Press, 1971). See also Oskar Negt, "Mass Media: Tools of Domination or Instruments of Emancipation? Aspects of the Frankfurt School's Communications Analysis," in Kathleen Woodward, ed., *The Myths of Information: Technology and Postindustrial Culture* (Madison, Wisc.: Coda Press, 1980), 65–87.
71. Whitebook: "The Problem," op. cit., 53.
72. Mark Poston, *Existential Marxism in Postwar France* (Princeton: Princeton University Press, 1975), 365.
73. Aronowitz, op. cit., 54. For Gorz's views, see "Technical Intelligence and the Capitalist Division of Labor," *Telos* 12: 27–41 (Spring 1972); and his *A Strategy for Labor: A Radical Proposal*, trans. by Martin A. Nicholas and Victoria Ortiz (Boston: Beacon Press, 1967); *Farewell to the Working Class: An Essay on Post-Industrial Socialism*, trans. by Michael Sonensher (Boston: Beacon Press, 1982); and *Paths to Paradise: On the Liberation from Work*, trans. by M. Imrie (London: Pluto Press, 1985).
74. Aronowitz, op. cit.
75. Poster, op. cit., 366–367.
76. *The Post-Industrial Society. Tomorrow's Social History: Classes, Conflicts, and Cultures in the Programmed Society* (New York: Random House, 1971).
77. Poster, op. cit., 363.
78. Ibid., 368, 369.
79. One exception among Communist party theorists was Roger Garaudy; ibid., 367.
80. Trans. by Patsy Vigderman and Jonathan Cloud (Boston: South End Press, 1980; orig. pub. in France, 1975). For a scathing critique, see Bookchin, *Toward an Ecological Society* (Montreal: Black Rose Books, 1980), 289–313. For a Marxist perspective, see Boris Frankel, *The Post-Industrial Utopians* (London: Polity Press, 1987), passim.
81. Bookchin, op. cit., 196.

8. Nature and Technology in Islam

1. *A Study of History*, vol. 12, *Reconsiderations* (London: Oxford University Press, 1961), 512.

2. *Machina Ex Deo*, op. cit., 85. See also Hilaire Belloc, *The Great Heresies* (New York: Sheed and Ward, 1935), 73–140.

3. One says, specifically referring to the contentions of White and Toynbee: "Another remarkable feature of the environmental controversy over the 'monotheistic debasement of nature' was that Islam—as usual—hardly figured in this discussion as if it were a religion on the moon and the living reality of one billion Muslims merely a statistical illusion. It was taken for granted that 'Islam, like Marxism, is a Judeo-Christian heresy,' and as such, it had hardly anything original to contribute"; S. Parvez Mansoor, "Environment and values: The Islamic perspective," in Ziaduddin Sardar, ed., *The Touch of Midas: Science, Values, and Environment in Islam and the West* (Manchester: Manchester University Press, 1984), 153.

4. On Islam, see Alfred Guillaume, *Islam* (Baltimore: Penguin Books, 1954); H.A.R. Gibb, *Mohammedanism: A Historical Survey* (New York: Mentor Books, 1955); and Mohammed Marmaduke Pickthall, *The Meaning of the Glorious Koran: An Explanatory Translation* (New York: Mentor Books, 1953).

5. Ziauddin Sardar, *Islamic Futures: The Shape of Ideas to Come* (London: Mansell Publishing Ltd., 1985), 173.

6. For an interesting perspective on this question, see Janet Abu-Lughod, *Before European History: The World System, 1250 A.D.–1350 A.D.* (New York: Oxford University Press, 1989).

7. White, *Medieval Religion*, op. cit., 77.

8. Sardar, introduction to *The Touch of Midas*, op. cit., 4.

9. Ismael al Faraqi, "Science and Traditional Values in Islamic Society," *Zygon: Journal of Religion and Science* 2:233, 235 (1967).

10. See Glyn Ford, "Rebirth of Islamic Science," in ibid., 26–39.

11. Giorgio de Santillana, preface to Seyyed Nossien Nasr, *Science and Civilization in Islam* (Cambridge: Harvard University Press, 1968), x.

12. Mansoor in Sardar, op. cit., 156–157. On stewardship in Islam, see also Black, *Dominion*, op. cit., 51. On a variant Muslim tradition, see Nasr, "The Ecological Problem in the Light of Sufism," in Jacob Needleman, A. K. Bierman, and James A. Gould, eds., *Religion for a New Generation*, 2d ed. (New York: Macmillan, 1977), 254–263.

13. Manzoor in Sardar, op. cit., 161. Quote is from Sir Mohammed Iqbal, *The Reconstruction of Religious Thought in Islam* (London: Routledge and Kegan Paul, 1934), 147.

14. Nasr, op. cit., 30.

15. Fazlur Rahman, *Islam and Modernity: Transformation of an Intellectual Tradition* (Chicago: University of Chicago Press, 1982), 34.

16. Ibid., 135. White is more explicit, saying that "the Arabic-speaking civilization knew what science was and was proficient in it. For four hundred years science was one of its major concerns. But a crystallization of other values occurred in the late eleventh century which shifted the whole focus of Islamic culture. Science was abandoned, and abandoned deliberately"; *Machina Ex Deo*, op. cit., 99.

17. See Saibo Mohamed Mauroof, "Elements for an Islamic Anthropology," in Professor Isma'il R. Al-Faruqi and Dr. Abdullah Omar Nasseef, eds., *Social and Natural Sciences: An Islamic Perspective* (Jeddah: King Abdulaziz University and Hodder and Stoughton, 1981), 136.

18. *Unbound Prometheus*, op. cit., 28.

19. *Islam: Essays in the Nature and Growth of a Cultural Tradition*, 2d ed. (London: Routledge and Kegan Paul, 1961), 114.

20. Needham, *The Grand Titration: Science and Society in East and West* (Toronto: University of Toronto Press, 1969), 189.
21. McNeill, *Rise of the West*, op. cit., 502–503.
22. Ibid., 503.
23. Ibid., 530n.
24. Gideon, op. cit., 34. For a historical overview, see Ahmed al-Hassan and Donald Hill, *Islamic Technology* (New York: Cambridge University Press, 1987).
25. Sardar, *Islamic Futures*, op. cit., 193.
26. White, *Machina Ex Deo*, op. cit., 126.
27. Rybczynski, op. cit., 177–180, 183–184.
28. McNeill, *The Pursuit of Power*, op. cit., 380, 774.
29. Rahman, op. cit., 50–51.
30. Nasseef, introduction to part 2 of Al-Faruqi and Nasseef, op. cit., 146.
31. Rahman, op. cit., 51, 131.
32. Ibid., 49.
33. Sardar, op. cit., 188,
34. Sardar, introduction to Sardar, *The Touch of Midas*, op. cit., 9.
35. Lloyd Timberlake, "The Emergence of Environmental Awareness in the West," in Sardar, op. cit., 132.
36. See generally Mansoor in ibid., 150–169.
37. Sardar, *Islamic Futures*, op. cit., 219, 220, 227.

9. Nazism

1. See George L. Mosse, *The Crisis of German Ideology: Intellectual Origins of the Third Reich* (New York: Grosset and Dunlap, 1964).
2. Peter Viereck, *Metapolitics: The Roots of the Nazi Mind* (New York: Capricorn Books, 1961), 114–120.
3. Hitler was close to the circle of Wagner's admirers and was once expected to marry the composer's son's widow; ibid., 132.
4. See Daniel Gasman, *The Scientific Origins of National Socialism* (London: Macmillan, 1971).
5. See A. James Gregor, *The Ideology of Fascism* (New York: Free Press, 1939), and *The Fascist Persuasion in Radical Politics* (Princeton: Princeton University Press, 1974).
6. Jeffrey Herf, *Reactionary Modernism: Technology, Culture, and Politics in Weimar and the Third Reich* (Cambridge: Cambridge University Press, 1984), 47.
7. Robert A. Pois, *National Socialism and the Religion of Nature* (New York: St. Martin's Press, 1968), 27. For commentary on Pois's views see the review essay, "Nazis as Nature Mystics?" by John Stroup, *This World: A Journal of Religion and Public Life* 23: 97–114 (Fall 1988).
8. On the concept of the *Volk*, see Mosse, op. cit., 13–51; and Viereck, op. cit., passim.
9. Herf, op. cit., 29.
10. Ibid., 18–19.
11. Ibid., 42. Mosse speaks of *Die Tat* as "at that time one of the principal organs of romantic-Volkish thought"; op. cit., 17.
12. Herf, op. cit., 45. Schmitt was to call himself "the Rousseau of the Nazi revolution and then repudiate it in 1945"; Viereck, op. cit, x.
13. Herf, op. cit., 120–121.
14. Mosse, op. cit., 283, 285.

15. Note that the party with the highest percentage of adherents among Germans with Ph.D. degrees, even before Hitler's accession to power, was the Nazis.

16. Herf, op. cit., 45–46. The book has been translated and edited by Karl F. Geiser as *A New Social Philosophy* (Princeton: Princeton University Press, 1937).

17. Herf, op. cit., 133, 146, 150. On Sombart and Nazism, see also Mosse, op. cit., 141–142.

18. German edition, 1818–1922. Viereck notes that the *Decline* was "one of the few 'heavy' books we know Hitler has read"; op. cit., 10.

19. Spengler, *The Decline of the West*. Abridged by Helmut Werner. English abridged edition prepared by Arthur Helps. Translated by Charles Francis Atherton (New York: Alfred A. Knopf, 1962), 413.

20. Herf, op. cit., 38, 55.

21. Herf says he "rejected political liberalism and endorsed technical rationality linked to the wilful self that knew no limits to its own celebration." Ibid.

22. Ibid., 24, 29, 68.

23. Ibid., 70, 95.

24. Ibid., 105, 107.

25. Quoted in ibid., 32, 33.

26. Ibid., 152.

27. Alex de Jonge, *The Weimar Chronicle: Prelude to Hitler* (New York: New American Library, 1978), 174.

28. On the Bauhaus, see Rybczynski, op. cit., 146–153; and Otto Friedrich, *Before the Deluge: A Portrait of Berlin in the 1920s* (New York: Avon Books, 1972), 185–188, 419–420.

29. See Mosse, op. cit., 16–19, 171–189.

30. Actually, the swastika, an ancient sun symbol, was popular among a wide variety of groups in Germany, especially the young. See Mosse, op. cit., 71, 77, 229, 267.

31. De Jonge, op. cit., 173.

32. Ibid, 176. See also Mosse, op. cit., 58–60, 71–72.

33. Ibid., 177. On Ludendorff, see also Viereck, op. cit., 296–298.

34. Winner, op. cit., 15.

35. Rybczynski, op. cit., 18.

36. Quoted in ibid., 19.

37. Ibid. Speer noted that the best-made German product of the time, the BMW engine, was actually produced on an assembly line "manned by unskilled Russian prisoners of war"; ibid.

38. Pacey, op. cit., 93. On the Nazi use of technology in general, see Landes, *Unbound Prometheus*, op. cit., 402–419.

39. See Alan Beyerchen, *Scientists under Hitler: Politics and the Physics Community in the Third Reich* (New Haven: Yale University Press, 1977).

40. The underlying tension in German society was illustrated by the lives of the inventor Rudolph Diesel and his son Eugene. Originally hopeful about technology, Rudolph came to despair of it and committed suicide. Eugene turned from technology to literature; see Donald E. Thomas, Jr., *Deisel: Technology and Society in Industrial Germany* (Tuscaloosa: University of Alabama Press, 1987), and Thomas, "Diesel, Father and Son, Social Philosophers of Technology," *Technology and Culture* 19:376–393 (1978).

41. See Walter Bluhm, *Theories of the Political System: Classics in Political Thought and Modern Political Analysis*. 3d ed. (Englewood Cliffs, N.J.: Prentice Hall, 1977), 493.

42. See especially Mosse, op. cit., 306–307.

43. Pois, op. cit, 3, 11. Even before Hitler's accession to power, a segment of the Nazi party

had been specially interested in nature, above all R. Walther Darre, the minister of agriculture from 1933 to 1942; see Anne Bramwell, *Blood and Soil: Richard Walther Darre and Hitler's 'Green Party'* (Bourne, U.K.: Kensal Press, 1985).

44. Pois, op. cit., 10.
45. Ibid., 41.
46. Ibid.
47. Ibid.
48. Ibid., 42.
49. Quoted by Herman Rauschning, *Hitler Speaks: A Series of Political Conversations with Adolf Hitler on His Real Aims* (London: Thornton and Butterworth, 1939), 242.
50. Quoted in Pois, op. cit., 64.
51. For contrasting views on Nazism and religion, see James M. Rhodes, *The Hitler Movement: A Modern Millennial Revolution* (Stanford: Stanford University Press, 1980), and Jean-Michel Angebert, *The Occult and the Third Reich: The Mystical Origins of Nazism and the Search for the Holy Grail*, trans. by Lewis A. M. Sumberg (New York: McGraw-Hill, 1974).
52. Rauschning, op.cit., 114.
53. Quoted in Pois, op.cit., 44.
54. Pois says of the Nazis that, despite disputes, "leading National Socialists *assumed* a fundamentally anti-Christian . . . stance" and were united in the belief that "National Socialism represented a new *Lebensphilosophie . . .* which . . . made it possible for science and religion to be brought together and the cleft between heaven and earth bridged"; ibid.
55. Quoted in ibid., 117.
56. Ibid., 110.

10. The Orient, Nature, and Technology

1. Frederick J. Taggert, *Rome and China: A Study of Correlations in Historical Events* (Berkeley: University of California Press, 1937).
2. Eliot Deutsch, "A Metaphysical Grounding for Natural Reverence: East-West," *Environmental Ethics* 8:296 (1986). See also J. Baird Callicott and Roger T. Ames, eds., *The Nature of Nature in Asian Traditions of Thought* (Albany: State University of New York Press, 1989); and Callicott, "Conceptual Resources for Environmental Ethics in Asian Traditions of Thought: A Propaedeutic," *Philosophy East and West* 37:115–130 (1987).
3. *The Meeting of East and West: An Inquiry Concerning World Understanding* (New York: Macmillan, 1946), 313, 375.
4. Thomas Welty, *The Asians: Their Heritage and Their Destiny* (Philadelphia: J. B. Lippincott, 1963), 58–59.
5. Lucille Schulberg, et al. *Historic India* (New York: Time-Life Books, 1968), 35.
6. On original Aryan religion, see ibid., 36–40.
7. John M. Koller, *Oriental Philosophies* (New York: Charles Scribner's Sons, 1970), 9.
8. Schulberg, op. cit., 114.
9. Alan Watts, *The Way of Zen* (New York: Vintage Books, 1957), 32.
10. This motif is masterfully dramatized in the last novel of Aldous Huxley, written while he was taking psychedelic drugs to ease the pain of illness; see *Island* (New York: Harper and Row, 1962).
11. See Nash, op. cit., 20
12. Welty, op. cit., 28–29.
13. On Buddhism generally, see Abraham Kaplan, *The New World of Philosophy* (New York:

Vintage Books, 1963), 237–266; on Buddhism in India, see E.W.F. Tomlin, *The Oriental Philosophers: An Introduction* (New York: Harper Colophon Books, 1963), 191–225; Welty, op. cit., 71–77; Watts, op. cit., 29–76; and Schulberg, op. cit., 57–70.

14. See Watts, op. cit., 43–44.
15. Koller, op. cit., 123.
16. See *A Guide for the Perplexed* (New York: Harper and Row, 1977).
17. "The central problem of Buddhism is overcoming suffering." Tomlin, op. cit., 4.
18. Koller, op. cit., 112.
19. Watts, op. cit., 47.
20. Ibid., 52.
21. See Kenneth K. Imada, "Environmental Problematics in the Buddhist Context," *Philosophy East and West* 37:135–149 (1987).
22. Daniel J. Kaluphana, "Man and Nature: Toward a Middle Path of Survival," *Environmental Ethics* 8:376 (1986).
23. So at least says a Western theologian. Carmody, op. cit., 126–127.
24. Koller, op. cit., 191.
25. Kaplan, op. cit., 202.
26. Schulberg, op. cit., 94.
27. Ibid., 97–98.
28. On this early Indian civilization, see McNeill, *Rise of the West*, op. cit., 170–188.
29. Pacey, *Maze*, op. cit., 187.
30. Murphy, op. cit., 16.
31. Karl Marx remarked on this as a special kind of economic development, "the Asiatic mode of production," not seen in the West. See Tucker, op. cit., 653–658, and Wittfogel, op. cit., passim.
32. Tucker, op. cit., 657–659.
33. On Gandhi's ideas, see Nirmal Kumar Bose, "Gandhi: Humanist and Socialist," in Fromm, op. cit., 98–106.
34. See F. A. Long, "Science, Technology and Industrial Development in India," *Technology in Society: An International Journal* 10:395–416 (1988).
35. See Taggert, op. cit., passim.
36. Wolfram Eberhard, *A History of China* (Berkeley: University of California Press, 1977), 224–229.
37. On pre-Confucian literature and ideas in China, see Wm. Theodore de Bary, Wing-tsit Chan, and Burton Watson, *Sources of Chinese Tradition*, vol. 1 (New York: Columbia University Press, 1960), 1–14. On the controversy over the historicity of Lao-tsu, see Watts, op. cit., 13.
38. Needham, op. cit., 218.
39. Welty, op. cit., 144.
40. Northrop, op. cit., 322. He discusses Buddhism, Taoism, and Confucianism in ibid., 322–358.
41. Watts, op. cit., 13. See *The I Ching*, trans. by James Legge, 2d ed. (New York: Dover, 1963). For a contrast between Western and Chinese culture based on the *I Ching*, see Capra, op. cit., 33–40.
42. Koller, op. cit., 200.
43. Watts, op. cit., 10.
44. On Confucianism, see Edward H. Shafer, *Ancient China* (New York: Time-Life Books, 1967), 60–61, 64–67, 82–84; Kaplan, op. cit., 275–287; and Koller, op. cit., 199–225.
45. Koller, op. cit., 199, 255.

46. Worster, op. cit., 46.
47. Kaplan, op. cit., 268.
48. Etienne Balasz, *Chinese Civilization and Bureaucracy: Variations on a Theme* (New Haven: Yale University Press, 1964), 22. A modern specialist on China holds that "on the whole the Confucian literati were consistently uninterested, and the intellectual affinities of science were mainly Taoist and unorthodox." Josoph Levinson, *Confucian China and Its Modern Fate*, vol. 1 (Berkeley: University of California Press, 1968), 13.
49. On Taoism, see Koller, op. cit., 231–248; Needham, op. cit., 158–170; and Schafer, op. cit., 62–66. On Taoist attitudes toward nature, see Roger T. Ames, "Taoism and the Nature of Nature," *Environmental Ethics* 8:317–350 (1986); Chgung-ying Cheng, "On the Environmental Ethics of the Tao and the Ch'i," *Environmental Ethics* 8:351–370 (1986); Russell Goodman, "Taoism and Ecology," *Environmental Ethics* 2:73–80 (1980); Po-Kueng Io, "Taoism and the Foundations of Environmental Ethics," *Environmental Ethics* 5: 335–343 (1983); Dolores La Chapelle, "Sacred Land, Sacred Sex," in Tobias, op. cit., 102–121; and Huston Smith, "Tao Now: An Ecological Testament," in Ian G. Barbour, ed., *Earth Might Be Fair: Reflections on Ethics, Religion, and Ecology* (Englewood Cliffs, N.J.: Prentice-Hall, 1972), 62–81.
50. Watts, op. cit., 25–27.
51. "The important difference between the Tao and the usual idea of God is that whereas God produces the world by making (*wei*), the Tao produces it by 'non-making' (*wu-wei*) which is approximately what we mean by 'growing.' . . . Because the natural universe works mainly according to the principles of growth, it would seem quite odd to the Chinese mind to ask how it was made." Ibid., 16–17.
52. Schafer, op. cit., 57–59.
53. Ibid., 105.
54. Ibid., 110, 146, 150.
55. Nash, op. cit., 21.
56. Kaplan, op. cit., 277.
57. Needham, op. cit., 158.
58. Worster, op. cit.
59. "Geopiety: A Theme in Man's Attachment to Nature and to Place," in David Lowenthal and Martin J. Bowden, eds., *Geographies of the Mind: Essays in Historical Geography in Honor of John Kirkland Wright* (New York: Oxford University Press, 1976), 11–39.
60. Ibid., 23.
61. Nash, op. cit., 20.
62. Pacey, op. cit., 287.
63. Watts has written, says Needham, "a brilliant, and, in other respects, a very perceptive book"; *The Grand Titration*, op. cit., 119. See also Needham's magisterial *Science and Civilization in China*, 5 vols. to date (Cambridge: Cambridge University Press, 1962–1987). On why China never experienced the industrial revolution, see as well Derek Bodde, *Chinese Thought, Society, and Science: The Intellectual and Social Background of Science and Technology in Pre-modern China* (Honolulu: University of Hawaii Press, 1991).
64. Mark Elvin, *The Pattern of the Chinese Past* (Stanford: Stanford University Press, 1973), 185.
65. Schafer, op. cit., 87.
66. Needham, op. cit., 85. Landes is less inclined to do so. See his discussion of Chinese clocks in *Revolution in Time*, op. cit., 17–36. On the paradoxical role of Confucian scholars in engineering see, Joseph Needham, Wang Ling, and Derek J. Price, *Heavenly*

Clockwork: The Great Astronomical Clocks of Medieval China (Cambridge: Cambridge University Press, 1960).

67. Needham, *Grand Titration*, op. cit., 33–34.

68. On traditional Chinese medicine and its philosophical aspects, see Capra, *The Turning Point*, op. cit., 312–320.

69. Needham, op. cit., 58–59.

70. Ibid., 17, 32, 35–36, 249.

71. Ibid., 249.

72. "It is extremely interesting that modern science, in so far as since the time of Laplace it has been found possible and even desirable to dispense completely with the hypothesis of a God as the basis of the Laws of Nature, has returned, in a sense, to the Taoist outlook. This is what accounts for the strangely modern ring in so much of the writings of that great school. But historically the question remains whether natural science could have reached its present stage of development without passing through a 'theological' stage"; ibid., 330. On the congruence between ancient Chinese and modern thought on physics, see Fritjof Capra, *The Tao of Physics: An Exploration of the Parallels between Modern Physics and Eastern Mysticism* (Boulder, Colo.: Shambala, 1975); and Gary Zukav, *The Dancing Wu Li Masters: An Overview of the New Physics* (New York: William Morrow, 1979).

73. Needham, op. cit., 39–40, 151, 153.

74. Volti, *Technology, Politics, and Society in China* (Boulder, Colo.: Westview Press, 1982), 15, 20, 23.

75. Levenson, op. cit., 8, 12.

76. Elvin, op. cit., 179–199.

77. Pacey, op. cit., 188, 288.

78. Levenson, op. cit., 62.

79. On modernization, see B. C. Hacker, "The Weapons of the West: Military Technology and Modernization in 19th-Century China and Japan," *Technology and Culture* 18:43–55 (1977).

80. Volti, op. cit., 26, 28.

81. On the Taipings, see De Bary et al., op. cit., 18–42.

82. For debates, see ibid., 151–195.

83. Rhoades Murphy, "Man and Nature in China," *Modern Asian Studies* 1:315–333 (1967).

84. Passmore, *Man's Responsibility*, op. cit., 26.

85. Levenson, op. cit., vol. 3, 99.

86. On Mao's policies, see Volti, op. cit., passim. On science and technology in the Maoist era, see John Sigurdson, *Technology and Science in the People's Republic of China* (Oxford: Pergamon Press, 1980). On recent years, see Denis Fred Simon and Merle Goldman, eds., *Science and Technology in Post-Mao China* (Cambridge: Harvard University Council on East Asian Studies, 1989). See also Lester Ross, *Environmental Policy in China* (Bloomington: Indiana University Press, 1988); and Vaclav Smil, *The Bad Earth: Environmental Degradation in China* (New York: M. E. Sharpe, 1984).

87. On Japanese religion, see H. Byron Earhart, *Japanese Religion: Unity and Diversity*, 3d ed. (Belmont, Calif.: Wadsworth, 1982); Earhart, *Religion in the Japanese Experience: Sources and Interpretations* (Encino, Calif.: Dickenson Publishing, 1974); and Welty, op. cit., 231–246.

88. Earhart, *Religion*, op. cit., 13. See also McNeill, *Rise of the West*, op. cit., 537.

89. Nash, op. cit., 21. On primitive Japanese religion, see also Earhart, *Japanese Religion*,

op. cit., 24–27; and J.E. Kidder, Jr., *Japan before Buddhism*, rev. ed. (New York: Praeger, 1966).

90. Earhart, *Religion*, op. cit., 8.
91. Ibid., 127. On its policy implications see Earhart, "The Ideal of Nature in Japanese Religion and Its Possible Significance for Environmental Concerns," *Contemporary Religions in Japan* 11:10–26 (1970).
92. Jonathan Norton Leonard, *Early Japan* (New York: Time- Life Books, 1968), 16–17.
93. Welty, op. cit., 77.
94. Earhart, *Religion*, op. cit., 127.
95. For a sketch of that influence from a Japanese perspective, see Daisetz T. Suzuki, *Zen and Japanese Culture* (New York: Pantheon Books, 1959); for one from that of an American aficionado, see Thomas Hoover, *Zen Culture* (New York: Random House, 1977).
96. Much of it was published in Tokyo, and in America by Charles E. Tuttle Company of Rutland, Vermont. For example, see Stewart W. Holmes and Chimyo Horioka, *Zen Art for Meditation* (Rutland, Vt.: Charles E. Tuttle, 1973).
97. See Dom Aelred Graham, O.S.B., *Zen Catholicism: A Suggestion* (New York: Harcourt, Brace and World, 1963).
98. See Devall and Sessions, op. cit., 11–15; and the appendix, "Gandhi, Dogon, and Deep Ecology," by the Zen master Robert Aitkin Roshi, 232–235.
99. See his "Buddhism and the Possibilities of a Planetary Culture," in ibid., 251–253; and "Buddhism and the Coming Revolution," in Gary Snyder, *Earth House Hold: Technical Notes & Queries to Fellow Dharma Revolutionaries* (New York: New Directions, 1969), 90–93.
100. Watts, *The Way*, op. cit., 80.
101. See the chapter, "Zen and the Samurai," in Suzuki, *Zen*, op. cit., 61–85.
102. Watts, op. cit., 107.
103. Suzuki, *Zen*, op. cit., 218.
104. See ibid., 32–33, 348.
105. Suzuki, "The Role of Nature in Zen Buddhism," in Earhart, *Religion*, op. cit., 134.
106. Suzuki, *Zen*, op. cit., 30.
107. G. B. Sansom, *Japan: A Short Cultural History*, rev. ed. (New York: D. Appleton-Century, 1943), 339.
108. Nobutaka Ike, *Japan: The New Superstate* (San Francisco: W. H. Freeman, 1973), 52.
109. Earhart, *Religion*, op. cit., 223.
110. Earhart, *Japanese Religion*, op. cit.
111. Maseo Watanabe, "The Conception of Nature in Japanese Culture," *Science* 183:282 (1974).
112. Junko Nakanishi, quoted in James D. Whitehill, "Ecological Consciousness and Values: Japanese Perspectives," in Schultz and Hughes, op. cit., 169. See also David Edward Shaner and R. Shannon Duval, "Conservation Ethics and the Japanese Intellectual Tradition," *Environmental Ethics* 11:197–214 (1989).
113. Whitehill in Schultz and Hughes, op. cit., 171.
114. Watanabe, op. cit.
115. Alan G. Grapard, "Nature and Culture in Japan," in Tobias, op. cit., 243, 254.
116. Ibid. See also Grapard, "Flying Mountains and Walkers in Emptiness: Toward a Definition of Sacred Space in Japanese Religion," *History of Religions* 20:195–221 (1982); and "Japan's Ignored Cultural Revolution: The Separation of Shinto and Buddhist Divinities in MEJII (*shimbutsu bunri*) and a Case Study: Yonomine," in ibid., 23:240–265 (1984).

117. Mayayoshi Sugimoto and David L. Swain, *Science and Culture in Traditional Japan, 600–1864* (Cambridge: MIT Press, 1978), 169, 197.
118. See, for example, Joseph S. Szylowitz, ed., *Technology and International Affairs* (New York: Praeger, 1981), 3.
119. Rybczynski, op. cit., 180–181.
120. See Delmar M. Brown, "The Impact of Firearms on Japanese Warfare," *Far Eastern Quarterly* 7:236–253 (1948).
121. Rybczynski, op. cit., 182.
122. Noel Perrin, *Giving Up the Gun: Japan's Reversion to the Sword, 1543–1879* (Boulder, Colo.: Shambala, 1980).
123. Rybczynski, op. cit., 185.
124. Sugimoto, op. cit., xxiii.
125. McNeill, *Pursuit*, op. cit., 720–721.
126. Sugimoto, op. cit., 348. For background, see Marion J. Levy, Jr., "Contrasting Factors in the Modernization of China and Japan," *Economic Development and Cultural Change* 2:161–197 (1953); and Michio Morishima, *Why Has Japan 'Succeeded': Western Technology and the Japanese Ethos* (Cambridge: MIT Press, 1962).
127. Sugimoto, op. cit., 293.
128. On the political history of Japan in the context of development, see Reinhard Bendix, *Kings or People: Power and the Mandate to Rule* (Berkeley: University of California Press, 1973), 61–87, 431–490.
129. Shigereu Nakayama, David L. Swain, and Yagi Eri, eds., *Science and Society in Modern Japan: Selected Historical Sources* (Cambridge: MIT Press, 1974), 13.
130. Sugimoto, op. cit., 395.
131. For an early attempt to survey Japanese culture from the outside, widely influential during World War II, see Ruth Benedict, *The Chrysanthemum and the Sword* (Boston: Houghton Mifflin, 1946). For a recent attempt by a longtime resident, see Robert C. Christopher, *The Japanese Mind* (New York: Fawcett Columbine, 1983).
132. See Sardar, op. cit., 182–183; and Frank C. Darling, *The Westernization of Japan* (Boston: Hall, 1979).
133. Swain in Nakayama et al., op. cit., xviii.
134. Yuasa Mitsomoto, "The Scientific Revolution and the Age of Technology," *Cahiers d'histoire mondiale* 9:187–207 (1965). See also Yabaouti Kiyosi, "The Pre-History of Modern Science in Japan: The Importation of Western Science during the Togugawa Period," ibid., 208–232; Hiroshige Tetu, "The Role of the Government in the Development of Science," ibid., 320–339; and Otohiko Hasumi, "The Impact of Science and Technology on Traditional Society in Japan," ibid., 363–379. See also James R. Bartholomew, "Why Was There No Scientific Revolution in Togugawa Japan?" *Japanese Studies in the History of Science* 15:111–125 (1976); and Bartholomew, "Science, Bureaucracy, and Freedom in Mejii and Taisho Japan," in Natsue Najita and J. Victor Korshmann, eds., *Conflict in Modern Japanese History: The Neglected Tradition* (Princeton: Princeton University Press, 1982), 295–341. On the role of the government in early Japanese industrial development, see Thomas C. Smith, *Political Change and Industrial Development in Japan: Government Enterprise, 1868–1880* (Stanford: Stanford University Press, 1955).
135. Moroshima, op. cit., 2, 6, 15, 16.
136. Ibid, 61.
137. Ibid., 136. On postwar science in Japan, see also Nakayama et al., op. cit., passim.
138. Moroshima, op. cit., 176–177.

139. Ibid., 177. See also Masao Watanabe, *The Japanese and Western Science*, trans. by Otto Theodor Benfey (Philadelphia: University of Pennsylvania Press, 1991).

140. Lee Smith, "Divisive Forces in an Inbred Nation," *Fortune*, March 30, 1987, 24–29.

141. Passmore notes that though it has no Christian tradition but rather a nature-venerating one, Japan has still managed largely to destroy its natural heritage; op. cit., 26.

142. See Ui Jun, "A Basic Theory of Kogai," in Nakayama et al., op. cit., 280–311. On the Japanese environmental movement, see Jim Griffith, "The Environmental Movement in Japan," *Whole Earth Review* 69:90–95 (Winter 1990).

11. Technological Cornucopianism

1. Critics view Wells differently. Though mainly an optimist, he had a strain of pessimism in him, to which he completely succumbed in his last work. On Wells as optimist, see especially W. Warren Wagar, *H. G. Wells and the World State* (New Haven: Yale University Press, 1961). On Wells's pessimism, see Jack Williamson, *H. G. Wells: Critic of Progress* (Baltimore: Mirage Press, 1973). Wells's ultimate pessimism is expressed in his last work, *Mind at the End of Its Tether* (London: Heinemann, 1945), wherein he writes: "The human story has come to an end and. . . . *Homo sapiens*, as he is pleased to call himself, is in his present form played out. . . "; quoted in Kumar, *Utopia*, op. cit., 351. See also W. Warren Wagar, "The Rebellion of Nature," in Eric S. Rabkin, Martin H. Greenberg, and Joseph D. Olander, eds., *The End of the World* (Carbondale: Southern Illinois University Press, 1983).

2. (New York: E. P. Dutton, 1924).

3. Reprinted (Lincoln: University of Nebraska Press, 1967). See also W. Warren Wagar, "Toward a World Set Free: The Vision of H. G. Wells," *Futurist* 17:24–31 (August 1983).

4. On futurism, see Marjorie Perloff, *The Futurist Movement: Avant-garde, Avant Guerre, and the Language of Rupture* (Chicago: University of Chicago Press, 1986). A recent retrospective exhibit of futurist artifacts and literature in Venice was one of the international artistic events of the year. See Robert Hughes, " 'Kill the Moonlight!' They Cried," *Time*, August 4, 1986; and John Golding, "The Futurist Past," *New York Review of Books*, August 14, 1986.

5. See Richard Guy Wilson, Dianne H. Pilgrim, and Dickran Tashjian, *The Machine Age in America, 1918–1941* (New York: Henry N. Abrams, 1986).

6. Robert Wuthnow, *The Restructuring of American Religion: Society and Faith since World War II* (Princeton: Princeton University Press, 1986), 286–287. Polls confirm this. See Robert Nisbet, "Utopia's Mores: Has the American Vision Dimmed?" *Public Opinion* 6:6–9 (April–May 1983).

7. About this paradox a philosopher writes that "such criticisms have not filtered down toward the masses in a technological society. Popular culture remains deferential toward technical experts even when it exhibits a receptive attitude toward negative stereotypes of those experts"; Durbin, *A Guide*, op. cit., xxiv.

8. Pacey, *Culture*, op. cit., 85–89.

9. Samuel C. Florman, *The Existential Pleasures of Engineering* (New York: St. Martin's Press, 1976), 101.

10. For overviews of this period, see William Braden, *The Age of Aquarius: Technology and the Cultural Revolution* (Chicago: Quadrangle Books, 1970); and Morris Dickstein, *Gates of Eden: American Culture in the Sixties* (New York: Basic Books, 1977).

11. Samuel C. Florman, *Blaming Technology: The Irrational Search for Scapegoats* (New York: St. Martin's Press, 1981), 8, 33, 91.

12. Ibid., 130.

13. For a useful introductory history of science fiction, especially in the United States, see Brian W. Aldiss, *Billion Year Spree: The True History of Science Fiction* (New York: Doubleday, 1973). For its relation to American literature generally, see David Ketterer, *New Worlds for Old: The Apocalyptic Imagination, Science Fiction, and American Literature* (Garden City, N.Y.: Anchor Books, 1974). Soviet science fiction—the largest school after American—was generally optimistic. See Darko Suvin, "The Utopian Tradition in Russian Science Fiction," *Modern Language Review* 66:139–159 (1971). More recently, science fiction in the Communist world had become pessimistic.

14. See his *Last and First Men* and *Last Men in London* (Baltimore: Penguin Books, 1972).

15. (New York: Bantam Books, 1964; rev. ed., New York: Popular Library, 1977).

16. (New York: Ballantine Books, 1963).

17. *Up-Wingers* (New York: Popular Library, 1967), 168. Estfandiary's ideas are also found in *Telespheres* (New York: Popular Library, 1977), and *Optimism One*, rev. ed. (New York: Popular Library, 1978). On his ideas, see "Optimism, Abundance, Universalism, and Immortality: The Philosophy of F. M. Estfandiary," a futurist interview by Mico Delianova in *Futurist* 12:185–189 (August 1978).

18. Quoted in Passmore, *Perfectability*, op. cit., 245.

19. Gordon Rattrey Taylor, *The Biological Time Bomb* (New York: World Publishing, 1968), 222.

20. Walter Truitt Anderson, "Food without Farms: The Biotech Revolution in Agriculture," *Futurist* 24:16–21 (January–February 1990).

21. Brian Stableford and David Langford, *The Third Millennium: A History of the World, A.D. 2000–3000* (New York: Alfred A. Knopf, 1985), 64–66, 132–133.

22. Robert H. Blank, *The Political Implications of Human Genetic Technology* (Boulder, Colo.: Westview Press, 1981), 70.

23. Quoted in ibid., 70–71.

24. *Fabricated Man: The Ethics of Genetic Control* (New Haven: Yale University Press, 1970), 95. See also generally E. J. Applewhite, *Paradise Mislaid: Birth, Death and the Human Predicament of Being Biological* (New York: St. Martin's Press, 1991).

25. For a feminist viewpoint, see Jalna Hanmer, "Reproductive Technology: The Future for Women?" in Joan Rothschild, ed., *Machina Ex Dea: Feminist Perspectives on Technology* (New York: Pergamon Press, 1983), 183–197.

26. See Daniel J. Rothman, *Strangers at the Bedside: A History of How Law and Bioethics Transformed Medical Decision Making* (New York: Basic Books, 1991).

27. On the background to the project, see Jerry E. Bishop and Michael Waldholz, *Genome: The Story of the Most Astonishing Scientific Adventure of Our Time—The Attempt to Map All the Genes in the Human Body* (New York: Simon and Schuster, 1991). See also Bernard D. Davis, ed., *The Genetic Revolution: Scientific Prospects and Public Perceptions* (Baltimore: Johns Hopkins University Press, 1991).

28. Quoted in Bishop and Waldholz, op. cit., 305. See also Robert A. Weinberg, "The Dark Side of the Genome," *Technology Review* 94: 44–51 (April 1991).

29. Edward A. Gargan, "Ultrasonic Tests Skew Ratio of Births in India," *New York Times*, December 13, 1991.

30. Examples would include Burnham Putnam Beckwith, *The Next 500 Years: Scientific Projections of Major Social Trends*, foreword by Daniel Bell (New York: Exposition Press, 1967); and Adrian Berry, *The Next Ten Thousand Years: A Vision of Man's Future in the Universe*, foreword by Robert Jastrow (New York: New American Library, 1974). On future studies, see Jib Fowles, ed., *Handbook of Futures Research* (Westport, Conn.: Greenwood

Press, 1978); and Victor C. Ferkiss, *Futurology: Promise, Performance, Prospects* (Beverly Hills, Calif.: Sage Publications, 1977).

31. (New York: Macmillan, 1962).

32. (New York: Alfred A. Knopf, 1971).

33. (New York: Macmillan, 1953).

34. See Joseph Wood Krutch, *The Measure of Man* (Indianapolis: Bobbs-Merrill, 1954); and David Spitz, "The Higher Reaches of the Lower Orders: A Critique of the Theories of B. F. Skinner," in Dante Germino and Klaus von Beyme, eds., *The Open Society in Theory and Practice* (The Hague: Martinus Nijoff, 1974), 237–275. For a defense of his ideas by Skinner himself, see his "Freeedom and the Control of Man," *American Scholar* 25:47–65 (Winter 1955–1956).

35. On its early years, see Kathleen Kinkade, *A Walden Two Experiment* (New York: William Morrow, 1973).

36. Skinner probably derives this belief more directly from the American naturalist tradition in philosophy, especially as promulgated by John Dewey. See Morton White, *Social Thought in America: The Revolt against Formalism* (Boston: Beacon Press, 1987), 212–219.

37. For Fuller's ideas, see his *Utopia or Oblivion: The Prospects for Humanity* (New York: Bantam Books, 1969); *No More Secondhand God and Other Writings* (Carbondale: Southern Illinois University Press, 1963); Fuller, Eric A. Walker, and James R. Killian Jr., *Approaching the Benign Environment*, preface by Taylor Littleton (New York: Collier Books, 1970); *Operating Manual for Spaceship Earth* (Carbondale: Southern Illinois University Press, 1969); *Nine Chains to the Moon* (Carbondale: Southern Illinois University Press, 1963); and, in collaboration with E. J. Applewhite, *Synergetics: Explorations in the Geometry of Thinking* (New York: Macmillan, 1975). See also Richard J. Brenneman, *Fuller's Earth: A Day with Bucky and the Kids* (New York: St. Martin's Press, 1984); and Lloyd Steven Sieden, *Buckminster Fuller's Universe: An Appreciation*, foreword by Norman Cousins (New York: Plenum Press, 1989). For discussion, see William Kuhns, *The Post-Industrial Prophets: Interpretations of Technology* (New York: Weybright and Talley, 1971), 220–246.

38. *Synergetics*, op. cit., 85, 631.

39. Jesco von Puttkamer, "Space: A Matter of Ethics—Toward a New Humanism," in Eugene M. Emme, ed., *Space Fiction and Space Futures Past and Present*, AAS History Series, vol. 5 (San Diego: American Astronautical Society, 1982), 206. For a closely related approach, see Earl Hubbard, *The Search Is On*, ed. by Barbara Hubbard (Los Angeles: Pace Publications, 1969). Earl Hubbard, the book notes, is a descendant of Thomas Sprat, Bishop of Rochester, a founder and first secretary of the Royal Society.

40. Von Puttkamer in Emme, op. cit., 208.

41. *On Thermonuclear War* (Princeton: Princeton University Press, 1962).

42. See his *Thinking about the Unthinkable* (New York: Horizon Press, 1962).

43. See his pioneering work, with Anthony Weiner, *The Year 2000: A Framework for Speculation on the Next Thirty-three Years* (New York: Macmillan, 1967).

44. Kahn and B. Bruce-Briggs, *Things to Come* (New York: Macmillan, 1972).

45. Kahn, William Brown, and Leon Martel, *The Next 200 Years* (New York: William Morrow, 1976). See also Kahn and William M. Brown, "The Optimistic Outlook," in Fowles, op. cit., 455–477. On Kahn generally, see Jerome Agel, *Herman Kahnsciousness: The Megaton Ideas of a One-Man Think Tank* (New York: Signet Books, 1973); and "Herman Kahn Remembered," *Futurist* 17:61–65 (October 1983). For discussion of his ideas on growth, see Barry B. Hughes, *World Futures: A Critical Analysis of Alternatives* (Baltimore: Johns

Hopkins University Press, 1985). See also "Kahn, Mead, and Thompson: A Three-Way Debate on the Future," *Futurist* 12: 229–232 (August 1978); and Herman Kahn and John B. Phelps, "The Economic Present and Future: A Chartbook for the Decades Ahead," *Futurist* 13:202–222 (June 1979).

46. *The Next 200 Years*, op. cit., 7.

47. Kahn discusses his methodology in "On Studying the Future," in Fred L. Greenstein and Nelson W. Polsby, eds., *Handbook of Political Science*, vol. 7, *Strategies of Inquiry* (Reading, Mass.: Addison-Wesley, 1975), 405–422.

48. See his "Space Colonies: The High Frontier," *Futurist* 10:25–33 (February 1976); *The High Frontier: Human Colonies in Space* (New York: Simon and Schuster, 1977); and *2081: A Hopeful View of the Future* (New York: Simon and Schuster, 1981).

49. See John Lewis and Ruth Lewis, *Space Resources* (New York: Columbia University Press, 1987).

50. Jesco von Puttkamer, "The Industrialization of Space: Transcending the Limits to Growth," *Futurist* 13:192–201 (June 1979); and "Space: The Long-Range Future," an interview with Jesco von Puttkamer, *Futurist* 19:36–38 (February 1985). See also G. Harry Stine, *The Third Industrial Revolution* (New York: Putnam, 1975); and Lelland A. C. Weaver, "Factories in Space: The Role of Robots," *Futurist* 21:29–34 (May–June 1987). For a negative view, see Daniel Duedney, "Space Industrialization: The Mirage of Abundance," *Futurist* 16:47–53 (December 1982).

51. Foreword to Frank White, *The Overview Effect: Space Exploration and Human Evolution* (Boston: Houghton Mifflin, 1987), xiv.

52. For a discussion of a similar problem, see Philip R. Harris, "Living on the Moon: Will Humans Develop an Unearthly Culture?" *Futurist* 19:30–35 (April 1985).

53. Note Magoroh Maruyama, "Designing a Space Community," *Futurist* 10:273–281 (1976).

54. See Paul L. Csonka, "Space Colonization: An Invitation to Disaster?" *Futurist* 11:285–290 (October 1977).

55. These are among the issues canvassed in Eugene C. Hargrove, ed., *Beyond Spaceship Earth: Environmental Ethics and the Solar System* (San Francisco: Sierra Club Books, 1987).

56. For a discussion of these issues, see Magoroh Maruyama and Arthur Harkins, *Cultures beyond the Earth*, foreword by Alvin Toffler (New York: Vintage Books, 1975); and Isaac Asimov, *Extraterrestrial Civilization* (New York: Crown, 1979).

57. How long this might take is examined in Robert Page Burruss, "Intergalactic Travel: The Long Voyage from Home," *Futurist* 21:29–33 (September–October 1987).

58. For the story of the American space program, see Walter A. McDougall, *The Heavens and the Earth: A Political History* (New York: Basic Books, 1985). For an imaginative philosophical interpretation, see Norman Mailer, *Of a Fire on the Moon* (New York: New American Library, 1971). On the future of the U.S. in space, see Richard S. Lewis, *Space in the 21st Century* (New York: Columbia University Press, 1990).

59. David Ehrenfeld, *The Arrogance of Humanism* (New York: Oxford University Press, 1978), 123.

60. Foreword to White, op. cit., xv.

61. And more than that: "However, if one sees humanism for what it is, a religion without God, then . . . space with its space stations and space inhabitants is just a replacement for heaven with its angels. Even the idea of immortality is there, fuzzy like everything else in this imaginary humanist domain."; Ehrenfeld, op. cit., 120.

62. For Teilhard's ideas, see his *The Phenomenon of Man*, introduction by Sir Julian Huxley, trans. by Bernard Wall (New York: Harper Torchbooks, 1961); *Le Milieu Divin: An Essay*

on the Interior Life, trans. by Bernard Wall (London: Fontana Books, 1962); and *The Future of Man*, trans. by Norman Denny (London: Collins, 1964). For a sample of the enormous secondary literature on him, see Benz, op. cit., passim; Philip Hefner, *The Promise of Teilhard: The Meaning of the Twentieth Century in Christian Perspective* (Philadelphia: J. B Lippincott, 1970); and Victor C. Ferkiss, *Technological Man: The Myth and the Reality* (New York: George Braziller, 1969), 92–99.

63. *The Phenomenon of Man*, op. cit., 220.
64. *The Future of Man*, op. cit., 147.

12. Contemporary Critics

1. William Barrett, *The Illusion of Technique: A Search for Meaning in a Technological Civilization* (Garden City, N.Y.: Anchor Books, 1979), 229.
2. On Heidegger as philosopher, see George Steiner, *Martin Heidegger* (New York: Viking Press, 1979); and Arne Naess, *Four Modern Philosophers: Carnap, Wittgenstein, Heidegger, Sartre*, trans. by Alistair Hannay (Chicago: University of Chicago Press, 1968), 173–264. Naess's views are especially interesting since he is one of the founders of "deep ecology." Heidegger's ideas on technology may be found in his *The Question Concerning Technology and Other Essays*, trans. and introduction by William Lovitt (New York: Harper Colophon Books, 1977). The major American cultural historian, Leo Marx, writes that this work "may well be the written work most directly concerned with the subject by a distinguished modern philosopher"; see "On Heidegger's Conception of 'Technology' and Its Historical Validity," *Massachusetts Review* 29:638 (1984). See also Michael E. Zimmerman, *Heidegger's Confrontation with Modernity: Technology, Politics, and Art* (Bloomington: Indiana University Press, 1990). On his relevance for environmental issues, see Canadian philosopher Neil Evernden's *The Natural Alien* (Toronto: University of Toronto Press, 1985), 60–72; Michael F. Zimmerman, "Toward a Heideggerian *Ethos* for Radical Environmentalism," *Environmental Ethics* 5:99–131 (1983); and Bruce V. Foltz, "On Heidegger and the Interpretation of Environmental Crisis," *Environmental Ethics* 6:323–338 (1984).
3. Barrett, op. cit., 148.
4. Ibid., 149. But see also W. R. Newell, "Politics and Progress in Heidegger's Philosophy of History," in Richard B. Day, Ronald Beiner, and Joseph Masciulli, eds., *Democratic Theory and Technological Society* (Armonk, N.Y.: M. E. Sharpe, 1988), 265.
5. Naess, op. cit., 175.
6. Steiner, op. cit., 34.
7. Barrett, op. cit., 220.
8. Lovitt, introduction to *The Question*, op. cit., xxxi.
9. Carl Mitcham, "Philosophy of Technology," in Durbin, A *Guide*, op. cit., 319. On Heidegger, see also ibid., 319–321.
10. Steiner, op. cit., 115.
11. William Barrett, *Irrational Man: A Study in Existential Philosophy* (London: Mercury Books, 1984), 194, 207, 209.
12. Steiner, op. cit., 55, 115.
13. Quoted in Barrett, *The Illusion*, op. cit., 233.
14. *Discourse on Thinking*, trans. by John M. Anderson and E. Hans Freund (New York: Harper and Row, 1966), 51.
15. Steiner, op. cit., 136.
16. *The Question*, op. cit., 13, 25.
17. Leo Marx, op. cit., 642.

18. Ibid., 645. See also *The Question*, op. cit., 25–26.
19. *The Question*, ibid., 28.
20. Ibid., 100.
21. Jung, op. cit., 110.
22. Barrett overly minimizes the significance of Heidegger's relationship with Naziism in *The Illusion*, op. cit., 149, 171–172. Steiner is more reliable, and more condemnatory in op. cit., 116–126. But see also Naess, op. cit., 179–186. On Heidegger's general political philosophy, see Richard Wolin, *The Politics of Being: The Political Thought of Martin Heidegger* (New York: Columbia University Press, 1990); George W. Ramoser, "Heidegger and Political Philosophy," *Review of Politics* 29:261–268 (1967); and Karsten Harries, "Heidegger as Political Thinker," in Michael Murray, ed., *Heidegger and Modern Philosophy: Critical Essays* (New Haven: Yale University Press, 1978), 304–328.
23. Victor Farias, *Heidegger and Nazism*, ed., with a foreword, by Joseph Margolis and Tom Rockmore (Philadephia: Temple University Press, 1989); James M. Markham, "Over Philosophy's Temple, Shadow of a Swastika," *New York Times*, February 4, 1988; and Thomas Sheehan, "Heidegger and the Nazis," *New York Review of Books*, June 16, 1988. See also Luc Ferry and Alain Renaut, *Heidegger and Modernity*, trans. by Franklin Philip (Chicago: University of Chicago Press, 1990).
24. Steiner, op. cit., 148.
25. His ideas on rootedness are compared by one admirer to those of primitive peoples, including American Indians. Paul Sheperd, "Homage to Heidegger," in Tobias, op. cit., 211.
26. Steiner, op. cit.
27. Leo Marx, op. cit., 652.
28. See, for example, the attention given his ideas in the preface to Mitcham and Mackey, *Philosophy and Technology*, 25–29; and Walter Hood, "The Aristotelian versus the Heideggerian Approach to Technology," in ibid., 347–363. But see also Anthony Gottlieb, "Heidegger for Fun and Profit," *New York Times Book Review*, January 7, 1990.
29. On his life and career, see his *Perspectives on Our Age. Jacques Ellul Speaks on His Life and Work*, ed. by William H. Vanderburg, trans. by Joachim Neugroschel (New York: Seabury Press, 1981); also his "Mirror of These Ten Years," *Christian Century* 87:200–204 (1970); and his *In Season and Out of Season: An Introduction to the Thought of Jacques Ellul*, trans. by Lani K. Niles (San Francisco: Harper and Row, 1982).
30. Carl Mitcham, "Jacques Ellul and His Contribution to Theology," *Cross Currents* 35:1 (1985).
31. Trans. by Olive Wyon (Philadelphia: Westminster Press, 1951).
32. Trans. by John Wilkinson, introduction by Robert K. Merton (New York: Alfred A. Knopf, 1964).
33. See, for example, Stover, op. cit.
34. Most directly related are *The Political Illusion*, trans. by Konrad Kellen (New York: Alfred A. Knopf, 1972); *Propaganda*, trans. by Konrad Kellen and Jean Lerner, foreword by Konrad Kellen (New York: Alfred A. Knopf, 1973); *The Technological System*, trans. by Joachim Neugroschel (New York: Continuum, 1980); *What I Believe*, trans. by Geoffrey W. Bromley (Grand Rapids, Mich.: Eerdmans, 1989); and *The Technological Bluff*, trans. by Geoffrey W. Bromley (Grand Rapids, Mich.: Eerdmans, 1990). Important articles include "The Technological Order," in Carl F. Stover, ed., *The Technological Order* (Detroit: Wayne State University Press, 1963), 10–38; "The Technological Order," *Technology and Culture* 3:391–421 (1962); "Technology and the Gospel," *International Review of Mission* 66:109–117 (1977); "Nature, Technique, and Artificiality," *Research in Philosophy and Technology* 3:263–283 (1980); and "The Technological Revolution and Its

Moral and Social Consequences," trans. by Simon King, in Johannes B. Metz, ed., *The Evolving World and Theology* (New York: Paulist Press, 1967), 97–107. See also the bibliography, "The Ellul Oeuvre," *Cross Currents* 35:107–108 (1985).

35. See, as examples, Christopher Lasch, *The World of Nations: Reflections on American History, Politics, and Culture* (New York: Alfred A. Knopf, 1973), 270–293; Paul T. Durbin, Daniel Cerezuelle, et al., "Symposium on Jacques Ellul," *Research in Philosophy and Technology* 3:161–262 (1980); Clifford J. Christians and Jay M. Van Hook, eds., *Jacques Ellul: Interpretative Essays* (Urbana: University of Illinois Press, 1981); W. H. Vanderberg, ed., *Perspectives on Our Age: Jacques Ellul Speaks of His Life and Work* (Toronto: CBC Enterprises, 1981); D. Menninger, "Politics or Technique: A Defense of Jacques Ellul," *Polity* 14:110–127 (1981); Menninger, "Political Dislocation in a Technological Universe," *Review of Politics* 42:73–91 (1980); and John B. Shaar, "Jacques Ellul: Between Babylon and the New Jerusalem," *Democracy: A Journal of Political Renewal and Radical Change* 2:102–118 (1982). An early critique is found in Kuhns, op. cit., 82–111. See also Carl Mitcham, "About Ellul," *Cross Currents* 35:105–107 (1985).

36. One critic does entitle his chapter on Ellul, "The New Manichaenism"; Kuhns, *The Post-Industrial Prophets*, op. cit. On the allegation that Ellul really believes this world to be the domain of Satan, see Winner, *Autonomous Technology*, op. cit., fn. 249; and Carl Mitcham, "Langdon Winner on Jacques Ellul: An Introduction to Alternative Political Critiques of Technology," *Technology Studies Resource Center Working Papers Series* 3:115–124 (June 1985).

37. Mumford, *The Pentagon of Power*, op. cit., 290–291.

38. Ian Miles, "The Ideologies of Futurists," in Fowles, op. cit., 77.

39. *Technology and the Human Condition* (New York: St. Martin's Press, 1977), 5.

40. Op.cit., passim.

41. *A Guide*, op. cit., 287.

42. *The Technological System*, op. cit., 47.

43. Ellul, "Technology and the Opening Chapters of Genesis," in Mitcham and Grote, *Theology and Technology*, op. cit., 125, 132.

44. Ellul, "From the Bible to a History of Non-Work," *Cross Currents* 35:44 (1985).

45. Ellul, "The Relationship between Man and Creation in the Bible," in Mitcham and Grote, op. cit., 139–155.

46. "Note to the Reader," *The Technological Society*, op. cit., xxv.

47. Ophuls, *Ecology*, op. cit., fn. 159.

48. Kuhns, *The Post-Industrial Prophets*, op. cit., 85, 90.

49. James Y. Holloway, ed., *Introducing Jacques Ellul* (Grand Rapids, Mich.: Eerdmans, 1970), 5.

50. *The Technological System*, op. cit., 33.

51. Ibid., 67.

52. *The Technological Society*, op. cit., 306; emphasis in original.

53. "The Technological Order," in Stover, op. cit., 10.

54. *The Technological System*, op. cit., 250.

55. Ibid., 38, 46.

56. Ibid., 89.

57. James M. Staudenmaier, S.J., *Technology's Storytellers: Reweaving the Human Fabric* (Cambridge: MIT Press, 1985), 136–137.

58. Ellul, *The Technological System*, op. cit., 149.

59. Ibid., 114.

60. "The Technological Order," op. cit., 18, 122.

61. *The Technological System*, op. cit., 95, 102.

62. Ibid., 59, 135.

63. "The Technological Order," op. cit., 16.

64. *The Technological Society*, op. cit., 258–259.

65. Ibid., 279.

66. *The Technological System*, op. cit., 37, 130.

67. Ibid., 112.

68. "The Technological Order," op. cit., 23–24.

69. Kuhns, *The Post-Industrial Prophets,*op. cit., 94.

70. "The Technological Order," op. cit., 26. As disciples of Ellul have argued, this would also involve the concomitant desacralization of the modern state; see William H. Vanderberg, "Political Imagination in a Technical Age," in Day et al., op. cit., 28.

71. *The Technological System*, op. cit., 114.

72. Ibid., 260, 282.

73. Ibid., 302–303.

74. Ibid., 235. See also George J. Graham, Jr., "Jacques Ellul— Prophetic or Apocalyptic Theologian of Technology?" *Political Science Reviewer* 13:213–239 (1983).

75. His former student and disciple Vanderberg speaks of him as a "deeply committed activist"; op. cit., 27.

76. *Post-Scarcity Anarchism* (Berkeley: Ramparts Press, 1971, 27, 40, 80). For Bookchin's latest views, see *Remaking Society: Pathways to a Green Future* (Boston: South End Press, 1990).

77. *Post-Scarcity Anarchism*, op. cit., 86.

78. This seems to be the message of ibid., 95–106.

79. Ibid., 109, 112, 118.

80. *Ecology of Freedom*, op. cit., 25, 122, 223, 241.

81. Ibid., 243, 251.

82. Ibid., 261–262.

83. Ibid., 261, 262.

84. Ibid., 274. The logic of Bookchin's ecological ethics is attacked from a biocentric position in Robyn Eckersley, "Divining Evolution: The Ecological Ethics of Murray Bookchin," *Environmental Ethics* 11:99–116 (1989).

85. Bookchin, *The Ecology of Freedom*, op. cit., 317, 324.

86. Preface to *Exploring New Ethics for Survival: The Voyage of the Spaceship Beagle* (Baltimore: Penguin Books, 1973), vii.

87. *Science* 162:1243–1248 (1968). For Hardin's views, see also *The Limits to Altruism* (Bloomington: Indiana University Press, 1977); *Stalking the Wild Taboo*, 2d ed. (Los Altos, Calif.: William Kaufman, 1978); and *Promethean Ethics: Living with Death, Competition, and Triage* (Seattle: University of Washington Press, 1980).

88. One critic holds that Hardin is an optimist and that the problem of population is insoluble even if one accepts his views; see Beryl L. Crowe, "The Tragedy of the Commons Revisited," *Science* 166:1103–1107 (1969).

89. Hardin, op. cit., 1243, 1244, 1245, 1247.

90. *Exploring*, op. cit., 65, 142, 144, 144–149.

91. The immediate target of their ire was a paper, "An Ecolate View of the Human Predicament," *Alternatives: A Journal of World Polity* 7:242–262 (1981). See Christian Bay, "On Ecolacy Sans Humanism," ibid., 395–402; Richard Falk, "On Advice to the Imperial Prince," ibid., 403–408; and Rajni Kothari, "On Eco-Imperialism," ibid., 383–394.

92. F. Berkes, D. Feeny, B. J. McCay, and J. M. Acheson, "The benefits of the Commons," *Nature* 340:91 (1989), 92, 93.

93. As one scholar notes, "Although Mumford has been lumped with four others as 'the leading antitechnologists,' by one protechnology author, this accusation oversimplifies Mumford's complicated and subtle thesis"; see Durbin, introduction to Durbin, op. cit., xxiii. The errant author is Florman, *Blaming Technology*, op. cit., 6. On Mumford, see Kuhns, op. cit., 32–64; Rybczynski, op. cit., 16–20; and Thomas P. Hughes and Agatha C. Hughes, eds., *Lewis Mumford: Public Intellectual* (New York: Oxford University Press, 1990).

94. (New York: Harcourt, Brace and World, 1934).

95. (New York: Columbia University Press, 1952).

96. (New York: Harcourt, Brace and World, 1967).

97. (New York: Harcourt Brace Jovanovich, 1970).

98. Kuhns, op. cit., 8.

99. Rybczynski, *Taming the Tiger*, op. cit., 16.

100. *Technics*, op. cit., 419–423.

101. Ibid., 295, 298, 299.

102. *The Myth*, vol. 1, op. cit., 11.

103. Ibid., vol. 2, 276, 395.

104. Ibid., 408.

105. Ibid., 413, 435.

13. Woman as Nature

1. Neumann, op. cit., 135.

2. For a survey of changes in gender dominance during prehistoric and early historic times, see Boulding, op. cit., 69–202. For a highly speculative feminist position on the origin of the sexes and humanity as a whole, see Elaine Morgan, *The Descent of Woman* (London: Souvenir Press, 1972). On gender and primitive religion, see also Marija Gumbutas, *The Language of the Goddess* (San Francisco: Harper, 1989); and Peter Steinfels, "Idyllic Theory of Goddesses Creates Storm," *New York Times*, February 13, 1990.

3. Neumann, op. cit., 212.

4. It would be interesting to speculate as to why the position of women in most non-Western societies has often been no higher than in the West, but that is outside of our concern here.

5. Merchant, op. cit., 13, 127.

6. "The ancient identity of nature as a nurturing mother links women's history with the history of the environment and ecological change. . . . The ecology movement has reawakened interest in the values and concepts associated historically with the premodern organic world"; ibid., xvi.

7. Easley, op. cit., 168.

8. Ibid., 214.

9. "In the seventeenth century mechanical philosophers not merely all but banished life conceptually from their cosmos but minimized the role of women in procreation"; ibid., 244, 245.

10. "Feminism and Ecology," appendix in Devall and Sessions, op. cit., 229–231. See also her "Earth Care: Women and the Environmental Movement," *Environment* 23:6–13, 38–40 (June 1981).

11. "Is Female to Male as Nature Is to Culture?" in Michelle Zimbalist Risaldo and Louise Lamphere, eds., *Women, Culture, and Society* (Stanford: Stanford University Press, 1974),

67–87. On ecofeminism, see Irene Diamond and Gloria Feman Orenstein, eds., *Reweaving the World: The Emergence of Ecofeminism* (San Francisco: Sierra Club Books, 1990); Judith Plant, ed., *Healing the Wounds: The Promise of Ecofeminism*, foreword by Petra Kelly (Santa Cruz, Calif.: New Society Publishers, 1989); Jim Cheney, "Eco-Feminism and Deep Ecology," *Environmental Ethics* 9:115–145 (1987); Ynestra King, "Ecological Feminism," *Zeta Magazine* 1:124–127 (July–August 1988); King,"Feminism and the Revolt of Nature," *Heresies: A Feminist Journal of Art and Politics* 13:12–16 (1981); and King, "Toward an Ecological Feminism and a Feminist Ecology," in Rothschild, op. cit., 118–129. For a recent critique, see Janet Biehl, *Rethinking Ecofeminist Politics* (Boston: South End Press, 1991).

12. "The Ecology of Feminism and the Feminism of Ecology," *Harbinger: A Journal of Social Ecology* 1:16 (1983).

13. (Wellesley, Mass.: Roundtable Press, 1979).

14. See, for example, *Women at the Edge of Time* (New York: Fawcett Crest, 1977).

15. See Ariel Kay Sallah, "Deeper Than Deep Ecology: The Ecofeminist Connection," *Environmental Ethics* 6:339–345 (1984).

16. Donald Davis, "Ecosophy: The Seduction of Sophia?" *Environmental Ethics* 8:151–162 (1986).

17. *Woman and Nature: The Roaring Inside Her* (New York: Harper and Row, 1978), xv.

18. *GYN/ECOLOGY: The Ethics of Radical Feminism* (Boston: Beacon Press, 1978), 9. For a similar if less extreme statement, see Andree Collard with Joyce Contrucci, *Rape of the Wild: Man's Violence against Animals and the Earth* (Bloomington: Indiana University Press, 1989).

19. *Beyond God the Father: A Philosophy of Womens' Liberation* (Boston: Beacon Press, 1973).

20. There has also been an extremist feminist group called WITCH, which did not stress—nor abjure—witchcraft rituals but whose "Covens" devote themselves to attacking the enemies of the cause. See "WITCH Documents" in Robin Morgan, *Sisterhood Is Powerful: An Anthology of Writings from the Women's Liberation Movement* (New York: Vintage Press, 1970), 538–553.

21. See her *New Woman/New Earth: Sexist Ideologies and Human Liberation* (New York: Seabury Press, 1975); and "Mother Earth and the Megamachine: A Theology of Liberation in Feminine, Somatic, and Ecological Perpsectives," in Rosemary Ruether, ed., *Liberation Theology: Human Hope Confronts Christian History and American Power* (New York: Paulist Press, 1972), 115–126.

22. Shulamith Firestone, *The Dialectic of Sex: The Case for Feminist Revolution* (New York: Bantam Books, 1971), 193, 198, 199.

23. Allison M. Jagger, *Feminist Politics and Human Nature* (Totowa, N.J.: Rowman and Allenheld, 1983), 93, 366, 372.

24. *In a Different Voice: Essays on Psychological Theory and Women's Development* (Cambridge: Harvard University Press, 1982).

25. See Jean Bethke Elshtain, *Public Man, Private Woman: Women in Social and Political Thought* (Princeton: Princeton University Press, 1981), 40.

26. On the gender gap, see Sandra Baxter and Marjorie Lansing, *Women and Politics: The Visible Majority* (Ann Arbor: University of Michigan Press, 1983).

27. *Life against Death: The Psychoanalytic Meaning of History* (New York: Vintage Books, 1959).

28. For a discussion see Balbus, op. cit., 292–302.

29. Ibid., 334–344.

30. Aronowitz, op. cit., 61, 62.

31. See especially Evelyn Fox Keller, *Reflections on Gender and Science* (New Haven: Yale University Press, 1985); "Feminism, Science, and Democracy," *Democracy: A Journal of Political Renewal and Radical Change* 3:50–58 (Spring 1983); and "Women, Science, and Popular Mythology," in Rothschild, op. cit., 130–146. For a summary of the discussion among feminists see Sandra Harding, *The Science Question in Feminism* (Ithaca: Cornell University Press, 1986). See also Harding and Merrill B. Hintikka, eds., *Discovering Reality: Feminist Perspectives in Epistemology, Methodology, and Philosophy of Science* (The Hague: D. Reidel, 1983). For a socialist perspective, see Evelyn Reed, *Sexism and Science* (New York: Pathfinder Press, 1978). A recent book holds that scientists influenced by the nature mysticism of John Muir sought to make science more "feminine." See Michael J. Smith, *Pacific Visions: California Scientists and the Environment, 1850–1915* (New Haven: Yale University Press, 1988).

32. See Margaret Alec, *Hypatia's Heritage: A History of Women in Science from Antiquity through the Nineteenth Century* (London: Womens' Press, 1986).

33. See, for example, Daniel Coleman, "Girls and Math: Is Biology Really Destiny?" *New York Times*, Education Life, August 2, 1987. See also Sally L. Hacker, "Mathematization of Engineering: Limits on Women and the Field," in Rothschild, op. cit., 38–58. But see also Constance Holden, "Female Math Anxiety on the Wane," *Science* 236:660–661 (1987); and Holden, "Is 'Gender Gap' Narrowing?" *Science* 253:959–960 (1991).

34. Quoted in David Dickson, *The New Politics of Science* (New York: Pantheon Books, 1984), 331.

35. Jagger, op. cit., 372.

36. Pacey, *The Culture of Technology*, op. cit., 105. He goes on to note that "women in their traditional roles and craftsmen with their special obligations had always to show their creativity in less egotistical ways; and their achievements are given rather limited recognition because, in technical and artistic terms, they were restrained in their originality by responsibility"; ibid., 111.

37. Florman, *Blaming Technology*, op. cit., 127–128, 130.

38. See Joan Rothschild, "A Feminist Perspective on Technology and the Future," *Women's International Quarterly* 4:65–74 (1981). On feminist utopias, see Charlotte Perkins Gilman, *Herland*, foreword by Ann J. Lane (New York: Pantheon Books, 1979; first published 1915); Howard Segal, "The Feminist Technological Utopia: Mary E. Bradley Lane's *Mizora* (1890)," *Alternative Futures: The Journal of Utopian Studies* 4:67–72 (1981); and Patrocinio Schweickart, "What If . . . Science and Technology in Feminist Utopias," in Rothschild, op. cit., 198–211.

39. Means, "Man and Nature: The Theological Vacuum," *Christian Century*, May 1, 1968.

40. See, as examples, Bruce Allsopp, *The Garden Earth: The Case for Ecological Morality* (New York: William Morrow, 1972); Richard Baer, "Ecology, Religion, and the Amerian Dream," *American Ecclesiastical Review* 165:43–59 (1971); Ian G. Barbour, *Technology, Environment, and Human Values* (New York: Praeger, 1980); Albert Fritsch, S.J., *Renew the Face of the Earth* (Chicago: Loyola University Press, 1989); Paulos Gregarios, *The Human Presence: An Orthodox View of Nature* (Geneva: World Council of Churches, 1978); Michael J. Himes and Kenneth J. Himes, "The Sacrament of Creation: Toward an Environmental Theology," *Commonweal*, January 20, 1990; Gordon D. Kaufman, "A Problem for Theology: The Concept of Nature," *Harvard Theological Review* 65:337–366 (1972); John Macquarrie, "The Idea of a Theology of Nature," *Union Seminary Quarterly Review* 30:69–75 (Winter–Summer 1975); Jay McDaniel, *Earth, Sky, Gods, and Mortals: Developing an Ecological Spirituality* (Mystic, Conn.: Twenty-Third Publications, 1989), and *Of God and Pelicans: A Theology of Reverence for Life*, foreword by John B. Cobb,

Jr. (Louisville, Ky.: Westminster/John Knox Press, 1989); Eric Rust, *Nature-Garden or Desert?* (Waco, Tex.: Word Books, 1971); H. Paul Santmire, *Brother Earth: Nature, God, and Ecology in a Time of Crisis* (New York: Thomas Nelson, 1970); Francis A. Schaeffer, *Pollution and the Death of Man: The Christian View of Ecology* (Wheaton, Ill.: Tyndal, 1970); and Joseph Sittler, *The Care of the Earth and Other University Sermons* (Philadelphia: Fortress Press, 1964). For a recent volume containing views from a variety of traditions, see Eugene C. Hargrove, ed., *Religion and Environmental Crisis* (Athens: University of Georgia Press, 1986).

41. *Ecology and Religion*, op. cit., 5.

42. Whitehead's views have also done much to inspire the American Jesuit John F. Haught; see *Nature and Purpose* (Washington, D.C.: University Press of America, 1980), and "The Emergent Environment and the Problem of Cosmic Purpose," *Environmental Ethics* 8:139–150 (1986).

43. "Ecology, Science, and Religion: Toward a Postmodern Worldview," in David Griffin, ed., *The Reenchantment of Science: Post-modern Proposals* (Albany: State University of New York Press, 1988), 91–113.

44. Carmody, op. cit., 126.

45. Ibid., 133, 135.

46. For summary statements, see Ian G. Barbour, "Attitudes toward Nature and Technology," in Barbour, *Earth Might Be Fair*, op. cit., 146–168; and Harold K. Schilling, "The Whole Earth Is the Lord's," in ibid., 100–122. See also the chapter, "The Greening of Religion," in Nash, *Rights of Nature*, op. cit., 87–120. On its resonance, see Pat Windsor, "Final WCC Assembly Report Leads with Ecology," *National Catholic Reporter*, March 1, 1991.

47. Barbour, "Attitudes," op. cit., 151.

48. For representative views, see Cobb, "Christian Existence in a World of Limits," *Environmental Ethics* 1:148–158 (1979); and Dieter T. Hessel, ed., *Energy Ethics: A Christian Response* (New York: Friendship Press, 1979).

49. See Ron Elsdon, *Bent World: A Christian Response to the Environmental Crisis* (Downers Grove, Ill.: InterVarsity Press, 1981).

50. Hart, *The Spirit of the Earth* (New York: Paulist Press, 1984), 58.

51. For a review of the Reagan record, see Jonathan Lash, Katherine Gillman, and David Sheridan, *A Season of Spoils: The Story of the Reagan Administration's Attack on the Environment* (New York: Pantheon Books, 1984); and Carl Pope, "The Politics of Plunder," *Sierra* 73:49–55 (November–December 1988).

52. On his convictions, see Susan Power Bratton, "The Ecotheology of James Watt," *Environmental Ethics* 5:225–236 (1983).

53. See Richard John Neuhaus, *In Defense of People: Ecology and the Seduction of Radicalism* (New York: Macmillan, 1971). It is interesting that since writing this book, Neuhaus has himself moved to the right in American politics.

54. Peter Heinegg, "Christian Ecology?" *Cross Currents* 33:376 (1983).

55. See the diverting account of Liz Harris, "Brother Sun, Sister Moon," *New Yorker*, April 27, 1987.

56. "Sollucitudo Rei Socialis," III, #27.

57. Pat Windsor, "Nestled in the Vatican Gardens," *National Catholic Reporter*, April 28, 1989.

58. Tim McCarthy, "Philippine Bishops Issue Ecology Pastoral," *National Catholic Reporter*, March 11, 1988. For text, see "Think Globally, Act Locally, Be Spiritually," *Amicus Journal* 10: 8–10 (Fall 1988). It was not until 1990 that an American bishop individually

issued an ecology pastoral; see Anthony M. Pilla, "Pastoral Letter Urges Reverence for Creation," *National Catholic Reporter*, November 16, 1990.

59. Quoted in Cindy Wooden, "Pope Says Environment's Destruction Is Threat to World Peace," *Catholic Standard* (Washington, D.C.), December 21, 1989. Excerpts are found in "Pope's New Year Message: Moral Roots of Ecology Crisis," *National Catholic Reporter*, December 29, 1989.

60. Quoted in Jennifer Parmalee, "Pope Says Environmental Misuse Threatens World Stability," *Washington Post*, December 6, 1989.

61. Peter Steinfels, "Praise for Catholic Bishops' Statement on Children," *New York Times*, November 17, 1991.

62. "Earth Summit Part of Religious Agenda," *National Catholic Reporter*, December 6, 1991.

63. *Original Blessing*, op. cit., 90.

64. *The Coming of the Cosmic Christ: The Healing of Mother Earth and the Birth of a Global Renaissance* (San Francisco: Harper and Row, 1988), 6. On his ideas, see also his recent *Creation Spirituality: Liberating Gifts for the Peoples of the Earth* (San Francisco: Harper and Row, 1991).

65. *Coming*, op. cit., 144.

66. Fox replied to his silencing in a tract, "Is the Catholic Church Today a Dysfunctional Family? A Pastoral Letter to Cardinal Ratzinger and the Whole Church," in which he bitterly attacked his critics; see *Creation* 4:23–38 (November–December 1988). See also Bill Kenkelen, "Some U.S. Theologians Protest Fox Silencing," *National Catholic Reporter*, November 4, 1988.

67. On Berry and his ideas, see Anne Lonergan and Caroline Richards, eds., *Thomas Berry and the New Cosmology* (Mystic, Conn.: Twenty-Third Publications, 1988). See also Stephen Dunn, C.P., and Anne Lonergan, eds., *Befriending the Earth: A Theology of Reconciliation between Humans and the Earth: Thomas Berry, C.P., in Dialogue with Thomas Clarke, S.J.* (Mystic, Conn.,: Twenty-Third Publications, 1991).

68. (San Francisco: Sierra Club Books, 1988).

69. Quoted in Michael J. Farrell, "Thomas Berry's Dream of Biocracy," *National Catholic Reporter*, November 13, 1987.

70. He is the author of *Buddhism* (New York: Hawthorn Books, 1966), and *Religions of India* (New York: Bruce- Macmillan, 1971).

71. In some sense, all religious traditions are saying the same thing. For Berry, "There is no definitive Christianity or Hinduism or Buddhism, but only an identifiable Christian process, Hindu process, or Buddhist process"; see *Dream*, op. cit., 116–117.

72. Quoted in Farrell, op. cit.

73. Quoted in ibid.

74. "Wonderworld as Wasteworld: The Earth in Deficit," *Cross Currents* 35:409 (Winter 1985–1986). Much of this article appears in only slightly different form in *Dream*, passim.

75. *Dream*, op. cit., 77, 80–81, 105.

76. "Wonderworld," op. cit., 418–419.

77. *Dream*, op. cit., 87.

78. "The New Story: Comments on the Origin, Identification and Transmission of Values," *Cross Currents* 37:187 (Summer–Fall 1987). See also Berry, "The New Story: Meaning and Value in the Technological World," in Mitcham and Grote, op. cit., 271–278.

79. Berry in Mitcham and Grote, op. cit., 274.

80. "The New Story: Comments," op. cit., 187, 191.

81. See Swimme, *The Universe Is a Green Dragon: A Cosmic Creation Story* (Santa Fe, N.M.:

Bear and Co., 1984). This widely popular book is dedicated to Berry, Swimme's former teacher, and Swimme has been a collaborator in creation spirituality with Matthew Fox.

82. Berry in Mitcham and Grote, op. cit., 276.

83. As Swimme puts it, "For Thomas Berry the universe is primary. He enters with no distracting agenda drawn from conciliar documents. He does not attempt to see the universe as a gloss on the Bible. From his point of view, to attempt to cram this stupendous universe into categories of thought fit for scriptural studies or systematic theology is to lose the very magnificence that has stunned us in the first place"; "Berry's Cosmology," *Cross Currents* 37:220 (Summer–Fall 1987).

84. Ibid., 222.

85. Ibid., 223. John Grim argues that Berry has not simply been influenced by scientists but by historians and philosophers as well, notably the biblical writers, Augustine in his creationist rather than his redemptive mood, Dante, Aquinas, Vico, Eric Vogelin, and Christopher Dawson; see "Time, History, Historians in Thomas Berry's Vision," *Cross Currents* 37:225–239 (Summer–Fall 1987).

86. Berry, "The New Story: Comments," op. cit., 194, 198.

87. "Our secular, rational, industrial society with its amazing scientific insight and technological skill has establshed the first radically anthropocentric society and has thereby broken the primary law of the universe"; see "The Dream of the Earth: Our Way into the Future," *Cross Currents* 37:205 (Summer–Fall 1987).

88. *Dream*, op. cit., 194, 195.

89. He discusses "The Historic Role of the American Indian" in ibid., 180–193. On American Indian religion and nature, see Catherine L. Albanese, *Nature Religion in America: From the Algonkian Indians to the New Age* (Chicago: University of Chicago Press, 1990).

90. "If there is revelation, it will not be singular but differentiated. If there is grace, it will be differentiated in its forms"; see "The Earth: A New Context for Religious Unity," in Lonergan and Richards, op. cit., 21.

91. "Our Future on Earth: Where Do We Go from Here?" in ibid., 103, 104, 105.

92. For clashing views, see James Farris, "Redemption: Fundamental to the Story," in ibid., 65–71; and Father Stephen Dunn, "Needed: A New Genre for Moral Theology," in ibid., 73–78.

93. Canadian theologian Gregory Baum argues that "there is hope for us only if the green concerns can be integrated into a socialist project"; see "The Grand Vision: It Needs Social Action," in ibid., 53.

94. Sean McDonagh, *To Care for the Earth: A Call to a New Theology* (London: Geoffrey Chapman, 1986), 3; also published in Santa Fe, N.M.: Bear and Co., 1987). See also his *The Greening of the Church* (Maryknoll, N.Y.: Orbis Books, 1990). On McDonagh's struggles in the Philippines, see Tim McCarthy, "Mission Fights to Answer Call of the Land," *National Catholic Reporter*, March 10, 1989.

95. McDonagh, *To Care for the Earth*, op. cit., 110.

14. *People Are Not above Nature*

1. See Bill Devall, "John Muir as Deep Ecologist," *Environmental Review* 6:63–86 (1982). For a recent attempt to reunite ecological science and values, see David Oates, *Earth Rising: Ecological Belief in an Age of Science* (Eugene: University of Oregon Press, 1989).

2. Devall and Sessions, *Deep Ecology*, op. cit., 65. The article appeared in a journal founded by Naess, *Inquiry* 16:95–100 (1973). Naess was born in 1912 and taught philosophy at the University of Oslo from 1939 until 1970, "when he resigned to devote himself more fully

to the urgent environmental problems facing mankind"; Devall and Sessions, ibid., 225. On Naess, see Nash, op. cit., 146–150. For his own views, see his *Ecology, Community, and Lifestyle: Outline of an Ecosophy* (Cambridge: Cambridge University Press, 1991); and "The Deep Ecological Movement: Some Philosophical Aspects," *Philosophical Inquiry* 8:10–31 (1986). See also Sessions, "Shallow and Deep Ecology: A Review of the Literature," in Schultz and Hughes, op. cit., 391–462.

3. Sessions, ibid., 399.
4. On the movement, see William B. Devall, "The Deep Ecology Movement," *Natural Resources Journal* 20:299–322 (1980); and Warwick Fox, "Deep Ecology: Toward a New Philosophy for our Time?" *The Ecologist* 14:194–204 (1984). On its practical implications, see Devall, *Simple in Means, Rich in Ends: Practicing Deep Ecology* (Salt Lake City: Peregrine Smith Books, 1989).
5. "The Shallow and the Deep, Long-Range Ecology Movement—A Summary," *Inquiry* 16:98, 99 (1973).
6. Devall and Sessions, op. cit., 67, 225.
7. Interview, ibid., 75.
8. Interview, ibid., 74. On his ideas see also "Appendix A, Ecosophy T, Arne Naess," ibid., 225–228.
9. Ibid., 10.
10. Richard A. Watson, "A Critique of Anti-Anthropocentric Biocentrism," *Environmental Ethics* 5:245–256 (1983). His fire is directed not only against Naess's paper on "The Shallow and Deep, Long- Range Ecology Movements," but also Sessions's "Spinoza and Jeffers on Man and Nature," *Inquiry* 20:481–528 (1977).
11. "A Defense of the Deep Ecology Movement," *Environmental Ethics* 6:265–270 (1984). See also Naess, "Deep Ecology and Ultimate Premises," *Ecologist* 18:128–131 (1988). On deep ecology generally, see Warwick Fox, *Toward a Transpersonal Ecology: Developing New Foundations for Environmentalism* (Boston: Shambala, 1990).
12. Snyder's most significant works include *The Back Country* (New York: New Directions, 1968); *Earth House Hold: Technical Notes and Queries to Fellow Dharma Revolutionaries* (New York: New Directions, 1989); *Regarding Wave* (New York: New Directions, 1970); *Turtle Island* (New York: New Directions, 1974); *The Old Ways* (San Francisco: City Lights Books, 1977); *The Real Work: Interviews and Talks, 1964–1979*, ed. and with an introduction by Wm. Scott McLean (New York: New Directions, 1980); *Riprap and Cold Mountain Poems* (San Francisco: North Point Press, 1990); and *The Practice of the Wild* (San Francisco: North Point Press, 1990). See also Geeta Dardick, "An Interview with Gary Snyder: 'When Life Starts Getting Interesting,' " *Sierra* 70:68–73 (September–October 1985); and "The Etiquette of Freedom," *Sierra* 74:74–77, 113–116 (September–October 1989)'bc.
13. *The Real Work*, op. cit., 93.
14. Shi, *In Search*, op. cit., 269.
15. Thomas Parkinson, "The Poetry of Gary Snyder," *Southern Review*, n.s., 4:616 (1968). On Snyder, see also William Everson, *Archetype West: The Pacific Coast as a Literary Region* (Berkeley: Oyez Press, 1976), 141–145; Bill McKibben, "The Mountain Hedonist," *New York Review of Books*, April 11, 1991; and Jon Halper, ed., *Gary Snyder: Dimensions of a Life* (San Francisco: Sierra Club Books, 1990).
16. Dan Ivan Janik, "Environmental Consciousness in Modern Literature: Four Representative Examples," in Schultz and Hughes, op. cit., 72.
17. *The Practice of the Wild*, op. cit., 93.
18. *Turtle Island*, op. cit., 107.

19. *The Practice*, op. cit., 6.
20. See "Buddhism and the Coming Revolution," *Earth House Hold*, op. cit., 90–93.
21. *Turtle Island*, op. cit., 104.
22. "Buddhism and the Coming Revolution," *Earth House Hold*, op. cit., 90.
23. "The *East-West* Interview," in *The Real Work*, op. cit., 109.
24. "The Bioregional Ethic," interview with Matt Helm in ibid., 147.
25. *The Practice*, op. cit., 41.
26. *Turtle Island*, introductory note, n.p.
27. "The Real Work," interview with Fred Geneson *The Real Work*, op. cit., 59.
28. Peter Borelli, "The Ecophilosophers," *Amicus Journal* 10: 34 (Spring 1988).
29. Feminist Carolyn Merchant has written that "the 'land ethic' of Aldo Leopold . . . the 'declarations of interdependence" of ecology action groups, and the nature religion of Callenbach's Ecotopians . . . present a community-oriented alternative to the homocentric ethics of ecosystem management"; op.cit., 252.
30. (New York: Bantam Books, 1977).
31. For Garreau the region includes also British Columbia and most of the coast of Alaska. See *The Nine Nations of North America* (New York: Avon Books, 1981), esp. 245–386. The term has been appropriated by the Times-Mirror corporation in their organization of poll data.
32. On the economy, see Richard Frye, "The Economics of *Ecotopia*," *Alternative Futures: The Journal of Utopian Studies* 3:71–81 (1980).
33. Sessions and Devall, "Ecotopia: The Vision Defined," in *Deep Ecology*, op. cit., 163.
34. Callenbach, op. cit., 19, 37, 46, 91–96.
35. Ibid., 46.
36. (Berkeley: Banyan Tree Books, 1981).
37. Callenbach, *Ecotopia*, op. cit., 46, 82, 107, 126, 129.
38. Ibid., 80, 81.
39. Ibid., 84. One should note in passing that Huxley, in his utopia, *Island*, op. cit., did not completely reject such an intent, and the conviction that an ideal society is simply a steppingstone to an ideal race is an old tradition in utopian literature, and is even found in the American classic, *Looking Backward*.
40. See pp. 21–24.
41. Kirkpatrick Sale, *Dwellers in the Land: The Bioregional Vision* (San Francisco: Sierra Club Books, 1985).
42. Borelli, op. cit.
43. Quoted in ibid. See also Berg, ed., *Reinhabiting a Separate Country: A Bioregional Anthology of Northern California* (San Francisco: Planet Drum Foundation, 1970); and Van Andruss, Christopher Plant, Judith Plant, and Eleanor Wright, eds., *Home! A Bioregional Reader*, foreword by Stephanie Mills (Santa Cruz, Calif.: New Society Publishers, 1990).
44. See "Shallow and Deep Ecology: A Review of the Literature," in Schultz and Hughes, op. cit., 419–420.
45. For an early formulation, see Lynn Margulis and James Lovelock, "The Atmosphere as Circulatory System of the Biosphere—The *Gaia* Hypothesis," *Co-Evolution Quarterly* 6:31–40 (Summer 1975). For the classic version, see Lovelock, *Gaia: A New Look at Life on Earth*, op. cit. Lovelock describes his unusual career in "The Independent Practice of Science," *Co-Evolution Quarterly* 25:22–30 (Spring 1980). On its relations to deep ecology, see George Sessions, "Deep Ecology, New Age, and Gain Consciousness," *Earth First!* 7(8):27–30 (1987). On the concept, see also Lawrence E. Joseph, *Gaia: The Growth of an*

Idea (New York: St. Martin's Press, 1991). On its political effects, see William Irwin Thompson, ed., *Gaia: A Way of Knowing. Political Implications of the New Biology* (Great Barrington, Mass.: Inner Traditions/Lindesfarne Press, 1987). For a similar view, see Dorian Sagan, *Biospheres: Metamorphosis of Planet Earth* (New York: McGraw-Hill, 1990).

46. Lovelock, *Gaia,* op. cit., 121.
47. See Jeanne McDermott, "Lynn Margulis—Unlike Most Microbiologists," *Co-Evolution Quarterly* 25:31–36 (Spring 1980).
48. See, for example, W. Ford Doolittle, "Is Nature Really Motherly?" *Co-Evolution Quarterly* 29:58–63 (Spring 1981); "James Lovelock Responds," *Co-Evolution Quarterly* 29:61–63 (Spring 1981); and "Lynn Margulis Responds," ibid., 63–65.
49. Quoted in Richard A. Kerr, "No Longer Willful, Gaia Becomes Respectable," *Science* 240:393 (1988).
50. See, for example, Norman Myers, *The Gaia Atlas of Future Worlds: Challenge and Opportunity in an Age of Change,* foreword by Kenneth E. Boulding (New York: Anchor Books, 1990).
51. See Paul Ehrlich, Barry Commoner, Francis Moore Lappe, et al., "The Population Bomb: An Explosive Issue for the Environmental Movement?" *Utne Reader: The Best of the Alternative Press* 27:78–87 (May–June 1988). The Ehrlichs have remained adamant in their position; see their review of Ben Wattenberg's *The Birth Dearth: What Happens When People in Free Countries Don't Have Enough Babies* (New York: Pharos Press, 1987) in *Amicus Journal* 10:40 (Winter 1988), wherein they say that "concerns about a birth dearth are not only anachronistic and misplaced but downright counterproductive. What is most needed is even lower birthrates around the world."
52. Excerpted in *Utne Reader,* op. cit., 83.
53. Quoted in Christopher Manes, *Green Rage: Radical Environmentalism and the Unmaking of Civilization* (Boston: Little, Brown, 1991), 70.
54. Dick Russell, "The Monkeywrenchers," *Amicus Journal* 9:31 (Fall 1987). On Earth First! see Trip Gabriel, "If a Tree Falls in the Forest, They Hear It," *New York Times Magazine,* November 4, 1990. On recent problems, see Thomas Goltz, "Earth First Meeting Reflects Gap Between Radicals, Mainstream," *Washington Post,* July 19, 1990. For a sampling of their ideas, see John Davis, ed., *The Earth First! Reader: Ten Years of Radical Environmentalism,* foreword by Dave Foreman (Salt Lake City: Peregrine Smith Books, 1991).
55. See Manes, op. cit., 175–190.
56. Brian Tokar, "What Kind of Environmentalism?" *Zeta Magazine* 4:46 (June 1991).
57. Manes, op. cit., 76.
58. Edward Abbey, *The Monkey Wrench Gang* (New York: Avon Books, 1983). See also his *Desert Solitaire* (Tucson: University of Arizona Press, 1988), and *The Journey Home: Some Words in Defense of the American West* (New York: Plume, 1991).
59. Alexander Cockburn, "Live Souls," *Zeta Magazine* 1:13–14 (February 1988). See also his followup debate with an irate member of Earth First! in "Legends of Anteus," *Zeta Magazine* 1:9–11 (April 1988).
60. *Desert Solitaire,* op. cit., 33.
61. For a partisan view of the trial, see Karen Pickett, "Arizona Conspiracy Trial Ends in Plea Bargain," *Earth First!,* September 13, 1991; and Pickett, "A Snitch on the Stand, and Other Revelations about the Pawns of the Evil Empire," ibid., 25.
62. (New York: Harmony Books, 1991). See also *Defending the Earth: A Dialog between Murray Bookchin and Dave Foreman,* edited and with an introduction by Steve Chase. Foreword by David Levine (Boston: South End Press, 1990).
63. Al Bravo, "Activists Seek Deal in Sabotage Case," *Albuquerque Journal,* August 14, 1991.

64. Charlene Smith, "Earth First! Vows More Sabotage after Guilty Pleas," *Albuquerque Journal*, December 8, 1991.
65. John Lancaster, "The Green Guerrilla," *Washington Post*, March 20, 1991.
66. Smith, op. cit.

15. *Reform or Revolution?*

1. Robert C. Paehlke, *Environmentalism and the Future of Progressive Politics* (New Haven: Yale University Press, 1989), 3.
2. Grant, op. cit., 17.
3. David M. Potter, *People of Plenty: Economic Abundance and the American Character* (Chicago: University of Chicago Press, 1982), passim.
4. In 1975 a Canadian political leader could write that "there is no doubt that a social ethics against exponential growth is developing in one way or another"; Maurice Lamontaigne, "The Loss of the Steady State," in Abraham Rotstein, ed., *Beyond Industrial Growth*, foreword by Robertson Davies (Toronto: University of Toronto Press, 1976), 17. On Canada, see also Paehlke, op. cit., passim.
5. Samuel P. Hays (in collaboration with Barbara B. Hays), *Beauty, Health, and Permanence: Environmental Politics in the United States, 1955–1975* (New York: Cambridge University Press, 1987), 3. On this change of emphasis, see also Roderick Nash, "Rounding Out the American Revolution," in Tobias, op. cit., 170–181.
6. Sagoff, op. cit., 154.
7. Hays, op. cit., 13. See also Hays, "From Conservation to Environment: Environmental Politics in the United States since World War II," *Environmental Review* 6:14–41 (1982).
8. *This Ugly Civilization* (New York: Simon and Schuster, 1929).
9. Fox, op. cit., 187.
10. Ibid., 217. But see also R. V. Young, Jr., "A Conservative View of Environmental Ethics," *Environmental Ethics* 1:241–254 (1979).
11. Sagoff, op. cit., 169.
12. See the chapter, "The Reagan Antienvironmental Revolution," in Hays, op. cit., 491–526.
13. "Environmental inquiry did not lead to a single system of thought such as social theorists might prefer, and it would be difficult to reduce its varied strands to a single pattern"; Hays, op. cit., 247.
14. For a brief, not wholly unsympathetic but critical history, see Winner, op. cit., 61–84. See also David Dickson, *The Politics of Alternative Technology* (New York: Universe Books, 1974); Franklin Long and Alexandra Oleson, eds., *Appropriate Technology and Social Values: A Critical Appraisal* (Cambridge: Ballinger, 1980); Kevin W. Willoughby, *Technology Choice: A Critique of the Appropriate Technology Movement* (Boulder, Colo.: Westview Press, 1990); and Winner, "The Political Philosophy of Alternative Technology," *Technology in Society* 1:75–86 (1979).
15. See above all his *Small Is Beautiful*, op. cit.
16. See as an introduction to his ideas, *Soft Energy Paths: Toward a Durable Peace* (San Francisco: Friends of the Earth International, 1977). For related discussions of the relationship between energy technologies and politics, see Victor Ferkiss, "Value Choices and Technological Options," *Space Solar Power Review* 4: 261–271 (1983); and Griffin Thompson, "Energy Technology and American Democratic Values," Ph.D. thesis, Department of Government, Georgetown University, Washington, D.C., 1988.
17. See philosopher Fred Ferre, "Toward a Postmodern Science and Technology," in David

Ray Griffin, ed., *Spirituality and Society: Postmodern Visions* (Albany: State University of New York Press, 1988), 133–141.

18. On technology assessment, see Stanley R. Carpenter, "Technoaxiology: Appropriate Norms for Technology Assessment," in Durbin and Rapp, op. cit., 115–136; Friedrich Rapp, "The Prospects for Technology Assessment," ibid., 141–150; Kristin Shrader-Frechette, "Technology Assessment and the Problem of Quantification," ibid., 151–164; Victor Ferkiss, "Technology Assessment and Appropriate Technology," *National Forum* 57:3–7 (Fall 1978); and John H. Gibbons and Holly L. Gwin, "Technology and Governance: The Development of the Office of Technology Assessment," in Michael E. Kraft and Norman J. Vig, eds., *Technology and Politics* (Durham, N.C.: Duke University Press, 1988), 98–122. Gibbons was head of OTA during the Reagan administration. See also Lynn White, Jr., "Technology Assessment from the Stance of a Medieval Historian," *American Historical Review* 79:1–13 (1974).

19. See Peter F. Drucker, "The New Technology: Predicting Its Impact," *New York Times*, April 8, 1973.

20. So also says Witold Rybczynski, *Paper Heroes: A Review of Technology Assessment* (Garden City, N.Y.: Anchor Books, 1980). On Schumacher, see also Hazel Henderson, "The Legacy of E. F. Schumacher," *Environment* 20:30–46 (1978). For his ideas in his own words, see *A Guide for the Perplexed*, op. cit.; "Buddhist Economics," in Theodore Roszak, ed., *Sources: An Anthology of Contemporary Materials Useful for Preserving Personal Sanity While Braving the Great Technological Wilderness* (New York: Harper Colophon Books, 1972), 259–272; "An Economics of Permanence," in ibid., 354–373; and "The Difference between Unity and Uniformity," *Co-Evolution Quarterly*, 7:52–59 (Fall 1975). See also Peter Gillingham, "Schumacher's Buddhism," *Co-Evolution Quarterly* 7:60 (Fall 1975). Schumacher, though sympathetic to Buddhism, was actually a Roman Catholic.

21. Schumacher was careful to stress that appropriate technology might in some instances be large-scale. Though normally this would not be true, all judgments had to be made on a case-by-case basis. For a critique of his ideas, see Charles T. Rubin, "E. F. Schumacher and the Politics of Technological Renewal," *Political Science Reviewer* 16:65–96 (1986).

22. Florman, op. cit., 90.

23. Alexander Cockburn in Cockburn and James Ridgeway, eds., *Political Ecology* (New York: Times Books, 1979), 347.

24. R. P. Turco, G. B. Toon, T. P. Ackerman, J. B. Pollack, and Carl Sagan, "Nuclear Winter: Global Consequences of Multiple Nuclear Explosions," *Science* 222:1283–1294 (1983); Paul R. Ehrlich, John Harte, Mark A. Harwell, Peter H. Raven, Carl Sagan, George M. Woodwell, Joseph Berry, Edward S. Ayelso, Anne R. Ehrlich, Thomas Norman Myers, David Pimentel, and John M. Teal, "Long-Term Biological Consequences of Nuclear War," *Science* 222:1293–1300 (1983); and Carl Sagan, "Nuclear War and Climatic Catastrophe: Some Policy Implications," *Foreign Affairs* 62:257–292 (Winter 1983–1984); R. Jeffrey Smith, "A Grim Portrait of the Postwar World," *Science* 229: 1245–1246 (1985); Eliot Marshall, "Armageddon Revisited," *Science* 236:1421–1422 (1987). But see also R. P. Turco, O. B. Toon, T. P. Ackerman, J. B. Pollack, and C. Sagan, "Climate and Smoke: An Appraisal of Nuclear Winter," *Science* 247:166–176 (1990); Malcolm W. Browne, "Nuclear Winter Theorists Pull Back," *New York Times*, January 23, 1990; and Stephen S. Rosenfeld, "Nuclear Winter: The Sun Peeks Through," *Washington Post*, January 26, 1990. See also Carl Sagan and Richard Turco, *A Path Where No Man Thought: Nuclear Winter and the End of the Arms Race* (New York: Random House, 1990).

25. Toulmin, op. cit., 262.

26. Actually, the first political party based on ecological values was the Values party of New

Zealand, founded in the late 1960s. But after flirting with socialism it had disappeared by 1981; see Fritjof Capra and Charlene Spretnak (in collaboration with Rudiger Lutz), *Green Politics* (New York: E. P. Dutton, 1984), 171–172.

27. Ibid., 14. On their intellectual origins, see also leftist German Werner Hulsberg, *The German Greens: A Social and Political Profile*, trans. by Gus Fagan (London: Verso, 1988), 2. According to an American unfriendly to the Greens, not only *The Limits to Growth* and *Small Is Beautiful* but Barry Commoner's *The Closing Circle* "had a tremendous impact on the critical intellectual community of West Germany"; see Kim R. Holmes, "The Origins, Development, and Composition of the Green Movement," in Robert L. Pfaltzgraff, Jr., Kim R. Holmes, Clay Clemens, and Werner Kaltefelter, *The Greens of West Germany: Origins, Strategies, and Transatlantic Implications* (Cambridge: Institute for Policy Analysis, 1983), 23–24. On the Greens, see also Wolfgang Rudig, ed., *Green Politics I 1990* (Carbondale: Southern Illinois University Press, 1991).

28. Holmes, op. cit., 25.

29. Emil Papadakis, *The Green Movement in West Germany* (London: Croom and Helm, 1984), 1. Inglehart's book is *The Silent Revolution: Changing Values and Political Styles among Western Publics* (Princeton: Princeton University Press, 1977).

30. Ibid., Papadakis, 3.

31. Capra and Spretnak, op. cit., 144. Ironically, given the Marxist origins of much German concern with industrial society, and the struggle for power within the Greens between leftists and others, the first impetus to Green politics came from the Right. Herbert Gruhl, author of the famous Green slogan, "We are neither left nor right, we are in front," was a member of the Christian Democratic Union; ibid., 15.

32. Diane Johnstone, "The Green Dilemma: To Be or Not to Be a Realo," *In These Times*, August 17–30, 1988.

33. Capra and Spretnak, op. cit., 53. Recently a publication, *Green Spirit: Bulletin for Spiritual Green Politics*, was founded and published in Vienna and dedicated to bringing together various viewpoints, including Christian, Buddhist, and pagan; *New Options*, December 28, 1987.

34. Spretnak, *The Spiritual Dimension of Green Politics* (Santa Fe, N.M.: Bear and Co., 1986), 22.

35. Ibid. She notes that since her original research trip in 1983, a group called Christians in the Greens had been formed in every West German state.

36. Capra and Spretnak, op. cit., 54–55.

37. Gary MacEoin, "Ratzinger Knocks Green Party, Dialogue with Jews," *National Catholic Reporter*, November 6, 1987. Since the pope's 1990 message, he has taken a somewhat different tack, saying that "with our ecological problems, we begin to understand that nature has its own moral message to give us"; Michael D. Schaffer, "Vatican's Defender Foresees More Tumult," *Washington Post*, January 20, 1990.

38. For her views, see Kelly, *Fighting for Hope*, introduction by Heinrich Boll, trans. by Marianne Howarth (Boston: South End Press, 1984).

39. Hulsberg, *The German Greens*, op. cit., 4–5.

40. Bahro, *From Red to Green: Interviews with New Left Review* (London: Verso, 1984), 62.

41. Ibid., 142, 143, 172, 185, 212, 213.

42. Ibid., 222.

43. John Sandford, introduction to Rudolf Bahro, *Building the Green Movement*, trans. by Mary Tyler (Philadelphia: New Society Publishers, 1986), 9.

44. Bahro, *From Red to Green*, op. cit., 210.

45. "This movement now has official political parties in 16 countries and members of parlia-

ment (MPs) in eight. The European Parliament itself now includes Green Party members"; Lester R. Brown, Christopher Flavin, and Sandra Postel, "A World at Risk," in Lester R. Brown et al., eds., *State of the World, 1989: A Worldwatch Institute Report on Progress toward a Sustainable Society* (New York: W. W. Norton, 1989), 7. See also Ferdinand Muller Rommell, ed., *New Politics in Western Europe: The Rise and Success of Green Parties and Alternate Lists* (Boulder, Colo.: Westview Press, 1989).

46. David S. Broder, "Haughey Fails in Bid for Majority," *Washington Post*, June 18, 1989.
47. "National Party Wins New Zealand Vote; Improved U.S. Ties Sought," *New York Times*, October 28, 1990.
48. Marlise Simons, "A Greens Mayor Takes On the Industrial Filth of Old Cracow," *New York Times*, March 25, 1990.
49. David Van Bieme, "Jeremy Rifkin Is a Little Worried about Your Future," *Washington Post Magazine*, January 17, 1988.
50. Diana Johnstone, "Red and White Fade to Green in EC Elections," *In These Times*, July 5–18, 1989.
51. Karen De Young, "Social Democrats Win Election in Sweden, Greens Gain Seats," *New York Times*, September 19, 1988.
52. Jack Mundey talking to Ellen Newman and David Gancher, "Labor and Environment— The Australian Green Bans," *Co-Evolution Quarterly* 13:40–45 (Spring 1977).
53. Jack Mundey, "Ecology, Capitalism, Communism," *Co-Evolution Quarterly* 13:46–48, (Spring 1977).
54. On the French Greens, see Sylvie Crossman, "The Greens (Les Vertes)," *Co-Evolution Quarterly* 16:94–99 (Winter 1977–1978).
55. D.D., " 'Greens' Challenge French Gene Research," *Science* 237:357 (1987).
56. Mark Hunter, "Greens Join Mainstream in France," *Washington Post*, March 22, 1989; and Diana Johnstone, "Spring Brings a Touch of Green to France," *In These Times*, April 5–11, 1989.
57. Johnstone, op. cit.
58. Ibid.
59. Jonathan C. Randal, " 'Greening' of Thatcher Surprises Many Britons," *Washington Post*, March 4, 1989; and Malcolm W. Browne, "Ozone Fading Fast, Thatcher Tells World Experts," *New York Times*, June 28, 1990.
60. Tim Golden, "Life's Gray for Mexico's Green Party," *New York Times*, August 14, 1991.
61. Serge Schmmann, "A Normal Germany," *New York Times*, December 4, 1990.
62. Marc Fisher, "Greens Fade from Germany's Political Spectrum," *Washington Post*, December 4, 1990.
63. Stephen Kinzer, "Kohl Loses State Election to Socialist-Green Coalition," *Washington Post*, January 22, 1991.
64. See, as an introduction, Capra and Spretnak, op. cit., passim.
65. Kirkpatrick Sale, "Greens Take Root in American Soil," *Utne Reader* 12:87 (October– November 1985).
66. Harry Boyte and Karen Lehmen, "Are You Curious (Green)?" *Utne Reader* 7:120–121 (December 1984–January 1985).
67. See Jay Walljasper, "Zeitgeist (Deep Ecology vs. Socialist Ecology?)" *Utne Reader* 30:135– 136 (November–December 1988). On periodicals covering the movement, see Walljasper, "Zeitgeist (The Politics of Ecology)," *Utne Reader* 28:126–127 (July–August 1988).
68. (San Pedro, Calif.: R and E Miles, 1987).
69. Walljasper, "Zeitgeist (Deep Ecology vs. Socialist Ecology?), op. cit., 135. See also Brian Tokar, "Ecological Radicalism," *Zeta Magazine* 1:84, 90–91 (December 1988).

70. See Phil Hill, "Left-Green Network Forms," *Potomac Valley Green Network*, May 1989, 8.

71. Quoted in Tokar, op. cit. In defense of deep ecology, see also Arne Naess, "Finding Common Ground," *Green Synthesis* 30:9–10 (March 1989); and Joanna Macy, "Deep Ecology Notes," ibid., 10.

72. Interview, "Murray Bookchin," *Whole Earth Review* 61:16 (Winter 1988).

73. For the source of his unhappiness, see Kirkpatrick Sale, "Deep Ecology and Its Critics," *Nation* 246:670–672+ (1988).

74. Bookchin, "Crisis in the Ecology Movement," *Zeta Magazine* 1:123 (July–August 1988). For Abbey's views, see interview, "Edward Abbey," *Whole Earth Review* 61:17 (Winter 1988).

75. Interview, "Murray Bookchin," op. cit.

76. Bookchin, "Crisis," op. cit., 121.

77. For a defense of spirituality in the movement, see Matthew Fox, "A Call for a Spiritual Renaissance," *Green Letter* 5:4, 16–17 (Spring 1989).

78. Spretnak, *Spiritual Dimension*, op. cit., 25. The book originated as a lecture given at the E. F. Schumacher Society of America. She describes her own religious orientation as being that of someone who left the Catholic church because it was not spiritual enough and now practices Buddhist Vipassana meditation but is "also drawn to the wisdom tradition of the Christian mystics and 'creation spirituality' "; ibid., 17. But see also Mike Wyatt, "Humanism and Ecology: The Social Ecology/Deep Ecology Schism," *Green Synthesis* 32:7–8 (October 1989).

79. See Dave Lindorff, "The One Issue That Could Unite the American Left," *In These Times*, November 16–22, 1988. See also Alexander Cockburn, "Socialist Ecology: What It Means, Why No Other Kind Will Do," *Zeta Magazine* 2:15–21 (February 1989); Cockburn, "Their Mullahs and Ours," ibid., 2:33–40 (April 1989); "Socialist Ecology," Letters from Bookchin and others, ibid., 2:3–5 (April 1989); and Tom Athanasiou, Janet Biehl, and Carl Boggs, "Is the Left-Green Network Really Green?" *Green Synthesis* 32:6, 10–14 (October 1989). Recently publication began in Santa Cruz of *Capitalism, Nature, Socialism: A Journal of Socialist Ecology*. On recent developments in the struggle, see Brian Tokar, "What Kind of Environmentalism?" *Zeta Magazine* 4:43–47 (June 1991). Another source of dissension is the attempt of "cults" to take over the Green movement in various nations, including the United States but especially Latin America; Joe Conason with Jill Weiner, "The Fake Greens," *Village Voice*, December 26, 1989.

80. Spretnak, op. cit., 38–39, 71.

81. Salim Muwakkil, "U.S. Green Movement Needs to Be Colorized," *In These Times*, May 2–8, 1990. He is referring not simply to the Greens but to the whole environmental movement.

82. Tim McCarthy, "Greens Going for a Green America," *National Catholic Reporter*, July 14, 1989; Jay Walljasper, "Can Green politics take root in the U.S.?" *Utne Reader* 35:140–143, (September–October 1989); Margo Adair and John Rensenbrink, "SPAKA: Democracy at Work. Determining Strategy and Policy Approaches in Key Areas," *Green Letter/Greener Times*, spec. ed., Autumn 1989, 3–5. The Draft Text of the Green Program USA is found in ibid., 6–41. See also Robert Boczkiewitz, "U.S. Greens Adopt Policy Platform but Delay Politicking," *National Catholic Reporter*, September 28, 1990; and John Rensenbrink, "Greens USA," *National Forum* 70:20–22 (Winter 1990).

83. Boczkiewitz, op. cit. See also Brian Tokar, "The Greens: To Party or Not?" *Zeta Magazine* 4:42–46 (October 1991).

84. Mark Satin, "You Don't Have to Be a Baby to Cry," *New Options*, September 24, 1990.

85. William K. Stevens, "Worst Fears on Acid Rain Unrealized," *New York Times*, February 20, 1990; Philip Shabecoff, "Acid Rain Report Unleashes a Torrent of Criticism," *New York Times*, March 20, 1990; and L.R., "How Bad Is Acid Rain?" *Science* 251:1303 (1991).

86. See John McCormick, *Reclaiming Paradise: The Global Environmental Movement* (Bloomington: Indiana University Press, 1989); and Lynton Keith Caldwell, *International Environmental Policy: Emergence and Dimensions*, 2d ed. (Durham, N.C.: Duke University Press, 1990). Environmental issues are now routinely part of the the agenda of summit conferences; see Edward Cody, "Another Environmental Summit Opens, Illustrating Issue's Currency," *Washington Post*, March 11, 1989. For general prescriptive propositions about environmental policy, see Lester F. Brown, Christopher Flavin, and Sandra Postel, *Saving the Planet: How to Shape an Environmentally Sustainable Global Economy* (New York: W. W. Norton, 1991).

87. Lester R. Brown, "Analyzing the Demographic Trap," in Brown et al., *State of the World, 1987: A Worldwatch Institute Report on Progress toward a Sustainable Society* (New York: W. W. Norton, 1987), 20–37; and Jodi Jacobson, "Planning the Global Family," in Lester R. Brown et al., *State of the World, 1988: A Worldwatch Institute Report on Progress toward a Sustainable Society* (New York: W.W. Norton, 1988), 151–169.

88. Lester R. Brown, "Conserving Soils," in Brown et al., *State of the World, 1984: A Worldwatch Institute Report on Progress toward a Sustainable Society* (New York: W.W. Norton, 1984), 53–73; Sandra Postel, "Protecting Forests," ibid., 74–94; Postel, "Halting Land Degradation," in Brown et al. op. cit. (1989), 21–40. See also Peter Warshall, "The Political Economy of Deforestation," *Whole Earth Review* 64: 68–75 (Fall 1989).

89. Sandra Postel, "Controlling Toxic Chemicals," in Brown et al., op. cit. (1988), 118–136; Nathaniel C. Nash, "Latin Nations Getting Others' Waste," *New York Times*, December 16, 1991; and Edward Cody, "105 Nations Back Treaty on Toxic-Waste Shipping," *Washington Post*, March 23, 1989. This is a domestic problem as well; see Robert Tomsho, "Dumping Grounds: Indian Tribes Contend with Some of the Worst of America's Pollution," *Wall Street Journal*, November 19, 1990.

90. See Kenneth Maxwell, "The Tragedy of the Amazon," *New York Review of Books*, March 7, 1991; and "The Mystery of Chico Mendes," ibid., March 28, 1991. Hope yet remains, however. See Julia Preston, "Destruction of Amazon Rain Forest Slowing," *Washington Post*, March 17, 1991.

91. On the charter, see Harold W. Wood, Jr., "The United Nations World Charter for Nature," *Ecology Law Quarterly* 12:977–996 (1985).

92. "The Heat Is On," *Time*, October 19, 1987.

93. The best single work on the basic implications of the greenhouse effect is Bill McKibben, *The End of Nature* (New York: Random House, 1989). See also Micheal Oppenheimer and Robert H. Boyle, *Dead Heat: The Race against the Greenhouse Effect* (New York: Basic Books, 1990); Jeremy Leggett, ed., *Global Warming: The Greenpeace Report* (New York: Oxford University Press, 1990); and Stephen H. Schneider, *Global Warming: Are We Entering the Greenhouse Century?* (San Francisco: Sierra Club Books, 1989). Much controversy exists, of course, about the exact nature and extent of the greenhouse, and a minority of scientists is skeptical. See Richard A. Kerr, "Hansen *vs* the World on the Greenhouse Threat," *Science* 244:1041–1043 (1989); William Booth, "Climate Study Halves Estimate of Global Warming," *Washington Post*, September 14, 1989; Booth, "New Models Chill Some Predictions of Severely Overheated Planet," *Washington Post*, January 29, 1990; Kerr, "New Greenhouse Report Puts Down Dissenters," *Science* 249:481–482 (1990); Kerr, "Global Temperatures Hit Record Again," *Science* 251:274 (1991); and Karen Young, "Heating the Global Warming Debate," *New York Times Magazine*, Feb-

ruary 3, 1991. The explosion of Mount Pinatubo in the Philippines in 1991 will of course lead to a temporary cooling; William K. Stevens, "The Earth Cools Off after Spell of Warmth," *New York Times*, December 24, 1991. Controversy also exists about its physical and social consequences; see Booth, "Projected Sea-Level Rise Scaled Back by Scientists," *Washington Post*, December 22, 1989; Shabecoff, "Draft Report on Global Warming Foresees Environmental Havoc in U.S.," *New York Times*, October 20, 1988; Leslie Roberts, "Is There Life after Climate Change?" *Science* 242:1010–1012 (1988); Roberts," "How Fast Can Trees Migrate?" *Science* 243:735–737 (1989); Booth, "Global Heating Could Benefit U.S. Farmers," *Washington Post*, May 17, 1990; Charles Campbell, "Science Committee Concludes Country Can Take Heat of Greenhouse Effect," *Albuquerque Journal*, September 7, 1991; and Peter Passell, "Warmer Globe, Greener Pastures?" *New York Times*, September 18, 1991.

94. Michawl Weisskopf, " 'Greenhouse Effect' Fueling Policy Makers," *Washington Post*, August 15, 1988; Stephen H. Schneider, "The Greenhouse Effect: Science and Policy," *Science* 243:771–781 (1989); William K. Stevens, "Governments Start Preparing for Global Warming Disasters," *New York Times*, November 14, 1989; William Booth, "Action Urged against Global Warming," *Washington Post*, February 2, 1990; Booth, "Carbon Dioxide Curbs May Not Halt Warming," *Washington Post*, March 10, 1990; " 'Global Marshall Plan' Urged for Environment," *Washington Post*, May 3, 1990; Leslie Roberts, "Academy Panel Split on Greenhouse Adaptation," *Science* 253:1206 (1991); and Weisskopf, "Strict Energy-Saving Urged to Combat Global Warming," *Washington Post*, April 11, 1991. See also Brian Tokar, "Politics under the Greenhouse," *Zeta Magazine* 2:90–96 (March 1989). The Bush administration has vacillated on the issue, in large measure because a key Bush aide, Governor John Sununu, an MIT Ph.D. graduate in physics, was skeptical; see Michael Weisskopf, "Reagan Aide Blocks 'Greenhouse' Rule," *Washington Post*, January 13, 1989; Philip Shabecoff, "Bush Asks Cautious Response to Threat of Global Warming," *New York Times*, February 6, 1990; William Drozdiak, "U.S. Refuses to Pledge Limit on Greenhouse Gases Emissions," *Washington Post*, November 6, 1990; and Richard A. Kerr, "U.S. Bites Greenhouse Bullet and Gags," *Science* 251:368 (1991). Sununu's replacement in December 1991 by the more pragmatic Samuel Skinner may lead to more flexibility in American policy. See also Robert Suro, "Europeans Accuse the U.S. of Balking on Plans to Combat Global Warming," *New York Times*, July 10, 1990. Some argue that it might be easier and less expensive to adjust to the greenhouse effect than to try to mitigate it; see Peter Passell, "Staggering Cost Is Foreseen to Curb Warming of Earth," *New York Times*, November 19, 1989.

95. Cass Peterson, "Skin Cancer Rises to Near Epidemic," *Washington Post*, March 10, 1987.

96. But skepticism remains; see Charles C. Mann, "Extinction: Are Ecologists Crying Wolf?" *Science* 253:736–738 (1991).

16. *Where Does Humanity Go from Here?*

1. Mark Sagoff, "The Greening of the Blue Collars," *Philosophy and Public Policy* 10:1–5 (Summer–Fall 1990). On the overwhelming general support for environmental concerns, at least in theory, see Rose Gutfeld, "Eight of 10 Americans Are Environmentalists, at Least So They Say," *Wall Street Journal*, August 2, 1991.

2. Herman E. Daly and John B. Cobb, Jr., with contributions by Clifford W. Cobb, *For The Common Good: Redirecting the Economy toward Community, the Environment, and a Sustainable Future* (Boston: Beacon Press, 1989), 37, 51, 55.

3. Bruce Piasecki and Peter Asmus. *In Search of Environmental Excellence: Moving Beyond Blame*, foreword by Jean-Michel Cousteau, introduction by Congresswoman Claudine

Schneider, afterword by the Honorable Robert K. Dawson (New York: Simon and Schuster, 1990), 128.

4. On Japan, see ibid., 134–135; on United States firms, see ibid., 18, 48, 129.

5. Bilateral pressure can often underline international consensus. See Stephen Weisman, "Japan Agrees to End the Use of Huge Fishing Nets," *New York Times*, November 27, 1991.

6. Some highly localized effects, in southern Chile, are already being noted. Nathaniel C. Nash, "Unease Grows under the Ozone Hole," *New York Times*, July 23, 1991. The United States has special problems. See William K. Stevens, "Summertime Harm to Shield of Ozone Detected Over U.S.," *New York Times*, October 23, 1991; and Michael Weisskopf, "Unusual Thinning Of Ozone Layer Found Over U.S.," *Albuquerque Journal*, October 23, 1991.

7. Quoted in Piasecki, op. cit., 124.

8. Jonathan Weiner, *The Next One Hundred Years: Shaping the Fate of Our Living Earth* (New York: Bantam Books, 1991), 215.

9. William K. Stevens, "At Meeting, U.S. Is Alone on Global Warming," *New York Times*, September 10, 1991; "U.S. Still Resists Curbs on Emissions," *New York Times*, September 22, 1991.

10. "Excerpts from Group of Seven's Declaration: Progress, Policies and Plans," *New York Times*, July 18, 1991.

11. Advertisement, "A Latin American Ecological Alliance," *New York Times*, July 22, 1991.

12. Daly, op. cit., 178.

13. William Pfaff, " 'New World Order' Bigger Than Bush," *Albuquerque Journal*, August 4, 1991.

14. Keith Bradsher, "U.S. and Mexico Draft Plan to Fight Border Pollution," *New York Times*, August 2, 1991.

15. See Diana Page, "Debt-for-Nature Swaps: Experience Gained, Lessons Learned," *International Environmental Affairs* 1:275–288 (1989).

16. Keith Bradsher, "U.S. Ban on Mexico Tuna Overruled," *New York Times*, August 23, 1991.

17. For an eloquent attack on free trade, see Daly, op. cit., 209–235.

18. Piasecki, op. cit., 108.

19. Ibid., 21.

20. William Booth, "Nuclear Power Debate Shifts," *Washington Post*, August 24, 1989; Weiner, op. cit., 223–226. It is argued that in any case it would still add to global warming; Piasecki, op. cit., 184, fn.18.

21. Ibid.,Piasecki, 74.

22. Ibid., 50–74.

23. Keith Schneider, "Military Has New Strategic Goal in Cleanup of Vast Toxic Waste," *New York Times*, August 5, 1991.

24. See Piasecki, op. cit., 90–91; Weiner, op. cit., 228; William Booth, "Johnny Appleseed and the Greenhouse," *Science* 242:19–20 (1988); Gregory Byrne, "Let 100 Million Trees Bloom," *Science* 242: 371 (1988); and Booth, "Forests May Combat Greenhouse Effect," *Washington Post*, March 23, 1990. Even the Bush administration was somewhat interested. Philip Shabecoff, "Bush Wants Billions of Trees For War Against Polluted Air," *New York Times*, January 8, 1990. Saving trees can take place at the grass roots anywhere. See Barbara Crossette, "Village Committees Learn to Guard Endangered Forest in Bangladesh," *New York Times*, August 6, 1991.

25. Weiner, op. cit., 227.

26. Daly, op. cit., 252–256. On the minimum amount of land necessary to preserve wilderness, see Weiner, op. cit., 163–190.
27. As most environmentalists realize. See Joe Alper, "Environmentalists: Ban the (Population) Bomb," *Science* 252:1247 (1991).
28. Paradoxically, economic growth may be necessary to check pollution. Keith Bradsher, "Lower Pollution Tied to Prosperity," *New York Times*, October 28, 1991.
29. Weiner, op. cit., 80.
30. Ibid., 234–239.
31. Keith Schneider, "Minorities Join to Fight Toxic Waste," *New York Times*, October 25, 1991. See also Felicity Barringer, "In Capital, No. 2 River Is a Cause," *New York Times*, December 1, 1991. The Highlander Folk School in Tennessee, whose training of Rosa Parks sparked the Civil Rights movement, now has STP (Stop the Poison) workshops to train environmental activists. See Carol Polsgrove, "Unbroken Circle," *Sierra* 77:130–133, 240 (January–February 1992).
32. Ibid. See also Gayle Geis, "Activists Lash 'Racist' Policies At EPA Banquet," *Albuquerque Journal*, August 1, 1991.
33. Keith Schneider, "Student Group Seeks Broader Agenda for Environmental Movement," *New York Times*, October 7, 1991. Not only students are concerned. On adult alliances, see Kent Paterson, "Expanding the Spectrum of U.S. Green Politics," *In These Times*, September 4–10, 1991.
34. See Katrin Snow, "Tribes' Activism Poses Hazard to Waste Industry's Health," *National Catholic Reporter*, December 9, 1991.
35. See Allie Kohn, *The Brighter Side of Human Nature* (New York: Basic Books, 1990); and Morton Hunt, *The Compassionate Beast: The Scientific Inquiry into Human Altruism* (New York: William Morrow, 1991).
36. Daly, op. cit., 166.
37. See Michael Walzer, "The Communitarian Critique of Liberalism," *Political Theory* 18:6–23 (1990). A new journal recently came into existence to push for communitarian values within a liberal context, *The Responsive Community*, founded and edited by Amitai Etzioni.
38. For a venture in this direction, see Jeremy Rifkin, *Biosphere Politics: A New Consciousness for a New Century* (New York: Crown Publishers, 1991).
39. Daly argues for a Christian theocentric basis; op.cit., 376–400.

BIBLIOGRAPHY

Books

Abbey, Edward. *Desert Solitaire*. Tucson: University of Arizona Press, 1988. Originally published in 1968.
———. *Hayduke Lives!* Boston: Little Brown, 1989.
———. *The Journey Home: Some Words in Defense of the American West*. New York: Plume, 1991.
———. *The Monkey Wrench Gang*. New York: Avon Books, 1983.
Abu-Lughod, Janet. *The World System, 1250 A.D.–1350 A.D.* New York: Oxford University Press, 1989.
Adams, Brooks. *The Law of Civilization and Decay*. Introduction by Charles A. Beard. New York: Vintage Books, 1955.
Adams, Henry. *Mont-Saint-Michel and Chartres*. Introduction by Ralph Adams Cram. Boston: Houghton Mifflin, 1933.
———. *The Education of Henry Adams*. Introduction by James Truslow Adams. New York: Modern Library, 1931.
Agel, Jerome. *Herman Kahnsciousness: The Megaton Ideas of a One-Man Think Tank*. New York: Signet Books, 1973.
Akin, William E. *Technocracy and the American Dream: The Technocratic Movement, 1900–1941*. Berkeley: University of California Press, 1977.
Albanese, Catherine L. *Nature Religion in America: From the Algonkian Indians to the New Age*. Chicago: University of Chicago Press, 1990.
Aldiss, Brian W. *Billion Year Spree: The True History of Science Fiction*. New York: Doubleday, 1973.
Alec, Margaret. *Hypatia's Heritage: A History of Women in Science from Antiquity through the Nineteenth Century*. London: Women's Press, 1986.
Al-Faruqi, Ismail R., and Abdullah Omar Naseef, eds. *Social and Natural Sciences: An Islamic Perspective*. Jeddah: King Abdulaziz University and Hodder and Stoughton, 1981.
Alford, C. Fred. *Science and the Revenge of Nature: Marcuse and Habermas*. Tampa: University Presses of Florida, 1985.
al-Hassan, Ahmed, and Donald Hill. *Islamic Technology*. New York: Cambridge University Press, 1987.

289

Allsopp, Bruce. *The Garden Earth: The Case for Ecological Morality*. New York: William Morrow, 1972.

Anderson, Walt. *A Place of Power: The American Episode in Human Evolution*. Santa Monica, Calif.: Goodyear, 1976.

Andruss, Van, Christopher Plant, Judith Plant, and Eleanor Wright, eds. *Home! A Bioregional Reader*. Foreword by Stephanie Mills. Santa Cruz, Calif.: New Society Publishers, 1990.

Angebert, Jean-Michel. *The Occult and the Third Reich: The Mystical Origins of Nazism and the Search for the Holy Grail*. Translated by Lewis A. M. Sumberg. New York: McGraw-Hill, 1974.

Applewhite, E. J. *Paradise Mislaid: Birth, Death and the Human Predicament of Being Biological*. New York: St. Martin's Press, 1991.

Arab-Ogly, E. *In the Forecasters' Maze*. Moscow: Progress, 1975.

Ardrey, Robert. *African Genesis*. New York: Atheneum, 1961.

Armstrong, Edward. *Saint Francis, Nature Mystic: The Derivation and Significance of the Nature Stories in the Franciscan Legend*. Berkeley: University of California Press, 1973.

Aronowitz, Stanley. *The Crisis in Historical Materialism: Classes, Politics, and Culture in Marxist Theory*. New York: Praeger, 1981.

Asimov, Isaac. *Extraterrestrial Civilization*. New York: Crown, 1979.

Attig, Tom, and Donald Scherer, eds. *Ethics and the Environment*. Englewood Cliffs, N.J.: Prentice Hall, 1983.

Bahro, Rudolf. *Building the Green Movement*. Introduction by John Sandford. Translated by Mary Tyler. Philadelphia: New Society Publishers, 1986.

———. *From Red to Green. Interviews with New Left Review*. London: Verso, 1984.

Bailyn, Bernard, David Brion Davis, David Herbert Donald, John L. Thomas, Robert L. Wiebe, and Gordon S. Wood. *The Great Republic: A History of the American People*. Lexington, Mass.: D. C. Heath, 1977.

———. *The Ideological Origins of the American Revolution*. Cambridge: Harvard University Press, 1967.

Balasz, Etienne. *Chinese Civilization and Bureaucracy: Variations on a Theme*. New Haven: Yale University Press, 1964.

Balbus, Isaac D. *Marxism and Domination: A Neo-Hegelian, Feminist, Psychoanalytic Theory of Sexual, Political, and Technological Liberation*. Princeton: Princeton University Press, 1982.

Baran, Paul A., and Paul M. Sweezy. *Monopoly Capitalism: An Essay on the American Social and Economic Order*. New York: Monthly Review Press, 1966.

Barbour, Ian, ed. *Earth Might Be Fair: Reflections on Ethics, Religion, and Ecology*. Englewood Cliffs, N.J.: Prentice-Hall, 1972.

———. *Technology, Environment, and Human Values*. New York: Praeger, 1980.

Barrett, William. *Irrational Man: A Study in Existential Philosophy*. London: Mercury Books, 1964.

———. *The Illusion of Technique: A Search for Meaning in a Technological Civilization*. Garden City, N.Y.: Anchor Books, 1979.

Baumer, Frederick L. *Modern European Thought: Continuity and Change in Ideas 1600–1950*. New York: Macmillan, 1977.

Baxter, Sandra, and Marjorie Lansing. *Women and Politics: The Visible Majority*. Ann Arbor: University of Michigan Press, 1983.

Beckwith, Bernham Putnam. *The Next 500 Years: Projections of Major Social Trends*. Foreword by Daniel Bell. New York: Exposition Press, 1967.

Beitzinger, A. J. *A History of American Political Thought*. New York: Dodd, Mead, 1972.

Bell, Daniel. *Marxist Socialism in the United States.* Princeton: Princeton University Press, 1967.

Bellamy, Edward. *Equality.* New York: D. Appleton, 1897.

————. *Looking Backward, 2000–1887.* Boston: Ticknor and Fields, 1888. Reprint. New York: Penguin Books, 1982.

Belloc, Hilaire. *The Great Heresies.* New York: Sheed and Ward, 1935.

Bendix, Reinhard. *Kings or People: Power and the Mandate to Rule.* Berkeley: University of California Press, 1978.

Benedict, Ruth. *The Chrysanthemum and the Sword: Patterns of Japanese Culture.* Boston: Houghton Mifflin, 1946.

Benz, Ernst. *Evolution and Christian Hope: Man's Concept of the Future from the Early Fathers to Teilhard de Chardin.* Garden City, N.Y.: Anchor Books, 1966.

Berg, Peter, ed. *Reinhabiting a Separate Country: A Bioregional Anthology of Northern California.* San Francisco: Planet Drum Foundation, 1970.

Berneri, Mary Louise. *Journey through Utopia.* Foreword by George Woodcock. New York: Schocken Books, 1950.

Bernstein, Barton J., ed. *Towards a New Past: Dissenting Essays in American History.* New York: Vintage Books, 1969.

Bernstein, Edward. *Cromwell and Communism: Socialism and Democracy in the Great English Revolution.* New York: Schocken Books, 1963.

Berry, Adrian. *The Next Ten Thousand Years: A Vision of Man's Future in the Universe.* Foreword by Robert Jastrow. New York: New American Library, 1974.

Berry, Thomas. *Buddhism.* New York: Hawthorn Books, 1966.

————. *Religions of India.* New York: Bruce-Macmillan, 1971.

————. *The Dream of the Earth.* San Francisco: Sierra Club Books, 1988.

Beyerchen, Alan. *Scientists under Hitler: Politics and the Physics Community in the Third Reich.* New Haven: Yale University Press, 1977.

Biehl, Janet. *Rethinking Ecofeminist Politics.* Boston: South End Press, 1991.

Biese, Alfred. *The Development of the Feeling for Nature in the Middle Ages and Modern Times.* London: Burt Franklin, 1964.

Bilsky, Lester J., ed. *Historical Ecology: Essays on Environment and Social Change.* Port Washington, N.Y.: Kennikat Press, 1980.

Bishop, Jerry E., and Michael Waldholz. *Genome: The Story of the Most Astonishing Scientific Adventure of Our Time— The Attempt to Map All the Genes in the Human Body.* New York: Simon and Schuster, 1991.

Black, John. *The Dominion of Man: The Search for Ecological Responsibility.* Edinburgh: Edinburgh University Press, 1970.

Blank, Robert H. *The Political Implications of Human Genetic Technology.* Boulder, Colo.: Westview Press, 1981.

Bloch, Marc. *Land and Work in Medieval Europe: Selected Papers.* Translated by J. E. Anderson. Berkeley: University of California Press, 1967.

Bluhm, Walter. *Theories of the Political System: Classics in Political Thought and Modern Political Analysis.* 3d ed. Englewood Cliffs, N.J.: Prentice Hall, 1977.

Bodde, Derek. *Chinese Thought, Society, and Science: The Intellectual and Social Background of Science and Technology in Pre-modern China.* Honolulu: University of Hawaii Press, 1991.

Boller, Paul F., Jr. *American Thought in Transition: The Impact of Evolutionary Naturalism, 1865–1900.* Chicago: Rand McNally, 1965.

Boman, Thorlieff. *Hebrew Thought Compared with Greek.* London: SCM Press, 1960.

Bookchin, Murray. *Defending the Earth: A Dialogue between Murray Bookchin and Dave*

Foreman. Edited with an introduction by Steve Chase. Foreword by David Levine. Boston: South End Press, 1990.

———. *The Ecology of Freedom: The Emergence and Dissolution of Hierarchy.* Palo Alto, Calif.: Cheshire Books, 1982.

———. *Post-Scarcity Anarchism.* Berkeley: Ramparts Press, 1971.

———. *Remaking Society: Pathways to a Green Future.* Boston: South End Press, 1990.

———. *Toward an Ecological Society.* Montreal: Black Rose Books, 1980.

Boorstin, Daniel J. *The Americans,* vol. 1, *The Colonial Experience.* New York: Vintage Books, 1958.

———. *The Americans,* vol. 2, *The National Experience.* New York: Vintage Books, 1965.

Borgmann, Albert. *Technology and the Character of Contemporary Life.* Chicago: University of Chicago Press, 1984.

Borsodi, Ralph. *This Ugly Civilization.* New York: Simon and Schuster, 1929.

Boulding, Elise. *The Underside of History: A View of Women through Time.* Boulder, Colo.: Westview Press, 1976.

Braden, William. *The Age of Aquarius: Technology and the Cultural Revolution.* Chicago: Quadrangle Books, 1970.

Bramwell, Anne. *Blood and Soil: Richard Walther Darre and Hitler's "Green Party".* Bourne, U.K.: Kensal Press, 1985.

Brenneman, Richard J. *Fuller's Earth: A Day with Bucky and the Kids.* New York: St. Martin's Press, 1984.

Brinkley, Alan. *Voices of Protest: Huey Long, Father Coughlin, and the Great Depression.* New York: Alfred A. Knopf, 1982.

Brown, Lester F., Christopher Flavin, and Sandra Postel. *Saving the Planet: How to Shape an Environmentally Sustainable Global Economy.* New York: W. W. Norton, 1991.

Brown, Lester F., et al., eds. *State of the World 1987: A Worldwatch Institute Report on Progress toward a Sustainable Society.* New York: W. W. Norton, 1987.

Brown, Lester F., et al., eds. *State of the World 1988: A Worldwatch Institute Report on Progress toward a Sustainable Society.* New York: W. W. Norton, 1988.

Brown, Lester F., et al., eds. *State of the World 1989: A Worldwatch Institute Report on Progress toward a Sustainable Society.* New York: W. W. Norton, 1989.

Brown, Lester F., et al., eds. *State of the World 1991: A Worldwatch Institute Report on Progress toward a Sustainable Society.* New York: W. W. Norton, 1991.

Brown, Norman O. *Life against Death: The Psychoanalytic Meaning of History.* New York: Vintage Books, 1959.

Bruchey, Stuart. *The Roots of American Economic Growth, 1607–1861: An Essay in Social Causation.* New York: Harper and Row, 1965.

Bugliarello, George, and Dean B. Doner, eds. *The History and Philosophy of Technology.* Introduction by Melvin Kranzberg. Urbana: University of Illinois Press, 1973.

Burlingame, Roger. *Machines That Built America.* New York: Harcourt, Brace, 1953.

Caldwell, Lynton Keith. *International Environmental Policy: Emergence and Dimensions.* 2d ed. Durham, N.C.: Duke University Press, 1990.

Callenbach, Ernest. *Ecotopia: The Notebook and Reports of William Weston.* New York: Bantam Books, 1977.

———. *Ecotopia Emerging.* Berkeley: Banyan Tree Books, 1981.

Callicott, J. Baird, ed. *Companion to a Sand County Almanac: Interpretative and Critical Essays.* Madison: University of Wisconsin Press, 1987.

Callicott, J. Baird, and Roger T. Ames, eds. *The Nature of Nature in Asian Traditions of Thought.* Albany: State University of New York Press, 1989.

Cantor, Milton. *The Divided Left: American Radicalism, 1900–1975*. New York: Hill and Wang, 1978.

Capra, Fritjof, and Charlene Spretnak, in collaboration with Rudiger Lutz. *Green Politics*. New York: E. P. Dutton, 1984.

———. *The Tao of Physics: An Exploration of the Parallels between Modern Physics and Eastern Mysticism*. Boulder, Colo.: Shambala, 1975.

———. *The Turning Point: Science, Society, and the Rising Culture*. New York: Bantam Books, 1983.

Carmody, John. *Ecology and Religion: Toward a New Christian Theology of Nature*. New York: Paulist Press, 1983.

Chenu, M. D., O.P. *Nature, Man, and Society in the Twelfth Century: Essays on New Theological Perspectives in the Latin West*. Selected and translated by Jerome Taylor and Lester K. Little. Chicago: University of Chicago Press, 1968.

Childe, V. Gordon. *Man Makes Himself*. New York: New American Library, 1951.

———. *What Happened in History*. Baltimore: Penguin Books, 1964.

Christians, Clifford J., and Jay M. Van Hook, eds. *Jacques Ellul: Interpretative Essays*. Urbana: University of Illinois Press, 1981.

Christopher, Robert C. *The Japanese Mind*. New York: Fawcett Columbine, 1983.

Cipolla, Carlo M. *Before the Industrial Revolution: European Society and Economy, 1000–1700*. New York: W. W. Norton, 1976.

———. *Clocks and Culture, 1300–1700*. New York: Walker, 1967.

Clarke, Arthur C. *Childhood's End*. New York: Ballantine Books, 1963.

———. *Profiles of the Future*. New York: Bantam Books, 1964. Rev. ed., New York: Popular Library, 1977.

Clecak, Peter. *Radical Paradoxes: Dilemmas of the American Left, 1945–1970*. New York: Harper Torchbooks, 1974.

Cochran, Thomas C., and William Miller. *The Age of Enterprise: A Social History of Industrial America*. New York: Harper Torchbooks, 1961.

Cockburn, Alexander, and James Ridgeway, eds. *Political Ecology*. New York: Times Books, 1979.

Cohen, Michael P. *The History of the Sierra Club*. San Francisco: Sierra Club Books, 1989.

———. *The Pathless Way: John Muir and American Wilderness*. Madison: University of Wisconsin Press, 1984.

Collard, Andree, with Joyce Contrucci. *Rape of the Wild: Man's Violence against Animals and the Earth*. Bloomington: Indiana University Press, 1989.

Collingwood, R. G. *The Idea of Nature*. Oxford: Clarendon Press, 1945.

Collins, James. *A History of Modern European Philosophy*. Milwaukee: Bruce Publishing, 1984.

———. *Spinoza on Nature*. Foreword by George Kimball Plochman. Carbondale: Southern Illinois University Press, 1984.

Conant, James B. *Science and Common Sense*. New Haven: Yale University Press, 1951.

Cooley, Mike. *Architect or Bee? The Human/Technology Relationship*. Introduction by David Noble. Compiled and edited by Shirley Cooley. Boston: South End Press, 1982.

Cooper, Chester L., ed. *Growth in America*. Westport, Conn.: Greenwood Press, 1976.

Copleston, Frederick, S. J. *A History of Philosophy*, vol. 1, *Greece and Rome*. Westminster, Md.: Newman Press, 1950.

———. *A History of Philosophy*, vol. 2, *Augustine to Scotus*. Westminster, Md.: Newman Press, 1952.

Cornford, Frances Macdonald. *Before and after Socrates*. Cambridge: Cambridge University Press, 1960.

Cox, Richard. *Locke on War and Peace*. Oxford: Clarendon Press, 1960.

Crunden, Robert M., ed. *The Superfluous Men: Conservative Critics of American Culture, 1900–1923*. Austin: University of Texas Press, 1976.

Daly, Herman E., and John B. Cobb, Jr., with contributions by Clifford W. Cobb. *For the Common Good: Redirecting the Economy toward Community, the Environment, and a Sustainable Future*. Boston: Beacon Press, 1989.

———. *Toward a Steady-State Economy*. San Francisco: W. H. Freeman, 1973.

Daly, Mary. *Beyond God the Father: A Philosophy of Women's Liberation*. Boston: Beacon Press, 1978.

———. *GYN/ECOLOGY: The Metaphysics of Radical Feminism*. Boston: Beacon Press, 1978.

Darling, Frank. *The Westernization of Japan*. Boston: Hall, 1979.

Davis, Bernard D., ed. *The Genetic Revolution: Scientific Prospects and Popular Perceptions*. Baltimore: Johns Hopkins University Press, 1991.

Davis, John, ed., with foreword by David Foreman. *The Earth First! Reader: Ten Years of Radical Environmentalism*. Salt Lake City: Peregrine Smith Books, 1991.

Dawson, Christopher. *Progress and Religion: An Historical Enquiry*. New York: Sheed and Ward, 1938.

Day, Richard B., Ronald Beiner, and Joseph Masciulli, eds. *Democratic Theory and Technological Society*. Armonk, N.Y.: M. E. Sharpe, 1988.

de Bary, Wm. Theodore, Wing-tsit Chan, and Burton Watson. *Sources of Chinese Tradition*, vols. 1 and 2. New York: Columbia University Press, 1960.

de Jonge, Alex. *The Weimar Chronicle: Prelude to Hitler*. New York: New American Library, 1978.

de Reincourt, Amaury. *Sex and Power in History*. New York: Delta Books, 1975.

De Rougemont, Denis. *The Future Is within Us*. Translated by Anthony J. C. Kerr. New York: Pergamon Press, 1983.

Devall, Bill, and George Sessions. *Deep Ecology*. Salt Lake City: Peregrine Smith Books, 1985.

———. *Simple in Means, Rich in Ends: Practicing Deep Ecology*. Salt Lake City: Peregrine Smith Books, 1989.

Diamond, Irene,and Gloria Feman Orenstein, eds. *Reweaving the World: The Emergence of Ecofeminism*. San Francisco: Sierra Club Books, 1990.

Dickson, David. *The New Politics of Science*. New York: Pantheon Books, 1984.

———. *The Politics of Alternative Technology*. New York: Universe Books, 1974.

Dickstein, Morris. *Gates of Eden: American Culture in the Sixties*. New York: Basic Books, 1977.

Diggins, John P. *The Bard of Savagery: Thorstein Veblen and Modern Social Theory*. New York: Seabury Press, 1978.

———. *The Lost Soul of American Politics: Virtue, Self-Interest, and the Foundations of Liberalism*. New York: Basic Books, 1984.

Dolbeare, Kenneth M., Patricia Dolbeare, and Jane Hadley. *American Ideologies: The Competing Political Beliefs of the 1970s*. Chicago: Rand McNally, 1973.

———. *Directions in American Political Thought*. New York: John Wiley, 1969.

Dorfman, Joseph. *The Economic Mind in American Civilization, 1606–1825*, vol. 1. New York: Viking Press, 1946.

Dunn, Stephen, C.P., and Anne Lonergan, eds. *Befriending the Earth: A Theology of Reconciliation between Humans and the Earth*. Thomas Berry, C.P., in Dialogue with Thomas Clarke, S.J. Mystic, Conn.: Twenty-Third Publications, 1991.

Durbin, Paul T., ed. *A Guide to the Culture of Science, Technology, and Medicine*. New York: Free Press, 1984.

Durbin, Paul T., and Friedrich Rapp, eds. *Philosophy and Technology*. Dordrecht, Neth.: D. Reidel, 1983.

Earhart, H. Byron. *Japanese Religion: Unity and Diversity*. 3d ed. Belmont, Calif.: Wadsworth, 1982.

———. *Religion in the Japanese Experience: Sources and Interpretations*. Encino, Calif.: Dickenson Publishing, 1974.

Easley, Brian. *Witch Hunting, Magic and the New Philosophy: An Introduction to Debates of the Scientific Revolution, 1450–1750*. Atlantic Highlands, N.J.: Humanities Press, 1980.

Eberhard, Wolfram. *A History of China*. Berkeley: University of California Press, 1977.

Economou, George. *The Goddess Natura in Medieval Literature*. Cambridge: Harvard University Press, 1972.

Ehrenfeld, David. *The Arrogance of Humanism*. New York: Oxford University Press, 1978.

Eisely, Loren. *Darwin's Century*. New York: Doubleday, 1958.

———. *The Firmament of Time*. New York: Atheneum, 1960.

Ekrich, Arthur A., Jr. *Man and Nature in America*. New York: Columbia University Press, 1963.

Elder, Frederick. *Crisis in Eden: A Religious Study of Man and Environment*. Nashville: Abingdon Press, 1970.

Ellul, Jacques. *In Season and Out of Season: An Introduction to the Thought of Jacques Ellul*. Translated by Lani K. Niles. San Francisco: Harper and Row, 1982.

———. *Perspectives on Our Age: Jacques Ellul Speaks on His Life and Work*. Edited by William H. Vanderberg. Translated by Joachim Neugroschel. New York: Seabury Press, 1981.

———. *Propaganda*. Translated by Konrad Kellen and Jean Lerner. Foreword by Konrad Kellen. New York: Alfred A. Knopf, 1973.

———. *The Political Illusion*. Translated by Konrad Kellen. New York: Alfred A. Knopf, 1972.

———. *The Presence of the Kingdom*. Translated by Olive Wyon. Philadelphia: Westminster Press, 1951.

———. *The Technological Bluff*. Translated by Geoffrey W. Bromley. Grand Rapids, Mich.: Eerdmans, 1990.

———. *The Technological Society*. Translated by John Wilkinson. Foreword by Robert K. Merton. New York: Vintage Books, 1967.

———. *The Technological System*. Translated by Joachim Neugroschel. New York: Continuum, 1980.

———. *What I Believe*. Translated by Geoffrey W. Bromley. Grand Rapids, Mich.: Eerdmans, 1989.

Elsdon, Ron. *Bent World: A Christian Response to the Environmental Crisis*. Downers Grove, Ill.: InterVarsity Press, 1981.

Elshtain, Jean Bethke. *Public Man, Private Woman: Women in Social and Political Thought*. Princeton: Princeton University Press, 1981.

Elsner, Henry, Jr. *The Technocrats: Prophets of Automation*. Syracuse: Syracuse University Press, 1967.

Elvin, Mark. *The Pattern of the Chinese Past*. Stanford: Stanford University Press, 1973.

Emme, Eugene M., ed. *Space Fiction and Space Futures Past and Present*. San Diego: AAS History Series, vol. 5, American Astronautical Society, 1982.

Engels, Friedrich. *Dialectics of Nature*. Translated and edited by Clemens Dutt. Preface and notes by J.B.S. Haldane. New York: International Publishers, 1940.

Estfandiary, E. F. *Optimism One*. Rev. ed. New York: Popular Library, 1978.

———. *Telespheres*. New York: Popular Library, 1977.

———. *Up-Wingers*. New York: Popular Library, 1977.

Evernden, Neil. *The Natural Alien*. Toronto: University of Toronto Press, 1985.

———, ed. *The Paradox of Environmentalism*. Toronto: York University Press, 1984.

Everson, William. *Archetype West: The Pacific Coast as a Literary Region*. Berkeley: Oyez Press, 1976.

Fairclough, Henry Rushton. *Love of Nature among the Greeks and Romans*. New York: Longmans, Green, 1930.

Faith-Man-Nature Group. *Christians and the Good Earth*. Alexandria, Va.: Faith-Man-Nature Group, n.d.

Fanfani, Amintore. *Catholicism, Protestantism, and Capitalism*. London: Sheed and Ward, 1949.

Farias, Victor. *Heidegger and Naziism*. Edited with a foreword by Joseph Margolis and Tom Rockmore. Philadelphia: Temple University Press, 1989.

Ferkiss, Victor C. *Futurology: Promise, Performance, Prospects*. Beverly Hills, Calif.: Sage Publications, 1977.

———. *Technological Man: The Myth and the Reality*. New York: George Braziller, 1969.

———. *The Future of Technological Civilization*. New York: George Braziller, 1974.

Ferre, Frederick. *Philosophy of Technology*. Englewood Cliffs, N.J.: Prentice Hall, 1989.

Ferry, Luc, and Alain Renaut. *Heidegger and Modernity*. Translated by Franklin Philip. Chicago: University of Chicago Press, 1990.

Firestone, Shulamith. *The Dialectic of Sex: The Case for Feminist Revolution*. New York: Bantam Books, 1981.

Flader, Susan L. *Thinking Like a Mountain: Aldo Leopold and the Evolution of an Ecological Attitude toward Deer, Wolves, and Forests*. Columbia: University of Missouri Press, 1974.

Fleron, Frederick J., ed. *Technology and Communist Culture: The Socio-cultural Impact of Technology under Socialism*. New York: Praeger, 1977.

Florman, Samuel C. *Blaming Technology: The Irrational Search for Scapegoats*. New York: St. Martin's Press, 1981.

———. *The Existential Pleasures of Engineering*. New York: St. Martin's Press, 1976.

Foreman, David. *Confessions of an Eco-Warrior*. New York: Harmony Books, 1991.

Fourier, Charles. *Design for Utopia: Selected Writings of Charles Fourier*. Introduction by Charles Gide. Foreword by Frank E. Manuel. Translated by Julian Franklin. New York: Schocken Books, 1971.

Fowles, Jib, ed. *Handbook of Futures Research*. Westport, Conn.: Greenwood Press, 1978.

Fox, Matthew, O.P. *Creation Spirituality: Liberating Gifts for the Peoples of the Earth*. San Francisco: Harper and Row, 1991.

———. *Original Blessing*. Santa Fe, N.M.: Bear and Co., 1983.

———. *The Coming of the Cosmic Christ: The Healing of Mother Earth and the Birth of a Global Renaissance*. San Francisco: Harper and Row, 1988.

———. *Western Spirituality: Historical Roots, Ecumenical Routes*. Santa Fe, N.M.: Bear and Co., 1981.

Fox, Stephen. *John Muir and His Legacy: The American Conservation Movement*. Boston: Little, Brown, 1981.

Fox, Warwick. *Toward a Transpersonal Ecology: Developing New Foundations for Environmentalism*. Boston: Shambala, 1990.

Frankel, Boris. *The Post-Industrial Utopians*. London: Polity Press, 1987.

Friedrich, Otto. *Before the Deluge: A Portrait of Berlin in the 1920s*. New York: Avon Books, 1972.

Fritsch, Albert, S. J. *Renew the Face of the Earth*. Chicago: Loyola University Press, 1989.

Fromm, Erich, ed. *Socialist Humanism: An International Symposium.* Garden City, N.Y.: Anchor Books, 1966.

Fuller, R. Buckminster. *Nine Chains to the Moon.* Carbondale: Southern Illinois University Press, 1963.

————. *No More Secondhand God and Other Writings.* Carbondale: Southern Illinois University Press, 1963.

————. *Operating Manual for Spaceship Earth.* Carbondale: Southern Illinois University Press, 1969.

————, in collaboration with E. J. Applewhite. *Synergetics: Explorations in the Geometry of Thinking.* New York: Macmillan, 1973.

————. *Utopia or Oblivion: The Prospects for Humanity.* New York: Bantam Books, 1969.

Fuller, R. Buckminster, Eric A. Walker, and James R. Killian, Jr. *Approaching the Benign Environment.* Preface by Taylor Littleton. New York: Collier Books, 1970.

Gabriel, Ralph Henry. *The Course of American Democratic Thought.* 2d ed. New York: Ronald Press, 1940.

Garreau, Joel. *The Nine Nations of North America.* New York: Avon Books, 1981.

Gasman, Daniel. *The Scientific Origins of National Socialism.* London: Macmillan, 1971.

Geikie, Sir Archibald, K.C.B. *The Love of Nature among the Greeks and the Romans: During the Later Decades of the Republic and the First Century of the Empire.* London: John Murray, 1912.

Gendron, Bernard. *Technology and the Human Condition.* New York: St. Martin's Press, 1977.

Germino, Dante, and Klaus von Beyme, eds. *The Open Society in Theory and Practice.* The Hague: Martinus Nijoff, 1974.

Giedion, Siegfried. *Mechanization Takes Command: A Contribution to Anonymous History.* New York: W. W. Norton, 1969.

Gilby, Thomas, O.P. *The Political Ideas of Thomas Aquinas.* Chicago: University of Chicago Press, 1958.

Gilligan, Carol. *In a Different Voice: Essays on Psychological Theory.* Cambridge: Harvard University Press, 1982.

Gilman, Charlotte Perkins. *Herland.* Introduction by Ann J. Lane. New York: Pantheon Books, 1979.

Gimbutas, Marija. *The Language of the Goddess.* San Francisco: Harper and Row, 1989.

Gimpel, Jean. *The Medieval Machine: The Industrial Revolution of the Middle Ages.* New York: Penguin Books, 1977.

Glacken, Clarence J. *Traces on the Rhodian Shore: Nature and Culture in Western Thought from Ancient Times to the End of the Eighteenth Century.* Berkeley: University of California Press, 1967.

Goldman, Marshall. *The Spoils of Progress.* Cambridge: MIT Press, 1972.

Goodwyn, Lawrence. *Democratic Promise: The Populist Movement in America.* New York: Oxford University Press, 1976.

Gorz, Andre. *A Strategy for Labor: A Radical Proposal.* Translated by Martin A. Nicholas and Victoria Ortiz. Boston: Beacon Press, 1967.

————. *Ecology as Politics.* Translated by Patsy Vigderman and Jonathan Cloud. Boston: South End Press, 1980.

————. *Farewell to the Working Class: An Essay on Post-Industrial Socialism.* Translated by Michael Sonensher. Boston: Beacon Press, 1982.

————. *Paths to Paradise: On the Liberation from Work.* Translated by M. Imrie. London: Pluto Press, 1985.

————. *The Post-Industrial Society: Tomorrow's Social History: Classes, Conflicts, and Cultures in the Programmed Society.* New York: Random House, 1971.

Graham, Dom Aelred, O.S.B. *Zen Catholicism.* New York: Harcourt, Brace and World, 1963.

Graham, Otis L. *An Encore for Reform: The Old Progressives and the New Deal.* New York: Oxford University Press, 1967.

Grant, George. *Technology and Empire. Perspectives on North America.* Toronto: House of Anansi, 1969.

Gray, Elizabeth Dodson. *Why the Green Nigger? Re- Mything Genesis.* Wellesley, Mass.: Roundtable Press, 1979.

Greenstein, Fred L., and Nelson W. Polsby, eds. *Handbook of Political Science, vol. 7, Strategies of Inquiry.* Reading, Mass.: Addison-Wesley, 1975.

Gregarios, Paulos. *The Human Presence: An Orthodox View of Nature.* Geneva: World Council of Churches, 1978.

Gregor, A. James. *The Fascist Persuasion in Radical Politics.* Princeton: Princeton University Press, 1974.

————. *The Ideology of Fascism.* New York: Free Press, 1939.

Griffin, David Ray, ed. *Spirituality and Society: Postmodern Values.* Albany: State University of New York Press, 1988.

————. *The Reenchantment of Science: Postmodern Proposals.* Albany: State University of New York Press, 1988.

Griffin, Susan. *Woman and Nature: The Roaring inside Her.* New York: Harper and Row, 1978.

Gutman, Herbert G. *Work, Culture, and Society in Industrializing America: Essays in American Working-Class and Social History.* New York: Vintage Books, 1977.

Habermas, Jürgen. *Communication and the Evolution of Society.* Translated by Thomas McCarthy. Boston: Beacon Press, 1979.

————. *Knowledge and Human Interests.* Translated by Jeremy Shapiro. Boston: Beacon Press, 1971.

————. *Legitimation Crisis.* Translated by Thomas McCarthy. Boston: Beacon Press, 1971.

————. *Toward a Rational Society: Student Protest, Science, and Politics.* Translated by Jeremy Shapiro. Boston: Beacon Press, 1970.

Hansot, Elizabeth. *Perfection and Progress Two Modes of Utopian Thought.* Cambridge: MIT Press, 1974.

Hardin, Garrett. *Exploring New Ethics for Sur ival: The Voyage of the Spaceship Beagle.* Baltimore: Penguin Books, 1973.

————. *Promethean Ethics: Living with Death, Competition, and Triage.* Seattle: University of Washington Press, 1980.

————. *Stalking the Wild Taboo,* 2d ed. Los Altos, Calif.: William Kaufmann, 1978.

————. *The Limits to Altruism.* Bloomington: Indiana University Press, 1977.

Harding, Sandra, and Merrill B. Hintikka, eds. *Discovering Reality: Feminist Perspectives in Epistemology, Methodology, and Philosophy of Science.* The Hague: D. Reidel, 1983.

————. *The Science Question in Feminism.* Ithaca: Cornell University Press, 1986.

Hargrove, Eugene C., ed. *Beyond Earth: Environmental Ethics and the Solar System.* San Francisco: Sierra Club Books, 1987.

————. *Foundations of Environmental Ethics.* Englewood Cliffs, N.J.: Prentice Hall, 1989.

————, ed. *Religion and Environmental Crisis.* Athens: University of Georgia Press, 1986.

Harrington, Michael. *Socialism.* New York: Saturday Review Press, 1972.

————. *The Twilight of Capitalism.* New York: Simon and Schuster, 1976.

Hart, John. *The Spirit of the Earth.* New York: Paulist Press, 1984.

Haught, John F. *Nature and Purpose.* Washington, D.C.: University Press of America, 1980.

Hayden, Dolores. *Seven American Utopias: The Architecture of Communitarian Socialism, 1790–1975.* Cambridge: MIT Press, 1976.

Hays, Samuel P., in collaboration with Barbara B. Hays. *Beauty, Health, and Permanence: Environmental Politics in the United States, 1955–1975.* New York: Cambridge University Press, 1987.

——. *Conservation and the Gospel of Efficiency: The Progressive Conservationists.* Cambridge: Harvard University Press, 1959.

Hefner, Philip. *The Promise of Teilhard de Chardin: The Meaning of the Twentieth Century in Christian Perspective.* Philadelphia: J. B. Lippincott, 1970.

Heidegger, Martin. *Discourse on Thinking.* Translated by John M. Anderson and E. Hans Freund. New York: Harper and Row, 1966.

——. *The Question Concerning Technology and Other Essays.* Translation and introduction by William Lovitt. New York: Harper Colophon Books, 1977.

Heilbroner, Robert L. *The Worldly Philosophers: The Lives, Times, and Ideas of the Great Economic Thinkers.* New York: Simon and Schuster, 1953.

Herf, Jeffrey. *Reactionary Modernism: Technology, Culture, and Politics in Weimar and the Third Reich.* Cambridge: Cambridge University Press, 1984.

Hessel, Dieter T., ed. *Energy Ethics: A Christian Response.* New York: Friendship Press, 1979.

Hindle, Brooke. *Technology in Early America: Needs and Opportunities for Study.* Chapel Hill: University of North Carolina Press, 1966.

Hodges, Henry. *Technology in the Ancient World.* New York: Alfred A. Knopf, 1970.

Hofstadter, Richard. *The Age of Reform: From Bryan to F.D.R.* New York: Vintage Books, 1955.

——. *The American Political Tradition and the Men Who Made It.* New York: Vintage Books, 1954.

——. *Social Darwinism in American Thought.* Rev. ed. Boston: Beacon Press, 1958.

Holloway, James Y., ed. *Introducing Jacques Ellul.* Grand Rapids, Mich.: Eerdmans, 1970.

Holmes, Stewart W., and Chimyo Horioka. *Zen Art for Meditation.* Rutland, Vt.: Charles E. Tuttle, 1973.

Hoover, Thomas. *Zen Culture.* New York: Random House, 1977.

Horwitz, Robert, ed. *The Moral Foundations of the American Republic.* Charlottesville: University Press of Virginia, 1977.

Hubbard, Earl. *The Search Is On.* Edited by Barbara Hubbard. Los Angeles: Pace Publications, 1969.

Hughes, J. Donald. *American Indian Ecology.* Preface by Jamake Highwater. El Paso: Texas Western Press, 1983.

——. *Ecology in Ancient Civilizations.* Albuquerque: University of New Mexico Press, 1975.

Hughes, Thomas P., and Agatha C. Hughes, eds. *Lewis Mumford: Public Intellectual.* New York: Oxford University Press, 1990.

Huizinga, Johan. *Homo Ludens: A Study in the Play- Element in Human Culture.* London: Temple Smith, 1970.

Hulsberg, Werner. *The German Greens: A Social and Political Profile.* Translated by Gus Fagan. London: Verso, 1988.

Hunt, Morton. *The Compassionate Beast: The Scientific Inquiry into Human Altruism.* New York: William Morrow, 1991.

Huth, Hans. *Nature and the American: Three Centuries of Changing Attitudes.* Berkeley: University of California Press, 1957.

Huxley, Aldous. *Island.* New York: Harper and Row, 1962.

The I Ching. Translated by James Legge. 2d ed. New York: Dover, 1963.

Ike, Nobutaka. *Japan: The New Superstate.* San Francisco: W. H. Freeman, 1973.

Inglehart, Ronald. *The Silent Revolution: Changing Values and Political Styles among Western Publics*. Princeton: Princeton University Press, 1977.

Iqbal, Sir Mohammed. *The Reconstruction of Religious Thought in Islam*. London: Routledge and Kegan Paul, 1934.

Jaeger, Werner. *Paideia: The Ideals of Greek Culture*, vol. 1. Translated by Gilbert Highet. New York: Oxford University Press, 1945.

Jagger, Allison M. *Feminist Politics and Human Nature*. Totowa, N.J.: Rowman and Allenheld, 1983.

Jefferson, Thomas. *Notes on the State of Virginia*. Edited and with an introduction by William Peden. Chapel Hill: University of North Carolina Press, 1955.

Jones, Howard Mumford. *O Strange New World. American Culture: The Formative Years*. New York: Viking Press, 1964.

Joranson, Philip N., and Ken Butigan, eds. *Cry of the Environment: Rebuilding the Christian Creation Tradition*. Santa Fe, N.M.: Bear and Co., 1984.

Joseph, Lawrence E. *Gaia: The Growth of an Idea*. New York: St. Martin's Press, 1991.

Jung, Hwa Yol. *The Crisis of Political Understanding: A Phenomenological Perspective in the Conduct of Political Inquiry*. Pittsburgh: Duquesne University Press, 1979.

Kahn, Herman. *On Thermonuclear War*. Princeton: Princeton University Press, 1962.

———. *Thinking about the Unthinkable*. New York: Horizon Press, 1962.

Kahn, Herman, William Brown, and Leon Martel. *The Next 200 Years*. New York: William Morrow, 1975.

Kahn, Herman, and R. Bruce-Briggs. *Things to Come*. New York: Macmillan, 1972.

Kahn, Herman, with Anthony Weiner. *The Year 2000: A Framework for Speculation on the Next Thirty-three Years*. New York: Macmillan, 1967.

Kallenberg, Arthur L., J. Donald Moon, and Donald R. Sabia Jr., eds. *Dissent and Affirmation: Essays in Honor of Mulford Q. Sibley*. Bowling Green, Ohio: Bowling Green University Popular Press, 1983.

Kammen, Michael. *People of Paradox: An Inquiry Concerning the Origins of American Civilization*. New York: Vintage Books, 1973.

Kasson, John F. *Civilizing the Machine: Technology and Republican Values in America, 1776–1900*. New York: Grossman, 1976.

Keller, Evelyn Fox. *Reflections on Gender and Science*. New Haven: Yale University Press, 1985.

Kelly, Petra. *Fighting for Hope*. Introduction by Heinrich Boll. Translated by Marianne Howarth. Boston: South End Press, 1984.

Ketterer, David. *New Worlds for Old: The Apocalyptic Imagination, Science Fiction, and American Literature*. Garden City, N.Y.: Anchor Books, 1974.

Kidder, J. E. Jr. *Japan before Buddhism*. Rev. ed. New York: Praeger, 1966.

Kiernan, Thomas. *The Road to Colossus: A Celebration of American Ingenuity*. New York: William Morrow, 1985.

King, Richard. *The Party of Eros: Radical Social Thought and the Realm of Freedom*. Chapel Hill: University of North Carolina Press, 1972.

Kinkade, Kathleen. *A Walden Two Experiment*. New York: William Morrow, 1973.

Kirk, Russell. *The Conservative Mind: From Burke to Santayana*. Chicago: Henry Regnery, 1953.

Kohn, Allie. *The Brighter Side of Human Nature*. New York: Basic Books, 1990.

Kolko, Gabriel. *The Triumph of Conservatism: A Reinterpretation of the Progressive Movement, 1900–1916*. New York: Free Press of Glencoe, 1963.

Koller, John M. *Oriental Philosophies*. New York: Charles Scribner's Sons, 1970.

Kouwenhoven, John A. *The Arts in Modern American Civilization*. New York: W. W. Norton, 1967.

Kraft, Micheal E., and Norman J. Vig, eds. *Technology and Politics*. Durham, N.C.: Duke University Press, 1988.

Kranzberg, Melvin, and Joseph Gies. *By the Sweat of Thy Brow: Work in the Western World*. New York: G. P. Putnam's Sons, 1975.

Krutch, Joseph Wood. *The Measure of Man*. Indianapolis: Bobbs-Merrill, 1954.

Kuhns, William. *Environmental Man*. New York: Harper and Row, 1969.

———. *The Post-Industrial Prophets: Interpretations of Technology*. New York: Weybright and Talley, 1971.

Kumar, Krishnan. *Prophecy and Progress: The Sociology of Industrial and Post-Industrial Society*. Harmondsworth, Eng.: Penguin Books, 1978.

———. *Utopia and Anti-Utopia in Modern Times*. London: Basil Blackwell, 1987.

Lakoff, Sanford A., ed. *Knowledge and Power: Essays on Science and Government*. New York: Free Press, 1966.

Landes, David S. *Revolution in Time: Clocks and the Making of the Modern World*. Cambridge: Belknap Press, 1983.

———. *The Unbound Prometheus: Technological Change and Industrial Development in Western Europe from 1750 to the Present*. Cambridge: Cambridge University Press, 1969.

Lapham, Lewis H., ed. *High Technology and Human Freedom*. Washington, D.C.: Smithsonian Institution Press, 1985.

Lasch, Christopher. *The World of Nations: Reflections on American History, Politics, and Culture*. New York: Alfred A. Knopf, 1973.

Lash, Jonathan, Katherine Gillman, and David Sheridan. *A Season of Spoils: The Story of the Reagan Administration's Attack on the Environment*. New York: Pantheon Books, 1984.

Lasky, Melvin J. *Utopia and Revolution: On the Origins of a Metaphor, or Some Illustrations of the Problem of Political Temperament and Intellectual Climate and How Ideas, Ideals, and Ideologies Have Been Historically Related*. Chicago: University of Chicago Press, 1976.

Lawson, R. Alan. *The Failure of Independent Liberalism, 1930–1941*. New York: Capricorn Books, 1971.

Layton, Edwin T., Jr. *The Revolt of the Engineers: Social Responsibility and the American Engineering Profession*. Cleveland: Case Western Reserve University Press, 1971.

Lears, T. Jackson. *No Place of Grace: Antimodernism and the Transformation of American Culture, 1880–1920*. New York: Pantheon Books, 1981.

Leggett, Jeremy, ed. *Global Warming: The Greenpeace Report*. New York: Oxford University Press, 1990.

Le Goff, Jacques. *Time, Work, and Culture in the Middle Ages*. Translated by Arthur Goldhammer. Chicago: University of Chicago Press, 1980.

Leiss, William. *The Domination of Nature*. New York: George Braziller, 1972.

Leonard, Jonathan Norton. *Early Japan*. New York: Time- Life Books, 1968.

Leopold, Aldo. *A Sand County Almanac and Sketches Here and There*. Illustrated by Charles W. Schwartz. New York: Oxford University Press, 1982.

Levenson, Joseph. *Confucian China and Its Modern Fate*. 3 vols. Berkeley: University of California Press, 1968.

Lewis, John, and Ruth Lewis. *Space Resources*. New York: Columbia University Press, 1987.

Lewis, Richard S. *Space in the 21st Century*. New York: Columbia University Press, 1990.

Lipow, Arthur. *Authoritarian Socialism in America: Edward Bellamy and the Nationalist Movement*. Berkeley: University of California Press, 1982.

Little, Daniel. *The Scientific Marx*. Minneapolis: University of Minnesota Press, 1986.

Long, Franklin, and Alexandra Oleson, eds. *Appropriate Technology and Social Values: A Critical Appraisal.* Cambridge: Ballinger, 1980.

Lonergan, Anne, and Caroline Richards, eds. *Thomas Berry and the New Cosmology.* Mystic, Conn.: Twenty-Third Publications, 1988.

Lora, Ronald. *Conservative Minds in America.* Chicago: Rand McNally, 1971.

Lovejoy, Arthur O., Gilbert Chinard, George Boas, and Ronald S. Crane. *A Documentary History of Primitivism and Related Ideas,* vol. 1. Baltimore: Johns Hopkins University Press, 1935.

———. *The Great Chain of Being: A Study in the History of an Idea.* Cambridge: Harvard University Press, 1936.

Lovelock, J. E. *Gaia: A New Look at Life on Earth.* New York: Oxford University Press, 1980.

Lovins, Amory. *Soft Energy Paths: Toward a Durable Peace.* San Francisco: Friends of the Earth International, 1987.

Lowenthal, David, and Martin J. Bowden, eds. *Geographies of the Mind: Essays in Historical Geography in Honor of John Kirkland Wright.* New York: Oxford University Press, 1976.

McCormick, John. *Reclaiming Paradise: The Global Environmental Movement.* Bloomington: Indiana University Press, 1989.

McDaniel, Jay. *Earth, Sky, Gods and Mortals: Developing an Ecological Spirituality.* Mystic, Conn.: Twenty-Third Publications, 1989.

———. *Of God and Pelicans: A Theology of Reverence for Life.* Foreword by John B. Cobb, Jr. Louisville, Ky.: Westminster/John Knox Press, 1989.

McDonagh, Sean. *The Greening of the Church.* Maryknoll, N.Y.: Orbis Books, 1990.

———. *To Care for the Earth: A Call to a New Theology.* London: Geoffrey Chapman, 1986; Santa Fe, N.M.: Bear and Co., 1987.

McDonald, Lee Cameron. *Western Political Theory: The Modern Age.* New York: Harcourt, Brace and World, 1962.

McDougall, Walter A. *The Heavens and the Earth: A Political History.* New York: Basic Books, 1985.

MacIntyre, Alisdair. *Herbert Marcuse: An Exposition and a Polemic.* New York: Viking Press, 1970.

McKibben, Bill. *The End of Nature.* New York: Random House, 1989.

McNeill, William. *The Pursuit of Power: Technology, Armed Force, and Society since A.D. 1000.* Chicago: University of Chicago Press, 1982.

———. *The Rise of the West: A History of the Human Community.* Chicago: University of Chicago Press, 1963.

Mailer, Norman. *Of a Fire on the Moon.* New York: New American Library, 1971.

Manes, Christopher. *Green Rage: Radical Environmentalism and the Unmaking of Civilization.* Boston: Little, Brown, 1991.

Manuel, Frank E., and Fritzie P. Manuel. *Utopian Thought in the Western World.* Cambridge: Belknap Press of Harvard University Press, 1979.

Marcuse, Herbert. *An Essay on Liberation.* Boston: Beacon Press, 1969.

———. *Eros and Civilization.* Boston: Beacon Press, 1955.

———. *One-Dimensional Man.* Boston: Beacon Press, 1969.

Martin, Calvin. *Keepers of the Game: Indian Animal Relationship and the Fur Trade.* Berkeley: University of California Press, 1978.

Maruyama, Magoroh, and Arthur Harkins. *Cultures beyond the Earth.* Foreword by Alvin Toffler. New York: Vintage Books, 1975.

Marx, Karl. *Capital: A Critique of Political Economy,* vol. 1, *The Process of Capitalist Production.*

Translated from the 3d German ed. by Samuel Moore and Edward Eveling. Edited by Friedrich Engels. New York: International Publishers, 1967.

Marx, Leo. *The Machine in the Garden: Technology and the Pastoral Ideal in America*. New York: Oxford University Press, 1964.

Matson, Floyd. *The Broken Image: Man, Science, and Society*. Garden City, N.Y.: Anchor Books, 1966.

Mead, Sidney E. *The Nation with the Soul of a Church*. New York: Harper and Row, 1975.

Medvedev, Zhores A. *Soviet Science*. New York: W. W. Norton, 1978.

Meek, Ronald L. *Social Science and the Ignoble Savage*. Cambridge: Cambridge University Press, 1976.

Meine, Curt. *Aldo Leopold: His Life and Work*. Madison: University of Wisconsin Press, 1987.

Merchant, Carolyn. *The Death of Nature: Women, Ecology, and the Scientific Revolution*. San Francisco: Harper and Row, 1979.

Metz, Johannes B., ed. *The Evolving World and Theology*. New York: Paulist Press, 1967.

Miller, Charles A. *Jefferson and Nature: An Interpretation*. Baltimore: Johns Hopkins University Press, 1988.

Mitcham, Carl, and Jim Grote, eds. *Theology and Technology: Essays in Christian Analysis and Exegesis*. Lanham, Md.: University Press of America, 1984.

Mitcham, Carl, and Robert Mackey, eds. *Philosophy and Technology: Readings in the Philosophical Problems of Technology*. New York: Free Press, 1983.

Mitrany, David. *Marx against the Peasant: A Study in Social Dogmatism*. Chapel Hill: University of North Carolina Press, 1951.

Morgan, Elaine. *The Descent of Woman*. London: Souvenir Press, 1972.

Morgan, Robin, ed. *Sisterhood Is Powerful: An Anthology of Writings from the Women's Liberation Movement*. New York: Vintage Books, 1972.

Morishima, Michio. *Why Has Japan 'Succeeded': Western Technology and the Japanese Ethos*. Cambridge: MIT Press, 1962.

Morison, Elting S. *From Know-How to Nowhere: The Development of American Technology*. New York: Basic Books, 1974.

Mosse, George L. *The Crisis of German Ideology: Intellectual Origins of the Third Reich*. New York: Grosset and Dunlap, 1964.

Mumford, Lewis. *Art and Technics*. New York: Columbia University Press, 1952.

———. *Technics and Civilization*. New York: Harcourt, Brace, 1934.

———. *The City in History: Its Origins, Its Transformations, and Its Prospects*. New York: Harcourt, Brace and World, 1961.

———. *The Myth of the Machine*, vol. 1, *Technics and Human Development*. New York: Harcourt, Brace and World, 1969.

———. *The Myth of the Machine*, vol. 2, *The Pentagon of Power*. New York: Harcourt, Brace and World, 1970.

Murphy, Earl Finbar. *Governing Nature*. Chicago: Quadrangle Books, 1967.

Murray, Michael, ed. *Heidegger and Modern Philosophy: Critical Essays*. New Haven: Yale University Press, 1978.

Myers, Norman. *The Gaia Atlas of Future Worlds: Challenge and Opportunity in an Age of Change*. Foreword by Kenneth E. Boulding. New York: Anchor Books, 1991.

Naess, Arne. *Ecology, Community, and Lifestyle: Outline of an Ecosophy*. Cambridge: Cambridge University Press, 1991.

———. *Four Modern Philosophers: Carnap, Wittgenstein, Heidegger, Sartre*. Translated by Alistair Hannay. Chicago: University of Chicago Press, 1968.

Najita, Natsue, and J. Victor Korshmann, eds. *Conflict in Modern Japanese History: The Neglected Tradition*. Princeton: Princeton University Press, 1982.

Nakayama, Shigereu, David L. Swain, and Yagi Eri, eds. *Science and Society in Modern Japan: Selected Historical Sources*. Cambridge: MIT Press, 1974.

Nash, George T. *The Conservative Intellectual Movement in America since 1945*. New York: Basic Books, 1979.

Nash, Roderick. *The Rights of Nature: A History of Environmental Ethics*. Madison: University of Wisconsin Press, 1989.

———. *Wilderness and the American Mind*. Rev. ed. New Haven: Yale University Press, 1973. 3d ed., 1982.

Nasr, Seyyed Nossien. *Science and Civilization in Islam*. Preface by Giorgio de Santillana. Cambridge: Harvard University Press, 1968.

Needham, Joseph, Wang Ling, and Derek J. Price. *Heavenly Clockwork: The Great Astronomical Clocks of Medieval China*. Cambridge: Cambridge University Press, 1960.

———. *Science and Civilization in China*. 5 vols. Cambridge: Cambridge University Press, 1962–1987.

———. *The Grand Titration: Science and Society in East and West*. Toronto: University of Toronto Press, 1969.

Needleman, Jacob, A. K. Bierman, and James A. Gould, eds. *Religion for a New Generation*. 2d ed. New York: Macmillan, 1977.

Nef, John U. *The Conquest of the Material World*. Cleveland: World Publishing, 1967.

———. *War and Human Progress: An Essay on the Rise of Industrial Civilization*. Cambridge: Harvard University Press, 1950.

Neihardt, John G., ed. *Black Elk Speaks*. Lincoln: University of Nebraska Press, 1932.

Neuhaus, Richard John. *In Defense of People: Ecology and the Seduction of Radicalism*. New York: Macmillan, 1971.

Neumann, Erich. *The Great Mother: An Analysis of the Archetype*. Princeton: Princeton University Press, 1955.

Nisbet, Robert A. *The Social Philosophers: Community and Conflict in Western Thought*. New York: Thomas A. Crowell, 1973.

———. *The Sociological Tradition*. New York: Basic Books, 1966.

Noble, David M. *The Eternal Adam and the New World Garden: The Central Myth of the American Novel since 1830*. New York: Grosset and Dunlap, 1971.

Noble, David W. *The Progressive Mind, 1890–1917*. Chicago: Rand McNally, 1970.

Novik, Ilya. *Society and Nature: Socio-Ecological Problems*. Translated by H. Campbell Creighton. Moscow: Progress Publishers, 1981.

Nye, Russell B. *Midwestern Progressive Politics: A Historical Study of Its Origin and Development, 1870–1950*. East Lansing: Michigan State University Press, 1951.

Oates, David. *Earth Rising: Ecological Belief in an Age of Science*. Eugene: University of Oregon Press, 1989.

Oelschlaeger, Max. *The Idea of Wilderness: From Prehistory to the Age of Ecology*. New Haven: Yale University Press, 1991.

Oliver, John W. *History of American Technology*. New York: Ronald Press, 1956.

Olson, Theodore. *Millennialism, Utopianism, and Progress*. Toronto: University of Toronto Press, 1982.

O'Neill, Gerard. *The High Frontier: Human Colonies in Space*. New York: Simon and Schuster, 1977.

———. *2081: A Hopeful View of the Future*. New York: Simon and Schuster, 1981.

O'Neill, William. *The Progressive Years: America Comes of Age*. New York: Dodd, Mead, 1975.

Ophuls, William. *Ecology and the Politics of Scarcity: Prologue to a Political Theory of the Steady State*. San Francisco: W. H. Freeman, 1977.

Oppenheimer, Michael, and Robert H. Boyle. *Dead Heat: The Race Against the Greenhouse Effect*. New York: Basic Books, 1990.

Orvitt, George. *The Restoration of Perfection: Labor and Technology in Medieval Culture*. New Brunswick: Rutgers University Press, 1987.

Pacey, Arnold. *Technology in World Civilization: A Thousand-Year History*. Cambridge: MIT Press, 1990.

———. *The Culture of Technology*. Cambridge: MIT Press, 1983.

———. *The Maze of Ingenuity: Ideas and Idealism in the Development of Technology*. New York: Holmes and Meier, 1975.

Paehlke, Robert C. *Environmentalism and the Future of Progressive Politics*. New Haven: Yale University Press, 1989.

Papadakis, Emil. *The Green Movement in West Germany*. London: Croom and Helm, 1984.

Parrington, Vernon L. *Main Currents in American Thought: An Interpretation of American Literature from the Beginnings to 1920*, vol. 1, *1620–1800: The Colonial Mind*. New York: Harcourt, Brace, 1927.

———. *Main Currents in American Thought: An Interpretation of American Literature from the Beginnings to 1920*, vol. 2, *1800–1860: The Romantic Revolution*. New York: Harcourt, Brace, 1927.

———. *Main Currents in American Thought: An Interpretation of American Literature from the Beginnings to 1920*, vol. 3, *1860–1920: The Beginnings of Critical Realism in America*. New York: Harcourt, Brace, 1930.

Parsons, Howard L., ed. and comp. *Marx and Engels on Ecology*. Westport, Conn.: Greenwood Press, 1977.

Passmore, John. *Man's Responsibility for Nature: Ecological Problems and Western Traditions*. New York: Charles Scribner's Sons, 1974.

———. *The Perfectability of Man*. London: Duckworth, 1970.

Patai, Daphne, ed. *Looking Backward, 1988–1888: Essays on Edward Bellamy*. Amherst: University of Massachusetts Press, 1988.

Perloff, Marjorie. *The Futurist Movement: Avant-garde, Avant-guerre, and the Language of Rupture*. Chicago: University of Chicago Press, 1986.

Perrin, Noel. *Giving Up the Gun: Japan's Reversion to the Sword, 1543–1879*. Boulder, Colo.: Shambala, 1980.

Petulla, Joseph M. *American Environmental History: The Exploitation and Conservation of Natural Resources*. San Francisco: Boyd and Fraser, 1977.

Pfaltzgraff, Robert L., Jr., Kim R. Holmes, Clay Clemens, and Werner Kaltefelter. *The Greens of West Germany: Origins, Strategies, and Transatlantic Implications*. Cambridge: Institute for Policy Analysis, 1983.

Piasecki, Bruce, and Peter Asmus. *In Search of Environmental Excellence: Moving beyond Blame*. Foreword by Jean-Michel Cousteau. Introduction by Congresswoman Claudine Schneider. Afterword by the Honorable Robert K. Dawson. New York: Simon and Schuster, 1990.

Pickthall, Mohammed Marmaduke. *The Meaning of the Glorious Koran: An Explanatory Translation*. New York: Mentor Books, 1953.

Piercy, Marge. *Women at the End of Time*. New York: Fawcett Crest, 1977.

Plant, Judith, ed. *Healing the Wounds: The Promise of Ecofeminism*. Foreword by Petra Kelly. Santa Cruz, Calif.: New Society Publishers, 1989.

Pocock, J. C. A. *The Machiavellian Moment: Florentine Political Thought and the Atlantic Republican Tradition*. Princeton: Princeton University Press, 1985.

Pois, Robert A. *National Socialism and the Religion of Nature*. New York: St. Martin's Press, 1968.

Polanyi, Karl. *The Great Transformation.* New York: Rinehart, 1944.

Pollack, Norman. *The Populist Response to Industrial America.* Cambridge: Harvard University Press, 1962.

Poston, Mark. *Existential Marxism in Postwar France.* Princeton: Princeton University Press, 1975.

Potter, David. *People of Plenty: Economic Abundance and the American Character.* Chicago: University of Chicago Press, 1982.

Pryde, Philip R. *Conservation in the Soviet Union.* Cambridge: Cambridge University Press, 1972.

Pursell, Carroll W., Jr. *Technology in America: A History of Individuals and Ideas.* Cambridge: MIT Press, 1981.

Rabkin, Eric S., Martin J. Greenberg, and Joseph D. Olander, eds. *The End of the World.* Carbondale: Southern Illinois University Press, 1983.

Rahman, Fazlur. *Islam and Modernity: Transformation of an Intellectual Tradition.* Chicago: University of Chicago Press, 1982.

Ramsey, Paul. *Prefabricated Man: The Ethics of Genetic Control.* New Haven: Yale University Press, 1970.

Rauschning, Herman. *Hitler Speaks: A Series of Political Conversations with Adolf Hitler on His Real Aims.* London: Thornton and Butterworth, 1939.

Reed, Evelyn. *Sexism and Science.* New York: Pathfinder Press, 1978.

Reich, Warren, ed. *Encyclopedia of Bioethics.* New York: Free Press, 1978.

Rexforth, Kenneth. *Communalism: From Its Origins to the Twentieth Century.* New York: Seabury Press, 1974.

Rhodes, James M. *The Hitler Movement: A Modern Millennial Revolution.* Stanford: Stanford University Press, 1980.

Richta, Radovan, et al. *Civilization at the Crossroads: Social and Human Implications of the Scientific and Technological Revolution.* Rev. ed. Prague: International Arts and Science Press, 1969.

Rifkin, Jeremy. *Biosphere Politics: A New Consciousness for a New Century.* New York: Crown, 1991.

Risaldo, Michelle Zimbalist, and Louise Lamphere, eds. *Women, Culture, and Society.* Stanford: Stanford University Press, 1974.

Rodgers, Daniel T. *The Work Ethic in Industrializing America, 1850–1920.* Chicago: University of Chicago Press, 1978.

Rommell, Ferdinard Muller, ed. *New Politics in Western Europe: The Rise and Success of Green Parties and Alternate Lists.* Boulder, Colo.: Westview Press, 1989.

Rosenthal, Bernard. *City of Nature: Journeys to Nature in the Age of American Romanticism.* Newark: University of Delaware Press, 1980.

Ross, Lester. *Environmental Policy in China.* Bloomington: Indiana University Press, 1988.

Rossi, Paolo. *Philosophy, Technology, and the Arts in the Early Modern Era.* Translated by Salvator Attanasio. Edited by Benjamin Nelson. New York: Harper and Row, 1970.

Rossiter, Clinton. *The Political Thought of the American Revolution.* New York: Harcourt, Brace and World, 1963.

Roszak, Theodore, ed. *Sources: An Anthology of Contemporary Materials Useful for Preserving Personal Sanity While Braving the Great Technological Wilderness.* New York: Harper Colophon Books, 1972.

Roth, John K. *American Dreams: Meditations on Life in the United States.* San Francisco: Chandler and Sharp, 1976.

Rothman, Daniel J. *Strangers at the Bedside: A History of How Law and Bioethics Transformed Medical Decision Making.* New York: Basic Books, 1991.

Rothschild, Joan, ed. *Machina Ex Dea: Feminist Perspectives on Technology.* New York: Pergamon Press, 1983.

Rotstein, Abraham, ed. *Beyond Industrial Growth.* Preface by Robertson Davies. Toronto: University of Toronto Press, 1976.

Rudig, Wolfgang, ed. *Green Politics I, 1990.* Carbondale: Southern Illinois University Press, 1991.

Ruether, Rosemary Radford, ed. *Liberation Theology: Human Hope Confronts Christian History and American Power.* New York: Paulist Press, 1972.

————. *New Woman/New Earth: Sexist Ideologies and Human Liberation.* New York: Seabury Press, 1975.

The Rule of St. Benedict. In Latin and English. Edited and translated by Abbot Justin McCann. Westminster, Md.: Newman Press, 1952.

Rust, Eric. *Nature-Garden or Desert?.* Waco, Tex.: Word Books, 1971.

Rybczynski, Witold. *Paper Heroes: A Review of Technology Assessment.* Garden City, N.Y.: Anchor Books, 1980.

————. *Taming the Tiger: The Struggle to Control Technology.* New York: Penguin Books, 1985.

Sagan, Carl, and Richard Turco. *A Path Where No Man Thought: Nuclear Winter and the End of the Arms Race.* New York: Random House, 1990.

Sagan, Dorion. *Biospheres: Metamorphosis of Planet Earth.* New York: Mc Graw-Hill, 1990.

Sagoff, Mark. *The Economy of the Earth: Philosophy, Law, and the Environment.* New York: Cambridge University Press, 1988.

Sakharov, Andrei P. *Progress, Coexistence, and Intellectual Freedom.* New York: W. W. Norton, 1968.

Sale, Kirkpatrick. *Dwellers in the Land: The Bioregional Vision.* San Francisco: Sierra Club Books, 1985.

————. *The Conquest of Paradise: Christopher Columbus and the Columbian Legacy.* New York: Alfred A. Knopf, 1990.

Samson, G. B. *Japan: A Short Cultural History.* Rev. ed. New York: D. Appleton-Century, 1943.

Sanders, N. K., ed. *The Epic of Gilgamesh.* Harmondsworth, Eng.: Penguin Books, 1972.

Sanford, Charles L. *The Quest for Paradise: Europe and the American Moral Imagination.* Urbana: University of Illinois Press, 1961.

Santmire, H. Paul. *Brother Earth: Nature, God, and Ecology in a Time of Crisis.* New York: Thomas Nelson, 1970.

————. *The Travail of Nature: The Ambiguous Ecological Promise of Christian Theology.* Philadelphia: Fortress Press, 1985.

Sardar, Ziaduddin, ed. *Islamic Futures: The Shape of Ideas to Come.* London: Mansell, 1985.

————. *The Touch of Midas: Science, Values, and Environment in Islam and the West.* Manchester: Manchester University Press, 1984.

Schafer, Edward H. *Ancient China.* New York: Time-Life Books, 1967.

Schaffer, Francis A. *Pollution and the Death of Man: A Christian View of Ecology.* Wheaton, Ill.: Tyndal, 1970.

Schlebecker, John T. *Whereby We Thrive: A History of American Farming, 1607–1972.* Ames: Iowa State University Press, 1975.

Schmidt, Alfred. *The Concept of Nature in Marx.* London: NLB, 1971.

Schneider, Stephen H. *Global Warming: Are We Entering the Greenhouse Century?* San Francisco: Sierra Club Books, 1989.

Schulberg, Lucille, et al. *Historic India.* New York: Time-Life Books, 1968.

Schultz, Robert C., and J. Donald Hughes, eds. *Ecological Consciousness: Essays from the Earthday X Colloquium, University of Denver, April 21–24, 1980.* Washington, D.C.: University Press of America, 1980.

Schumacher, E. F. *A Guide for the Perplexed.* New York: Harper and Row, 1977.

———. *Small Is Beautiful: Economics as If People Mattered.* New York: Harper and Row, 1973.

Scully, Vincent. *The Earth, The Temple, and the Gods.* New Haven: Yale University Press, 1962.

Segal, Howard P. *Technological Utopianism in American Culture.* Chicago: University of Chicago Press, 1985.

Shi, David E. *In Search of the Simple Life: American Voices, Past and Present.* Salt Lake City: Gibbs M. Smith, 1986.

———. *The Simple Life: Plain Living and High Thinking in American Culture.* New York: Oxford University Press, 1985.

Sibley, Mulford Q. *Nature and Civilization: Some Implications for Politics.* Itaska, Ill.: F. E. Peacock, 1977.

Sieden, Lloyd Steven. *Buckminster Fuller's Universe: An Appreciation.* Foreword by Norman Cousins. New York: Plenum Press, 1989.

Sigurdson, John. *Technology and Science in the People's Republic of China.* Oxford: Pergamon Press, 1980.

Simon, Denis Fred, and Merle Goldmnan, eds. *Science and Technology in Post-Mao China.* Cambridge: Harvard University Council on East Asian Studies, 1989.

Singleton, Fred, ed. *Environmental Problems in the Soviet Union and Eastern Europe.* Boulder, Colo.: Lynne Reinner Publishers, 1987.

Sittler, Joseph. *The Care of the Earth and Other University Sermons.* Philadelphia: Fortress Press, 1964.

Skinner, B. F. *Beyond Freedom and Dignity.* New York: Alfred A. Knopf, 1971.

———. *Science and Human Behavior.* New York: Macmillan, 1953.

———. *Walden Two.* New York: Macmillan, 1962.

Smil, Vaclav. *The Bad Earth: Environmental Degradation in China.* New York: M. E. Sharpe, 1984.

Smith, Henry Nash. *Virginland: The American West as Symbol and Myth.* Cambridge: Harvard University Press, 1950.

Smith, Michael J. *Pacific Visions: California Scientists and the Environment.* New Haven: Yale University Press, 1988.

Smith, Neil. *Uneven Development: Nature, Capital, and the Production of Space.* London: Basil Blackwell, 1984.

Smith, Thomas C. *Political Change and Industrial Development in Japan: Government Enterprise, 1868–1880.* Stanford: Stanford University Press, 1955.

Snyder, Gary. *Earth House Hold: Technical Notes and Queries to Fellow Dharma Revolutionaries.* New York: New Directions, 1969.

———. *Regarding Wave.* New York: New Directions, 1970.

———. *Riprap and Cold Mountain Poems.* San Francisco: North Point Press, 1990.

———. *The Back Country.* New York: New Directions, 1968.

———. *The Old Ways.* San Francisco: City Lights Books, 1977.

———. *The Practice of the Wild.* San Francisco: North Point Press, 1990.

———. *The Real Work: Interviews and Talks, 1964–1979.* Edited and with an introduction by Wm. Scott McLean. New York: New Directions, 1980.

Sombart, Werner. *A New Social Philosophy.* Translated and edited by Karl F. Geiser. Princeton: Princeton University Press, 1937.

Spengler, Oswald. *The Decline of the West.* Abridged by Helmut Werner. English abridged edition prepared by Arthur Helps. Translated by Charles Francis Atherton. New York: Alfred A. Knopf, 1962.

Spitz, David, ed. *Political Theory and Social Change.* New York: Atherton Press, 1967.

Spragens, Thomas A., Jr. *The Irony of Liberal Reason.* Chicago: University of Chicago Press, 1981.

Spretnak, Christine. *The Spiritual Dimension of Green Politics.* Santa Fe, N.M.: Bear and Co., 1986.

Spring, David, and Ellen Spring, eds. *Ecology and Religion in History.* New York: Harper Torchbooks, 1974.

Stableford, Brian, and David Langford. *The Third Millennium: A History of the World, A.D. 2000–3000.* New York: Alfred A. Knopf, 1985.

Stanlis, Peter J. *Edmund Burke and the Natural Law.* Ann Arbor: University of Michigan Press, 1958.

Stapledon, Olaf. *Last and First Men* and *Last Men in London.* Baltimore: Penguin Books, 1972.

Staudenmaier, James M., S.J. *Technology's Storytellers: Reweaving the Human Fabric.* Cambridge: MIT Press, 1985.

Steiner, George *Martin Heidegger.* New York: Viking Press, 1979.

Stine, G. Harry. *The Third Industrial Revolution.* New York: Putnam, 1975.

Stone, Glenn C., ed. *A New Ethic for a New Earth.* Andover, Conn.: Faith-Man-Nature Group, n.d.

Stover, Carl F., ed., with a foreword by William Benton. *The Technological Order: Proceedings of the Encyclopaedia Britannica Conference.* Detroit: Wayne State University Press, 1963.

Strauss, Leo, and Joseph Cropsey, eds. *History of Political Philosophy.* Chicago: Rand McNally, 1963.

———. *Natural Right and History.* Chicago: University of Chicago Press, 1973.

———. *Thoughts on Machiavelli.* Seattle: University of Washington Press, 1969.

Sugimoto, Mayayoshi, and David L. Swain. *Science and Culture in Traditional Japan, 600–1864.* Cambridge: MIT Press, 1978.

Sutton, Francis X., Seymour E. Harris, Carl Kaysen, and James Tobin. *The American Business Creed.* Cambridge: Harvard University Press, 1956.

Suzuki, Daisetz T. *Zen and Japanese Culture.* New York: Pantheon Books, 1959.

Swimme, Brian. *The Universe Is a Green Dragon: A Cosmic Creation Story.* Santa Fe, N.M.: Bear and Co., 1984.

Szylowitz, Joseph S. *Technology and International Affairs.* New York: Praeger, 1981.

Taggert, Frederick J. *Rome and China: A Study of Correlations in Historical Events.* Berkeley: University of California Press, 1937.

Tannahill, Reay. *Sex in History.* New York: Stein and Day, 1981.

Taylor, Gordon Rattrey. *The Biological Time Bomb.* New York: World, 1968.

Teilhard de Chardin, Pierre. *Le Milieu Divin: An Essay on the Interior Life.* Translated by Bernard Wall. London: Fontana Books, 1962.

———. *The Future of Man.* Translated by Norman Denny. London: Collins, 1964.

———. *The Phenomenon of Man.* Introduction by Sir Julian Huxley. Translated by Bernard Wall. New York: Harper Torchbooks, 1961.

Thomas, Donald E., Jr. *Diesel: Technology and Society in Industrial Germany.* Tuscaloosa: University of Alabama Press, 1987.

Thomas, Keith. *Man and the Natural World: A History of the Modern Sensibility.* New York: Pantheon Books, 1983.

————. *Religion and the Decline of Magic.* New York: Charles Scribner's Sons, 1971.

Thompson, E. P. *William Morris: Romantic to Revolutionary.* New York: Pantheon Books, 1976.

Thompson, Griffin. "Energy Technology and American Values." Ph.D. diss., Department of Government, Georgetown University, 1988.

Thompson, John B. *Habermas: Critical Debates.* Cambridge: MIT Press, 1982.

Thompson, William Irwin. *At the Edge of History.* New York: Harper and Row, 1971.

————. *Gaia, A Way of Knowing. Political Implications of the New Biology.* Great Barrington, Mass.: Inner Traditions/Lindesfarne Press, 1987.

Thoreau, Henry David. *Walden, or Life in the Woods.* Boston: Houghton Mifflin, 1906.

Tichi, Cecelia. *New World, New Earth: Environmental Reform in American Literature from the Puritans through Whitman.* New Haven: Yale University Press, 1979.

————. *Shifting Gears.* Chapel Hill: University of North Carolina Press, 1987.

Tokar, Brian. *The Green Alternative: Creating an Ecological Future.* San Pedro, Calif.: R & E Miles, 1987.

Tomlin, E. W. F. *The Oriental Philosophers: An Introduction.* New York: Harper Colophon Books, 1963.

Toulmin, Stephen. *The Return to Cosmology: Postmodern Science and the Theology of Nature.* Berkeley: University of California Press, 1982.

Toynbee, Arnold. *A Study of History,* vol. 12, *Reconsiderations.* London: Oxford University Press, 1961.

Trotsky, Leon. *Literature and Revolution.* New York: International Publishers, 1925.

Tucker, Robert C. *Philosophy and Myth in Karl Marx.* Cambridge: Cambridge University Press, 1961.

————. *The Lenin Anthology.* New York: W. W. Norton, 1975.

————. *The Marx-Engels Reader.* 2d ed. New York: W. W. Norton, 1972.

Turner, Frederick. *Beyond Geography: The Western Spirit against the Wilderness.* New York: Viking Press, 1980.

————. *Rediscovering America: John Muir in His Time and Ours.* San Francisco: Sierra Club Books, 1985.

Twain, Mark. *A Connecticut Yankee in King Arthur's Court.* Edited by Bernard L. Stein. Berkeley: University of California Press, 1983.

Twelve Southerners. *I'll Take My Stand: The South and the Agrarian Tradition.* Baton Rouge: Louisiana State University Press, 1977.

Van Creveld, Martin. *Technology and War: From 2000 B.C. to the Present.* New York: Free Press, 1989.

Vanderberg, W. H., ed. *Perspectives on Our Age: Jacques Ellul Speaks on His Life and Work.* Toronto: CBC Enterprises, 1981.

Van Sertima, Ivan. *They Came before Columbus.* New York: Random House, 1972.

Veblen, Thorstein. *The Engineers and the Price System.* New York: B.W. Huebsch, 1921.

————. *The Theory of the Leisure Class: An Economic Study of Institutions.* Foreword by C. Wright Mills. New York: Mentor Books, 1953.

Venable, Vernon. *Human Nature: The Marxian View.* Cleveland: World, 1966.

Viereck, Peter. *Metapolitics: The Roots of the Nazi Mind.* New York: Capricorn Books, 1961.

Volti, Rudy. *Technology, Politics, and Society in China.* Boulder, Colo.: Westview Press, 1982.

Von Grunebaum, G. E. *Islam: Essays in the Nature and Growth of a Cultural Tradition.* 2d ed. London: Routledge and Kegan Paul, 1961.

Wagar, W. Warren. *H. G. Wells and the World State.* New Haven: Yale University Press, 1961.

Wahlgren, Eric. *The Vikings and America.* New York: Thames and Hudson, 1986.

Wallace, Anthony F. C. *The Social Context of Innovation: Bureaucrats, Families, and Heroes in the Early Industrial Revolution As Foreseen in Bacon's New Atlantis.* Princeton: Princeton University Press, 1982.

Wallace-Hadrill, D. S. *The Greek Patristic View of Nature.* Manchester: Manchester University Press, 1976.

Ware, Norman. *The Industrial Worker, 1840–1860: The Reaction of American Industrial Society to the Advance of the Industrial Revolution.* Chicago: Quadrangle Books, 1964.

Watanabe, Maseo. *The Japanese and Western Science.* Translated by Otto Theodor Benfey. Philadelphia: University of Pennsylvania Press, 1991. First published in Japanese in Tokyo in 1976.

Wattenberg, Ben. *The Birth Dearth: What Happens When People in Free Countries Don't Have Enough Babies?* New York: Pharos Press, 1987.

Watts, Alan. *The Way of Zen.* New York: Vintage Books, 1957.

Weiner, Douglas R. *Models of Nature: Ecology, Conservation, and Cultural Revolution in Soviet Russia.* Bloomington: Indiana University Press, 1988.

Weiner, Jonathan. *The Next One Hundred Years: Shaping the Fate of Our Living Earth.* New York: Bantam Books, 1991.

Weinstein, James. *Ambiguous Legacy: The Left in American Politics.* New York: Frederick Watts, 1975.

———. *The Corporate Ideal in the Liberal State, 1900–1918.* Boston: Beacon Press, 1969.

Weisberger, Bernard. *The New Industrial Society.* New York: John Wiley, 1969.

Wells, H. G. *A Modern Utopia.* Lincoln: University of Nebraska Press, 1967. First published 1905.

———. *Mind at the End of Its Tether.* London: Heinemann, 1945.

———. *The Outline of History.* New York: E. P. Dutton, 1924.

Welty, Thomas. *The Asians: Their Heritage and Their Destiny.* Philadelphia: J. B. Lippincott, 1963.

White, Frank. *The Overview Effect: Space Exploration and Human Evolution.* Boston: Houghton Mifflin, 1987.

White, Lynn, Jr. *Machina Ex Deo: Essays in the Dynamics of Western Culture.* Cambridge: MIT Press, 1968.

———. *Medieval Religion and Technology: Collected Essays.* Berkeley: University of California Press, 1978.

White, Morton. *Social Thought in America: The Revolt against Formalism.* Boston: Beacon Press, 1957.

White, Morton, and Lucia White. *The Intellectual versus the City: From Thomas Jefferson to Frank Lloyd Wright.* New York: Mentor Books, 1964.

Wiebe, Robert W. *Businessmen and Reform: A Study of the Progressive Movement.* Cambridge: Harvard University Press, 1967.

———. *The Search for Order, 1877–1920.* New York: Hill and Wang, 1967.

Williams, Charles. *Witchcraft.* Cleveland: Meridian Books, 1959.

Williamson, Jack. *H. G. Wells: Critic of Progress.* Baltimore: Mirage Press, 1973.

Willoughby, Kevin W. *Technology Choice: A Critique of the Alternative Technology Movement.* Boulder, Colo.: Westview Press, 1990.

Wilson, Richard Guy, Dianne H. Pilgrim, and Dickran Tashjian. *The Machine Age in America, 1918–1941*. New York: Henry N. Abrams, 1986.

Winner, Langdon. *Autonomous Technology: Technics- out-of-Control as a Theme in Political Thought*. Cambridge: MIT Press, 1977.

———. *The Whale and the Reactor: A Search for Limits in an Age of High Technology*. Chicago: University of Chicago Press, 1986.

Wittfogel, Karl A. *Oriental Despotism*. New Haven: Yale University Press, 1957.

Wood, Gordon S. *The Creation of the American Republic, 1776–1787* . Chapel Hill: University of North Carolina Press, 1969.

Woodward, Kathleen, ed. *The Myths of Information: Technology and Postindustrial Culture*. Madison, Wisc.: Coda Press, 1980.

Wolfe, Donald M. *The Image of Man in America*. 2d ed. New York: Thomas Y. Crowell, 1961.

Wolin, Richard. *The Politics of Being: The Political Thought of Martin Heidegger*. New York: Columbia University Press, 1990.

Worster, Donald. *American Environmentalism: The Formative Period, 1860–1915*. New York: John Wiley, 1973.

———. *Nature's Economy: The Roots of Ecology*. Garden City, N.Y.: Anchor Books, 1979.

———. *Rivers of Empire: Water, Aridity, and the Growth of the American West*. New York: Pantheon Books, 1985.

Wuthnow, Robert. *The Reconstruction of American Religion: Society and Faith since World War II*. Princeton: Princeton University Press, 1986.

Ziegler, Charles E. *Environmental Policy in the USSR*. Amherst: University of Massachusetts Press, 1987.

Zimmerman, Michael E. *Heidegger's Confrontation with Modernity: Technology, Politics, and Art*. Bloomington: Indiana University Press, 1990.

Zukav, Gary. *The Dancing Wu Li Masters: An Overview of the New Physics*. New York: William Morrow, 1979.

Periodicals

Adair, Margo, and John Rensinbrink. "SPAKA: Democracy at Work. Determining Strategy and Policy Approaches in Key Areas." *Green Letter/Greener Times*, spec. ed., 3–5, Autumn 1989.

Advertisement. "A Latin American Ecological Alliance." *New York Times*, July 22, 1991.

al-Faraqi, Ismail. "Science and Traditional Values in Islamic Society." *Zygon: Journal of Religion and Science* 2:231–246 (1967).

Alper, Joe. "Environmentalists: Ban the (Population) Bomb." *Science* 252:1247 (1991).

Ames, Roger T. "Taoism and the Nature of Nature." *Environmental Ethics* 8:317–350 (1986).

Anderson, Walter Truett. "Food without Farms: The Biotech Revolution in Agriculture." *Futurist* 24:16–21 (January–February 1990).

Athanasiou, Tom, Janet Biehl, and Carl Boggs. "Is the Left-Green Network Really Green?" *Green Synthesis* 32:6, 10–14 (October 1989).

Baer, Richard. "Ecology, Religion, and the American Dream." *American Ecclesiastical Review* 165:45–59 (971).

Bailes, Kendall. "The Politics of Technology: Stalin and Technocratic Thinking among Soviet Engineers." *American Historical Review* 79:445–469 (1974).

Barringer, Felicity. "In Capital, No. 2 River Is a Cause." *New York Times*, December 1, 1991.

Bartholomew, James R. "Why Was There No Scientific Revolution in Togugawa Japan?" *Japanese Studies in the History of Science* 15:111–125 (1976).

Battista, Mary. "Eastern Europe Faces Vast Environmental Blight." *Washington Post*, March 20, 1990.

Bay, Christian. "On Ecolacy Sans Humanism." *Alternatives: A Journal of World Polity* 7:385–402 (1981–1982).

Berkes, F., D. Feeny, B. J. McKay, and J. M. Acheson. "The Benefits of the Commons." *Nature* 340:91–93 (1989).

Berry, Thomas. "The Dream of the Earth: Our Way Into the Future." *Cross Currents* 37:200–215 (Summer/Fall 1987).

———. "The New Story: Comments on the Origin, Identification and Transmission of Values." *Cross Currents* 37:187–199 (Summer/Fall 1987).

———. "Wonderworld as Wasteworld: The Earth in Deficit." *Cross Currents* 35:408–422 (Winter 1985–1986).

Boczkiewitz, Robert. "U.S. Greens Adopt First Policy Platform but Delay Politicking." *National Catholic Reporter*, September 28, 1990.

Bookchin, Murray. "Crisis in the Ecology Movement." *Zeta Magazine* 1:121–123 (July–August 1988).

Booth, William. "Action Urged against Global Warming." *Washington Post*, February 2, 1990.

———. "Carbon Dioxide Curbs May Not Halt Warming." *Washington Post*, March 10, 1990.

———. "Climate Study Halves Estimate of Global Warming." *Washington Post*, September 14, 1989.

———. "Forest May Combat Greenhouse Effect." *Washington Post*, March 23, 1990.

———. "Global Heating Could Benefit U.S. Farmers." *Washington Post*, May 17, 1990.

———. "Johnny Appleseed and the Greenhouse." *Science* 242:19–20 (1988).

———. "Nuclear Power Debate Shifts." *Washington Post*, August 28, 1989.

Borelli, Peter. "The Ecophilosophers." *Amicus Journal* 10:30–39 (Spring 1988).

Boyte, Harry, and Karen Lehmen. "Are You Curious (Green)?" *Utne Reader* 7:120–121 (December 1984/January 1985).

Bradsher, Keith. "Lower Pollution Tied to Prosperity." *New York Times*, October 28, 1991.

———. "U.S. and Mexico Draft Plan to Fight Border Pollution." *New York Times*, August 2, 1991.

———. "U.S. Ban on Mexico Tuna Is Overruled." *New York Times*, August 23, 1991.

Bratton, Susan Power. "The Ecotheology of James Watt." *Environmental Ethics* 5:225–236 (1983).

Broder, David S. "Haughey Fails in Bid for Majority." *Washington Post*, June 18, 1989.

Brown, Delmer M. "The Impact of Firearms on Japanese Warfare." *Far Eastern Quarterly* 7:236–253 (1948).

Browne, Malcolm W. "Ozone Fading Fast, Thatcher Tells World Experts." *New York Times*, June 28, 1990.

Burgess, Robert Page. "Intergalactic Travel:The Long Voyage From Home." *Futurist* 21:29–33 (September–October 1987).

Bush, Donald J. "Futurama: World's Fair as Utopia." *Alternative Futures: The Journal of Utopian Studies* 2:3–20 (Fall 1979).

Byrne, Gregory. "Let 100 Million Trees Bloom." *Science* 242:371 (1988).

Callicott, J. Baird. "Conceptual Resources for Environmental Ethics in Asian Traditions of Thought: A Propaedeutic." *Philosophy East and West* 37:115–130 (1987).

———. "Traditional American Indian and Western European Attitudes Toward Nature: An Overview." *Environmental Ethics* 4:293–318 (1982)

Campbell, Charles. "Science Committee Concludes Country Can Take Heat of Greenhouse Effect." *Albuquerque Journal*, September 7, 1991.

Cheney, Jim. "Eco-Feminism and Deep Ecology." *Environmental Ethics* 9:115–145 (1987).

Cheng, Chgung-ying. "On the Environmental Ethics of the Tao and the Ch'i." *Environmental Ethics* 8:351–370 (1986).

Cherfas, Jeremy. "East Germany Struggles to Clean Up Its Air and Water." *Science* 248:295–296 (1990).

Clark, John P. "Marx's Organic Body." *Environmental Ethics* 11: 243–258 (1989).

Cobb, John, Jr. "Christian Existence in a World of Limits." *Environmental Ethics* 1:149–158 (1979).

Cockburn, Alexander. "Legends of Antreus." *Zeta Magazine* 1:9–11 (April 1988).

———. "Live Souls." *Zeta Magazine* 1:13–14 (February 1988).

———. "Socialist Ecology: What It Means, Why No Other Kind Will Do." *Zeta Magazine* 2:15–21 (February 1989).

———. "Their Mullahs and Ours." *Zeta Magazine* 2: 33–40 (April 1988).

Cody, Edward. "Another Environmental Summit Opens, Illustrating Issue's Currency." *Washington Post*, March 11, 1989.

———. "105 Nations Back Treaty on Toxic-Waste Shipping." *Washington Post*, March 23, 1989.

Coleman, Daniel. "Girls and Math: Is Biology Really Destiny?" *New York Times*, Education Life, August 2, 1987.

Conason, Joe, and Jill Weiner. "The Fake Greens." *Village Voice*, December 26, 1989.

Cooperman, Alan. "Nuke Debate Rekindled." *Albuquerque Journal*, October 13, 1991.

Crossette, Barbara. "Village Committees Learn to Guard Endangered Forest in Bangladesh." *New York Times*, August 6, 1991.

Crossman, Sylvie. "The Greens (Les Vertes)." *Co-Evolution Quarterly* 16:94–99 (Winter 1977–1978).

Crowe, Beryl L. "The Tragedy of the Commons Revisited." *Science* 166:1103–1107 (1969).

Csonka, Paul L. "Space Colonization: An Invitation to Disaster?" *Futurist* 11:285–290 (October 1977).

D.D. " 'Greens' Challenge French Gene Research." *Science* 237:357 (1987).

Dardick, Geeta. "An Interview with Gary Snyder: 'When Life Starts Getting Interesting.' " *Sierra* 70:68–73 (September–October 1985).

Davis, Donald. "Ecosophy: The Seduction of Sophia?" *Environmental Ethics* 8:151–162 (1986).

Delianova, Mico. "Optimism, Abundance, Universalism, and Immortality: The Philosophy of E. F. Estfandiary." Interview in *Futurist* 12:185–189 (June 1978).

Deutsch, Eliot. "A Metaphysical Grounding for Natural Reverence: East-West." *Environmental Ethics* 8:293–299 (1986).

Devall, Bill. "John Muir as Deep Ecologist." *Environmental Review* 6:63–86 (1982).

Devall, William B. "The Deep Ecology Movement." *Natural Resources Journal* 20:299–322 (1980).

De Young, Karen. "Social Democrats Win Election in Sweden, Greens Gain Seats." *Washington Post*, September 19, 1988.

Drozdiak, William. "U.S. Refuses to Pledge Limit on Greenhouse Gas Emissions." *Washington Post*, November 8, 1990.

Drucker, Peter F. "The New Technology: Predicting Its Impact." *New York Times*, April 8, 1973.

Duedney, Daniel. "Space Industrialization: The Mirage of Abundance." *Futurist* 16:47–53 (December 1982).

Durbin, Paul T., Daniel Cerezuelle, et al. "Symposium on Jacques Ellul." *Research in Philosophy and Technology* 3:161–262 (1980).

Earhart, H. Byron. "The Ideal of Nature in Japanese Religion and Its Possible Significance for Environmental Concerns." *Contemporary Religions in Japan* 11:10–26 (1970).

Eckersly, Robyn. "Divining Evolution: The Ecological Ethics of Murray Bookchin." *Environmental Ethics* 11:99–116 (1989).

"Edward Abbey." Interview, *Whole Earth Review* 61:17 (Winter 1988).

Ehrenfeld, David, and Philip J. Bentley. "Judaism and the Practice of Stewardship." *Judaism* 34:301–311 (1985).

Ehrlich, Paul, and Anne Ehrlich. Review of *The Birth Dearth*. *Amicus Journal* 10:40 (Winter 1988).

Ehrlich, Paul, Anne Ehrlich, Barry Commoner, Francis Moore Lappe, et al. "The Population Bomb: An Explosive Issue for the Environmental Movement." *Utne Reader* 27:77–87 (May/June 1988).

Ehrlich, Paul, Anne Ehrlich, John Harte, Mark A. Hartwell, Peter H. Raven, Carl Sagan, George M. Woodwell, Joseph Berry, Edward S. Ayelso, Anne M. Ehrlich, Thomas Eisner, Stephen J. Gould, Herbert D. Grover, Rafael Herrera, Robert M. May, Ernst Mayr, Christopher P. McKay, Harold A. Mooney, Norman Myers, David Pimentel, and John M. Teal. "Long-Term Biological Consequences of Nuclear War." *Science* 222: 1293–1300 (1983).

Ellul, Jacques. "From the Bible to a History of Non-Work." *Cross Currents* 35:43–48 (1985).

———. "Mirror of These Ten Years." *Christian Century* 87:200–204 (1970).

———. "Nature, Technique, and Artificiality." *Research in Philosophy and Technology* 3:263–283 (1980).

———. "The Technological Order." *Technology and Culture* 3:394–421 (1962).

———. "Technology and the Gospel." *International Review of Mission* 66: 109–117 (1977).

"The Ellul Oeuvre." *Cross Currents* 35:107–108 (1985).

"Excerpts from Group of Seven's Declaration: Progress, Policies and Plans." *New York Times,* July 18, 1991.

Falk, Richard. "On Advice to the Imperial Prince." *Alternatives: A Journal of World Polity* 7:403–408 (1981–1982).

Farrell, Michael J. "Thomas Berry's Dream of Biocracy." *National Catholic Reporter,* November 13, 1987.

Fedoseev, P. N. "The Social Significance of the Scientific and Technological Revolution." *International Social Science Journal,* 27:152–162 (1975).

Ferkiss, Victor. "Daniel Bell's Concept of Post-Industrial Society: Theory, Myth, Ideology." *Political Science Reviewer* 9:87–100 (1979).

———. "Populism: Myth, Reality, Common Danger." *Western Political Quarterly* 14:737–740 (1961).

———. "Populist Influences on American Fascism." *Western Political Quarterly* 10:350–373 (1957).

———. "Technology Assessment and Appropriate Technology." *National Forum* 57:3–7 (Fall 1978).

———. "Values, Choices and Technological Options." *Space Solar Power Review* 4:261–271 (1983).

Fertig, Fred. "Child of Nature: The American Indian as Ecologist." *Sierra Club Bulletin* 55:4–7 (1976).

Finley, M. I. "Technical Innovations and Economic Progress in the Ancient World." *Economic History Review* 18:29–45 (1965).

Fisher, Marc. "Greens Fade from Germany's Political Spectrum," *Washington Post,* December 4, 1990.

Foltz, Bruce V. "On Heidegger and the Interpretation of Environmental Crisis." *Environmental Ethics* 6:323–338 (1984).

Fox, Matthew. "A Call for a Spiritual Renaissance." *Green Letter* 5:4, 16–17 (Spring 1989).

———. "Is the Catholic Church Today a Dysfunctional Family? A Pastoral Letter to Cardinal Ratzinger and the Whole Church." *Creation* 4:23–38 (November/December 1988).

Fox, Warwick. "Deep Ecology: Towards a New Philosophy for Our Time?" *Ecologist* 14:194–204 (1984).

Frye, Richard. "The Economics of Ecotopia." *Alternative Futures: The Journal of Utopian Studies* 3:71–81 (1980).

Fussell, G. E. "Farming Systems in the Classical Era." *Technology and Culture* 21:16–44 (1961).

Gabriel, Trip. "If a Tree Falls in the Forest, They Hear It." *New York Times Magazine*, November 4, 1990.

Gargan, Edward A. "Ultrasonic Tests Skew Ratio of Births in India." *New York Times*, December 13, 1991.

Geis, Gayle. "Activists Lash 'Racist' Policies At EPA Banquet." *Albuquerque Journal*, August 1, 1991.

Gillingham, Peter. "Schumacher's Buddhism." *Co-Evolution Quarterly.* 7:60 (Fall 1975).

" 'Global Marshall Plan' Urged for Environment." *Washington Post*, May 3, 1990.

Golden, Tim. "Life's Gray for Mexico Green Party." *New York Times*, August 14, 1991.

Golding, John. "The Futurist Past." *New York Review of Books*, August 14, 1986.

Goltz, Thomas. "Earth First Meeting Reflects Gap between Radicals, Mainstream." *Washington Post*, July 19, 1990

Goodman, Russell. "Taoism and Ecology." *Environmental Ethics* 2:73–80 (1980).

Gorz, Andre. "Technical Intelligence and the Capitalist Division of Labor." *Telos* 12:27–41 (1972).

Gottlieb, Anthony. "Heidegger for Fun and Profit." *New York Times Book Review*, January 7, 1990.

Graham, George J., Jr. "Jacques Ellul—Prophetic or Apocalyptic Theologian of Technology." *Political Science Reviewer* 13:213–239 (1983).

Grapard, Alan G. "Flying Mountains and Walkers in Emptiness: Toward a Definition of Sacred Space in Japanese Religion." *History of Religions* 20:195–221 (1982).

———. "Japan's Ignored Cultural Revolution: The Separation of Shinto and Buddhist Divinities in MEJII (*shimbatsu bunri*) and a Case Study:Yonomine." *History of Religions* 23:240–265 (1984).

Griffith, Jim. "The Environmental Movement in Japan." *Whole Earth Review* 69:90–96 (Winter 1990).

Grim, John. "Time, History, Historians in Thomas Berry's Vision." *Cross Currents* 37:225–239 (Summer/Fall 1987).

Gutfeld, Rose. "Eight of 10 Americans Are Environmentalists, at Least So They Say." *Wall Street Journal*, August 2, 1991.

Guthrie, Daniel A. "Primitive Man's Relationship to Nature." *Bioscience* 21:721–723 (1971).

Hacker, B.C. "The Weapons of the West: Military Technology and Modernization in 19th Century China and Japan." *Technology and Culture* 18:43–55 (1977).

Hardin, Garrett. "An Ecolate View of the Human Predicament." *Alternatives: A Journal of World Polity* 7:242–262 (1981).

———. "The Tragedy of the Commons." *Science* 162:1243–1248 (1968).

Harris, Liz. "Brother Sun, Sister Moon." *New Yorker*, April 27, 1987.

Harris, Philip R. "Living on the Moon: Will Humans Develop an Unearthly Culture?" *Futurist* 19:30–35 (April 1985).

Hasumi, Otohiko. "The Impact of Science and Technology on Traditional Society in Japan." *Cahiers d'histoire mordiale* 9:363–379 (1965).

Haught, John F. "The Emergent Environment and the Problem of Cosmic Purpose." *Environmental Ethics* 8:139–150 (1986).

Hays, Samuel P. "From Conservation to Environment: Environmental Politics in the United States since World War II." *Environmental Review* 6:14–41 (1982).

"The Heat Is On." *Time,* October 19, 1987.

Heffernan, James D. "The Land Ethic: A Critical Appraisal." *Environmental Ethics* 4:235–247 (1982).

Heinegg, Peter. "Christian Ecology?" *Cross Currents* 33:376, 1983.

Henderson, Hazel. "The Legacy of E. F. Schumacher." *Environment* 20:30–46, 1978.

"Herman Kahn Remembered." *Futurist* 17:61–65, October 1983.

Hiers, Richard H. "Ecology, Biblical Theology, and Methodology: Biblical Perspectives on the Environment." *Zygon: A Journal of Religion and Science* 19:43–49 (1984).

Hill, Phil. "Left-Green Network Emerges." *Potomac Valley Green Network* 8 (May 1989).

Hilton, Rodney, and P. H. Sawyer. "Technical Determinism: The Stirrup and the Plough." *Past and Present* 24:90–100 (1963).

Himes, Michael J., and Kenneth R. Himes. "The Sacrament of Creation: Toward an Environmental Theology." *Commonweal* 117:42–49 (January 20, 1990).

Hoffert, Robert W. "The Scarcity of Politics: Ophuls and Western Political Thought." *Environmental Ethics* 8:5–32 (1986).

Hoffman, Eric P. "Soviet Views of 'The Scientific-Technological Revolution,' " *World Politics* 30:615–644 (1978).

———. "The Scientific Management of Society." *Problems of Communism* 26:59–67 (May–June 1977).

Holbo, Paul S. "Wheat or What? Populism and American Fascism." *Western Political Quarterly* 14:727–736 (1961).

Holden, Constance. "Female Math Anxiety on the Wane." *Science* 236:660–661 (1987).

———. "Is 'Gender Gap' Narrowing?" *Science* 253:959–960 (1991).

Hughes, J. Donald. "Ecology in Ancient Greece." *Inquiry,* 18:115–125 (1975).

Hughes, Robert. " 'Kill the Moonlight!' They Cried." *Time,* August 4, 1986.

Hunter, Mark. "Greens Join Mainstream in France." *Washington Post,* March 22, 1989.

Imada, Kenneth K. "Environmental Problematics in the Buddhist Context." *Philosophy East and West* 37:135–149 (1987).

Io, Po-Kueng. "Taoism and the Foundations of Environmental Ethics." *Environmental Ethics* 5:335–343 (1983).

Johnstone, Diane. "Red and White Fade to Green in EC Elections." *In These Times,* July 5–18, 1989.

———. "Spring Brings a Touch of Green to France." *In These Times,* April 5–11, 1989.

———. "The Green Dilemma: To Be or Not to Be a Realo." *In These Times,* August 17–30, 1988.

Kahn, Herman, and John B. Phelps. "The Economic Present and Future: A Chartbook for the Decades Ahead." *Futurist* 13:202–222 (June 1979).

"Kahn, Mead, and Thompson. A Three-Way Debate on the Future." *Futurist* 12:229–232 (August 1978).

Kateb, George. "The Political Thought of Herbert Marcuse." *Commentary* 49:48–63 (January 1970).

Kaufman, Gordon D. "A Problem for Theology: The Concept of Nature." *Harvard Theological Review* 65:337–366 (1972).

Kay, Jeanne. "Comments on 'The Unnatural Jew,' " *Environmental Ethics* 7:189–191 (1985).
———. "Concepts of Nature in the Hebrew Bible." *Environmental Ethics* 10:309–327 (1988).
Keller, Bill. "Developers Turn Aral Sea Into a Catastrophe." *New York Times*, December 20, 1988.
Keller, Evelyn Fox. "Feminism, Science, and Democracy." *Democracy: A Journal of Political Renewal and Social Change* 3:50–58 (Spring 1983).
Kenkelen, Bill. "Some U.S. Theologians Protest Fox Silencing." *National Catholic Reporter*, November 4, 1988.
Kerr, Richard A. "Global Temperature Hits Record Again." *Science* 251:374 (1991).
———. "Hansen vs. the World on the Greenhouse Effect." *Science* 244:1041–1043 (1989).
———. "New Greenhouse Report Puts Down Dissenters." *Science* 249:481–482 (1990).
———. "No Longer Willful, Gaia Becomes Respectable." *Science* 240:393–395 (1988).
———. "U.S. Bites Greenhouse Bullet and Gags." *Science* 251:868 (1991).
King, Ynestra. "Ecological Feminism." *Zeta Magazine* 1:124–127 (July–August 1988).
———. "Feminism and the Revolt of Nature." *Heresies: A Feminist Journal of Art and Politics* 13:12–16 (1981).
———. "The Ecology of Feminism and the Feminism of Ecology." *Harbinger: A Journal of Social Ecology* 1:16 (1983).
Kinzer, Stephen. "Kohl Loses State Election to Socialist-Green Coalition." *New York Times*, January 22, 1991.
Kiyoshi, Yabaouti. "The Pre-History of Modern Science in Japan: The Importation of Modern Science During the Togugawa Period." *Cahiers d'histoire mondiale* 9:208–232 (1965).
Kolack, Shirley. "The Impact of Technology on Democratic Values: The Case of Lowell, Massachusetts." *Bulletin of Science, Technology, and Society* 8:405–410 (1988).
Kothari, Rajni. "On Eco-Imperialism." *Alternatives: A Journal of World Polity* 7:383–394 (1981–1982).
L.R. "How Bad Is Acid Rain?" *Science* 251:1303 (1991).
Lancaster, John. "The Green Guerilla." *Washington Post*, March 20, 1991.
Laud, Robin. "Post-Industrial Society: East and West." *Survey* 21: 1–17 (1975).
Layton, Edwin T., Jr. "American Ideologies of Science and Engineering." *Technology and Culture* 16:48–66 (1975).
Lee, Donald C. "On the Marxian View of the Relationship between Man and Nature." *Environmental Ethics* 2:3–16 (1980).
———. "Toward a Marxian Environmental Ethic." *Environmental Ethics* 4:339–343 (1982).
Levy, Marion J., Jr. "Contrasting Factors in the Modernization of China and Japan." *Economic Development and Cultural Change* 2:161–197 (1953).
Lindorff, David. "The One Issue That Could Unite the American Left." *In These Times*, November 16–22, 1988.
Long, F. A. "Science, Technology and Industrial Development in India." *Technology in Society: An International Journal* 10:395–416 (1988).
Lovelock, James. "James Lovelock Responds." *Co-Evolution Quarterly* 29:62–16 (Spring 1981).
———. "The Independent Practice of Science." *Co-Evolution Quarterly* 25:22–30 (Spring 1980).
Lowry, S. T. "The Classical Greek Theory of National Resource Economics." *Land Economics* 41:203–208 (1966).
McCarthy, Tim. "Greens Going for a Green America." *National Catholic Reporter*, July 14, 1989.
———. "Mission Fights to Answer Cry of the Land." *National Catholic Reporter*, March 10, 1989.

————. "Philippine Bishops Issue Ecology Pastoral." *National Catholic Reporter*, March 11, 1988.

McDermott, Jeanne. "Lynn Margulis—Unlike Most Microbiologists." *Co-Evolution Quarterly* 25:31–36 (Spring 1980).

MacEoin, Gary. "Ratzinger Knocks Green Party, Dialogue with Jews." *National Catholic Reporter*, November 6, 1987.

Macquarrie, John. "The Idea of a Theology of Nature." *Union Seminary Quarterly Review* 30:69–75 (Winter–Summer 1975).

Macy, Joanna. "Deep Ecology Notes." *Green Synthesis* 30:10 (March 1989).

Mandel, William. "The Soviet Ecology Movement." *Science and Society* 36:385–416 (1972).

Marcuse, Herbert. "Some Implications of Modern Technology." *Studies in Philosophy and Social Science* 9:414–439 (1941).

Margulis, Lynn. "Lynn Margulis Responds." *Co-Evolution Quarterly* 29:63–65 (Spring 1981).

Margulis, Lynn, and James Lovelock. "The Atmosphere as Circulatory System of the Biosphere—The Gaia Hypothesis." *Co- Evolution Quarterly* 6:31–40 (Summer 1975).

Markham, James M. "Over Philosophy's Temple, the Shadow of a Swastika." *New York Times*, February 4, 1988.

Marshall, Eliot. "Armageddon Revisited." *Science* 236:1421–1422 (1987).

————. "Nuclear Winter Debate Heats Up." *Science* 235:271–273 (1987).

Maruyama, Magorah. "Designing a Space Community." *Futurist* 10:273–281 (October 1976).

Marx, Leo. "On Heidegger's Conception of 'Technology' and Its Historical Validity." *Massachusetts Review* 29:637–652 (1984).

Massey, Marshall. "Carrying Capacity and the Greek Dark Ages." *Co-Evolution Quarterly* 40:29–35 (Winter 1983).

Matthiessen, Peter. "The Blue Pearl of Siberia." *New York Review of Books*, February 16, 1991.

Maxwell, Kenneth. "The Mystery of Chico Mendes." *New York Review of Books*, March 28, 1991.

————. "The Tragedy of the Amazon." *New York Review of Books*, March 7, 1991.

Means, Richard L. "Man and Nature: The Theological Vacuum." *Christian Century* 85:579–581 (May 1, 1968).

Meier, Hugo A. "Technology and Democracy, 1800–1860." *Mississippi Valley Historical Review* 43:618–640 (1957).

Meiksins, Peter. "The 'Revolt of the Engineers' Reconsidered." *Technology and Culture* 29:219–246 (1988).

Menninger, D. "Political Dislocation in a Technological Universe." *Review of Politics* 42:73–91 (1980).

————. "Politics or Technique: A Defense of Jacques Ellul." *Polity* 14:110–127 (1981).

Merchant, Carolyn. "Earth Care: Women and the Environment." *Environment* 23:6–13, 38–40 (June 1981.)

Mitcham, Carl. "About Ellul." *Cross Currents* 35:105–107 (1985).

————. "Jacques Ellul and His Contribution to Theology." *Cross Currents* 35:1–8 (1985).

————. "Langdon Winner on Jacques Ellul: An Introduction to Alternative Political Critiques of Technology." *Technology Studies Resource Center Working Papers Series* 3:115–124 (June 1985).

Mitsotomo, Yuasa. "The Scientific Revolution and the Age of Technology." *Cahiers d'histoire mondiale* 9:187–207 (1965).

Moline, Jon D. "Aldo Leopold and the Moral Community." *Environmental Ethics* 8:99–120 (1986).

Montgomery, David. "On Goodwyn's Populists." *Marxist Perspectives* 1:166–173 (1978).

Mundey, Jack. "Ecology, Capitalism, Communism." *Co-Evolution Quarterly* 13:46–48 (Spring 1977).

Mundey, Jack, talking to Ellen Newman and David Gancher. "Labor and Environment—The Australian Green Bans." *Co-Evolution Quarterly* 13:40–45 (Spring 1977).

Murphy, Rhoades. "Man and Nature in China." *Modern Asian Studies* 1:315–333 (1967).

"Murray Bookchin." Interview in *Whole Earth Review*.61:16 (Winter 1988).

Muwakkil, Salim. "U.S. Green Movement Needs to be Colorized." *In These Times*, May 2–8, 1990.

Naess, Arne. "A Defense of the Deep Ecology Movement." *Environmental Ethics* 6:265–270 (1984).

———. "Finding Common Ground." *Green Synthesis* 30, 9–10 (March 1989).

———. "Self-realization in Mixed Communities of Humans, Bears, Sheep, and Wolves." *Inquiry* 22:231–241 (1979)

———. "The Deep Ecological Movement: Some Philosophical Aspects." *Philosophical Inquiry* 8:10–31 (1986).

———. "The Shallow and the Deep, Long-Range Ecology Movement— A Summary." *Inquiry* 16:95–100 (1973).

Nash, Nathaniel C. "Latin Nations Getting Others' Waste." *New York Times*, December 16, 1991.

———. "Unease Grows under the Ozone Hole." *New York Times*, July 23, 1991.

"National Party Wins New Zealand, Improved U.S. Ties Sought." *New York Times*, October 28, 1990.

Nisbet, Robert. "Utopia's Mores: Has the American Vision Dimmed?" *Public Opinion* 6:6–9 (April–May 1983).

Norr, John. "German Social Theory and the Hidden Face of Technology." *European Journal of Sociology* 15:323–336 (1974).

Oates, Joyce Carol. "The Mysterious Mr. Thoreau." *New York Times Book Review*, May 1, 1988.

O'Neill, Gerard. "Space Colonies: The High Frontier." *Futurist* 10:25–33 (February 1976).

Page, Diana. "Debt-for-Nature Swaps: Experience Gained, Lessons Learned." *International Environmental Affairs* 1:275–288 (1989).

Parkinson, Thomas. "The Poetry of Gary Snyder." *Southern Review* 4, n.s.:616–632 (1968).

Parmalee, Jennifer. "Pope Says Environmental Misuse Threatens World Stability." *Washington Post*, December 6, 1989.

Partridge, Ernest. "Are We Ready for an Environmental Morality?" *Environmental Ethics* 4:175–190 (1982).

Passell, Peter. "Staggering Cost Is Foreseen to Curb Warming of Earth." *New York Times*, November 19, 1989.

———. "Warmer Globe, Greener Pastures?" *New York Times*, September 18, 1991.

Paterson, Kent. "Expanding the Spectrum of U.S. Green Politics." *In These Times*, September 4–10, 1991.

Peterson, Cass. "Skin Cancer Rate Rises to 'Near Epidemic,' " *Washington Post*, March 10, 1987.

Pfaff, William. " 'New World Order' Bigger Than Bush." *Albuquerque Journal*, August 4, 1991.

Pickett, Karen. "Arizona Conspiracy Trial Ends in Plea Bargain." *Earth First!* 11:5 (1991).

———. "A Snitch on the Stand and Other Revelations about the Pawns of the Evil Empire." *Earth First!* 11:25 (1991).

Pilla, Anthony M. "Pastoral Letter Urges Reverence for Creation." *National Catholic Reporter*, November 16, 1990.

Polsgrove, Carol. "Unbroken Circle." *Sierra* 77:130–133, 140 (January–February 1992).

Pope, Carl. "The Politics of Plunder." *Sierra* 73:49–55 (November/December 1988).

"Pope's New Year Message: Moral Roots of Ecology Crisis." *National Catholic Reporter*, December 29, 1989.

Preston, Julia. "Destruction of Amazon Rain Forest Slowing." *Washington Post*, March 17, 1991.

Ramoser, George W. "Heidegger and Political Philosophy." *Review of Politics* 29:261–268 (1967).

Randall, Afrian J. "The Philosophy of Luddism: The Case of the West of England Wollen Workers, ca. 1790–1809." *Technology and Culture* 27:1–17 (1986).

Randall, Jonathan C. " 'Greening' of Thatcher Surprises Many Britons." *Washington Post*, March 4, 1989.

Rensenbrink, John. "Greens USA." *National Forum* 70:20–22 (Winter 1990).

Reynolds, Terry S. "Medieval Roots of the Industrial Revolution." *Scientific American* 251:122–130 (July 1984).

Roberts, Leslie. "Academy Panel Split on Greenhouse Adaptation." *Science* 253:1206 (1991).

———. "How Fast Can Trees Migrate?" *Science* 243:735–737 (1989).

———. "Is There Life After Climate Change?" *Science* 242:1010–1012 (1988).

Rodman, John. "The Other Side of Ecology in Ancient Greece: Comments on Hughes." *Inquiry* 19:108–112 (1976).

Ross, Stephen J. "Political Economy for the Masses: Henry George." *Democracy: A Journal of Political Renewal and Social Change* 2: 125–134 (July 1982).

Rothschild, Joan. "A Feminist Perspective on Technology and the Future." *Women's International Quarterly* 4:65–74 (1981).

Routley, Val. "On Karl Marx as Environmental Hero." *Environmental Ethics* 3:237–244 (1981).

Rubin, Charles T. "E. F. Schumacher and the Politics of Technological Renewal." *Political Science Reviewer* 16:65–96 (1986).

Russell, Dick. "The Monkeywrenchers." *Amicus Journal* 9:28–42 (Fall 1987).

Sagan, Carl. "Nuclear War and Climatic Catastrophe: Some Policy Implications." *Foreign Affairs* 62:257–292 (Winter 1983–84).

Sagoff, Mark. "The Greening of the Blue Collars." *Philosophy and Public Policy* 10:1–5 (Summer/Fall 1990).

Sale, Kirkpatrick. "Deep Ecology and Its Critics." *Nation* 246:670–672+ (1988).

———. "Greens Take Root in American Soil." *Utne Reader* 12:86–89 (October/November 1985).

Sallah, Ariel Kay. "Deeper than Deep Ecology: The Eco-Feminist Connection." *Environmental Ethics* 6:339–345 (1984).

Satin, Mark. "You Don't Have to Be a Baby to Cry." *New Options*, September 24, 1990.

Schaar, John B. "Jacques Ellul:Between Babylon and the New Jerusalem." *Democracy: A Journal of Political Renewal and Social Change* 2:102–118 (1982).

Schemann, Serge. "A Normal Germany." *New York Times*, December 4, 1990.

Schneider, Keith. "Military Has New Strategic Goal in Cleanup of Vast Toxic Waste." *New York Times*, August 5, 1991.

———. "Minorities Join to Fight Toxic Waste." *New York Times*, October 25, 1991.

———. "Student Group Seeks Broader Agenda for Environmental Movement." *New York Times*, October 7, 1991.

Schneider, Stephen H. "The Greenhouse Effect: Science and Policy." *Science* 243:771–781 (1989).

Schoenfeld, Clay. "Aldo Leopold Remembered." *Audubon* 80:79 (March 1978).

Schumacher, E. F. "The Difference between Unity and Uniformity." *Co-Evolution Quarterly* 7:52–59 (Fall 1975).

Schwartzschild, Steven S. "The Unnatural Jew." *Environmental Ethics* 6:347–362 (1984).

Scott, Peter Dale. "Peace, Power, and Revolution: Peace Studies, Marxism, and the Academy." *Alternatives: A Journal of World Policy* 11:351–372 (1983–84).

Segal, Howard. "The Feminist Technological Utopia: Mary E. Bradley Lane's *Mizora* (1890)." *Alternative Futures: The Journal of Utopian Studies* 4:67–72 (1981).

Sessions, George. "Deep Ecology, New Age, and Gain Consciousness." *Earth First!* 7(8):27–30 (1987).

———. "Spinoza and Jeffers on Man and Nature." *Inquiry* 20:481–528 (1977).

Shabekoff, Philip. "Acid Rain Report Unleashes a Torrent of Criticism." *New York Times*, March 20, 1990.

———. "Bush Asks Cautious Response to Threat of Global Warming." *New York Times*, February 6, 1990.

———. "Bush Wants Billions of Trees for War Against Polluted Air." *New York Times*, January 28, 1990.

———. "Draft Report on Global Warming Foresees Environmental Havoc in U.S.." *New York Times*, October 28, 1988.

Shammas, Carole. "How Efficient Was Early America?" *Journal of Social History* 14:3–24 (1980).

Shaner, David Edward, and R. Shannon Duval. "Conservation Ethics and the Japanese Intellectual Tradition." *Environmental Ethics* 11:197–214 (1989).

Sheehan, Thomas. "Heidegger and the Nazis." *New York Review of Books*, June 16, 1988.

Simons, Marlise. "A Greens Mayor Takes On the Industrial Filth of Old Cracow." *New York Times*, March 25, 1990.

Skinner, B. F. "Freedom and the Control of Man." *American Scholar* 25:47–65 (Winter 1966–1956).

Smith, Charlene. "Earth First! Vows More Sabotage after Guilty Pleas." *Albuquerque Journal*, December 8, 1991.

Smith, Lee. "Divisive Forces in an Inbred Nation." *Fortune* 115:24–29, March 30, 1987.

Smith, R. Jeffrey. "A Grim Portrait of the Postwar World." *Science* 229:1245–1246 (1985).

Snow, Katrin. "Tribes' Activism Poses Hazard to Waste Industry's Health." *National Catholic Reporter*, December 6, 1991.

Snyder, Gary. "The Etiquette of Freedom." *Sierra* 74:74–77,113–116 (September/October 1989).

"Space: The Long-Range Future." Interview with Jesco von Puttkamer, *Futurist* 19:36–38 (February 1985).

Squadrito, Kathleen M. "Locke's View of Dominion." *Environmental Ethics* 1:255–262 (1979).

Steinfels, Peter. "Idyllic Theory of Goddesses Creates Storm." *New York Times*, February 13, 1990.

Stevens, William K. "At Meeting, U.S. Is Alone on Global Warming." *New York Times*, September 10, 1991.

———. "Summertime Harm to Shield of Ozone Detected Over U.S.." *New York Times*, October 23, 1991.

———. "The Earth Cools Off after Spell of Warmth." *New York Times*, December 24, 1991.

———. "Worst Fears of Acid Rain Unrealized." *New York Times*, February 20, 1990.

Stroup, John. "Nazis as Nature Mystics?" *This World: A Journal of Religion and Public Life* 23:97–114 (Fall 1988).

Subo, Roberto. "Europeans Accuse the U.S. of Balking on Plans to Combat Global Warming." *New York Times*, July 10, 1990.

Sun, Marjorie. "Environmental Awakening in the Soviet Union." *Science* 241:1033–1035 (1988).

Suvin, Darko. "The Utopian Tradition in Russian Science Fiction." *Modern Language Review* 66:139–159, 1971.

Swimme, Brian. "Berry's Cosmology." *Cross Currents* 27:218–224 (Summer/Fall 1987).

Tetu, Hirosige. "The Role of the Government in the Development of Science." *Cahiers d'histoire mondiale* 9:320–339 (1965).

"Think Globally, Act Locally, Be Spiritually." *Amicus Journal* 10:8–10 (Fall 1988).

Thomas, Donald E., Jr. "Diesel, Father and Son, Social Philosophers of Technology." *Technology and Culture* 19:376–393 (1978).

Tokar, Brian. "Ecological Radicalism." *Zeta Magazine* 1:84, 90–91 (December 1988).

———. "Politics under the Greenhouse." *Zeta Magazine* 2:90–96 (March 1989).

———. "The Greens: To Party or Not?" *Zeta Magazine* 4:42–46 (October 1991).

———. "What Kind of Environmentalism?" *Zeta Magazine* 4:43–47 (June 1991).

Tolman, Charles. "Karl Marx, Alienation, and the Mastery of Nature." *Environmental Ethics* 3:63–74 (1981).

Tomsho, Robert. "Dumping Grounds: Indian Tribes Contend with Some of Worst of America's Pollution." *Wall Street Journal*, November 29, 1990.

Turco, R. P., G. B. Toon, T. P. Ackerman, J. B. Pollack, and Carl Sagan. "Nuclear Winter: Global Consequences of Multiple Nuclear Explosions." *Science* 222:1283–1294 (1983).

"U.S. Still Resists Curbs on Emissions." *New York Times*, September 22, 1991.

Van Bieme, David. "Jeremy Rifkin Is a Little Worried About Your Future." *Washington Post Magazine*, January 17, 1988.

Von Puttkamer, Jesco. "The Industrialization of Space: Transcending the Limits to Growth." *Futurist* 13:192–201 (June 1979).

Wagar, W. Warren. "The Steel Grey Savior." *Alternative Futures: The Journal of Utopian Studies* 2:38–54 (1979).

———. "Toward a World Set Free: The Vision of H. G. Wells." *Futurist* 17:24–31 (August 1983).

Walljasper, Jay. "Can Green Politics Take Root in the U.S." *Utne Reader* 35:140–143 (September/October 1989).

———. "Zeitgeist (Deep Ecology vs. Socialist Ecology?" *Utne Reader* 30:135–136 (November/December 1988).

———. "Zeitgeist (The Politics of Ecology)." *Utne Reader* 28:126–127 (July/August 1988).

Warren, Mark. "The Marx-Darwin Question: Implications for the Critical Aspects of Marx's Social Theory." *International Sociology* 2:251–269 (1987).

Warshall, Peter. "The Political Economy of Deforestation." *Whole Earth Review* 64:68–75 (Fall 1989).

Watanabe, Maseo. "The Conception of Nature in Japanese Culture." *Science* 183:279–282 (1974).

Watson, Richard A. "A Critique of Anti-Anthropocentric Biocentrism." *Environmental Ethics* 5:245–256 (1983).

Weaver, Lelland A. C. "Factories in Space: The Role of Robots." *Futurist* 16:47–53, December 1982.

Webking, Robert. "Liberalism and the Environment." *Political Science Reviewer* 11:223–249 (1981).

———. "Plato's Response to Environmentalism." *Intercollegiate Review* 20:27–35 (Winter 1984).

Weinberg, Robert A. "The Dark Side of the Genome." *Technology Review* 94:44–51 (April 1991).

Weisman, Steven R. "Japan Agrees to End the Use of Huge Fishing Nets." *New York Times*, November 27, 1991.

Weisskopf, Michael. " 'Greenhouse Effect' Fueling Policy Makers." *Washington Post*, August 15, 1988.

———. "Reagan Aide Blocks 'Greenhouse' Rule." *Washington Post*, January 13, 1989.

———. "Strict Energy-Saving Urged to Combat Global Warming." *Washington Post*, April 11, 1991.

———. "Unusual Thinning of Ozone Layer Found Over U.S.." *Albuquerque Journal*, October 23, 1991.

Wells, Ken. "Earth First! Group Manages to Offend Nearly Everybody." *Wall Street Journal*, June 19, 1990.

White, Lynn, Jr. "Technology Assessment from the Stance of a Medieval Historian." *American Historical Review* 79:1–13 (1974).

Whitebook, Joel. "The Problem of Nature in Habermas." *Telos* 40:41–79 (Summer 1979).

Windsor, Pat. "Final WCC Assembly Report Leads with Ecology." *National Catholic Reporter*, March 1, 1991.

———. "Nestled in the Vatican Gardens." *National Catholic Reporter*, April 28, 1989.

Winner, Langdon. "The Political Philosophy of Alternative Technology." *Technology in Society: An International Journal* 1:75–86 (1979).

Wood, Harold W., Jr. "The United Nations World Charter for Nature." *Ecology Law Quarterly* 12:977–996 (1985).

Woodbury, Robert W. "The Legend of Eli Whitney and Interchangeable Parts." *Technology and Culture* 1:235–253 (1960).

Woodin, Cindy. "Pope Says Environment's Destruction Is Threat to World Peace." *Catholic Standard* (Washington), December 21, 1989.

Wright, Karen. "Heating the Global Warming Debate." *New York Times Magazine*, February 3, 1991.

Wyatt, Mike. "Humanism and Ecology: The Social Ecology/Deep Ecology Schism." *Green Synthesis* 32:7–8 (October 1989).

Young, R. V., Jr. "A Conservative View of Environmental Ethics." *Environmental Ethics* 1:241–254 (1979).

Zimmerman, Michael F. "Toward a Heideggerian *Ethos* for Radical Environmentalism." *Environmental Ethics* 5:99–131 (1983).

INDEX